Transport and Chemical Transformation of Pollutants in the Troposphere

Series editors (Volumes 2 to 8 and 10): Peter Borrell, Patricia M. Borrell, Tomislav Cvitaš, Kerry Kelly and Wolfgang Seiler

Series editors (Volumes 1 and 9): Peter Borrell, Patricia M. Borrell and Pauline Midgley

Springer
Berlin
Heidelberg
New York
Barcelona
Hong Kong
London
Milan
Paris
Singapore
Tokyo

Transport and Chemical Transformation
of Pollutants in the Troposphere

Volume 9

Exchange and Transport of Air Pollutants over Complex Terrain and the Sea

Field Measurements and Numerical Modelling;
Ship, Ocean Platform and Laboratory Measurements

Søren Larsen (Roskilde), **Franz Fiedler** (Karlsruhe)
and **Peter Borrell** (Garmisch-Partenkirchen)
Editors

EUROTRAC-2 International Scientific Secretariat,
GSF-Forschungszentrum für Umwelt und Gesundheit, München

Springer

Prof. Søren E. Larsen
Riso National Laboratory
Department of Meteorology
DK-4000 Roskilde, Denmark

Prof. Franz Fiedler
Universität Karlsruhe/Forschungszentrum Karlsruhe
Institut für Meteorologie und Klimaforschung
Postfach 6980
D-76128 Karlsruhe, Germany

Dr. Peter Borrell
P&PMB Consultants
Ehrwalder Strasse 9
D-82467 Garmisch-Partenkirchen, Germany

With 124 Figures, two in Colour

The cover picture shows a view of the Rhine Valley taken from one of the measuring aircraft in the TRACT campaign. The pollution trapped between the inversion layer can be easily seen. An account of the campaign is given in chapter 5. The photograph was kindly supplied by Dr. Meinolf Kossmann from the Forschungszentrum, Karlsruhe.

ISBN 3-540-67438-1 Springer-Verlag Berlin Heidelberg New York

Library of Congress Cataloging-in-Publication Data
Exchange and transport of air pollutants over complex terrain and the sea: field measurements and numerical modelling: ship, ocean platform and laboratory measurements/Søren Larsen, Franz Fiedler, and Peter Borrell, editors
p. cm. – (Transport and chemical transformation of pollutants in the troposphere; v. 9)
Includes bibliographical references and index.
ISBN 3540674381 (alk. paper)
1. Air-pollution – Europe – Mathematical models. 2. Ocean-atmosphere interaction. 3. Atmospheric deposition – Measurement. 4. Marine pollution – Europe. 5. Air-Pollution – Alpine regions. I. Larsen, Søren. II. Fiedler, Franz. III. Borrell, Peter. IV. Series
TD883.7.E85 E93 2000
628.5'3–dc21 00-038813

Springer-Verlag Berlin Heidelberg New York
a member of BertelsmannSpringer Science+Business Media GmbH

Printed in Germany

Cover Design: Struve & Partner, Heidelberg

SPIN 10514823 30/3130xz-5 4 3 2 1 0 – Printed on acid-free paper

Transport and Chemical Transformation of Pollutants in the Troposphere

Series editors (Volumes 2 to 8 and 10): Peter Borrell, Patricia M. Borrell, Tomislav Cvitaš, Kerry Kelly and Wolfgang Seiler

Series editors (Volumes 1 and 9): Peter Borrell, Patricia M. Borrell and Pauline Midgley

Volume 1 Transport and Chemical Transformation of Pollutants in the Troposphere
An overview of the work of EUROTRAC

Volume 2 Heterogeneous and Liquid Phase Processes
Laboratory studies related to aerosols and clouds

Volume 3 Chemical Processes in Atmospheric Oxidation
Laboratory studies of chemistry related to troposphere ozone

Volume 4 Biosphere-Atmosphere Exchange of Pollutants and Trace Substances
Experimental and theoretical studies of biogenic emission and of pollutant deposition

Volume 5 Cloud Multi-phase Processes and High Alpine Air and Snow Chemistry
Ground-based cloud experiments and pollutant deposition in the high Alps

Volume 6 Tropospheric Ozone Research
Tropospheric ozone in the regional and sub-regional context

Volume 7 Tropospheric Modelling and Emission Estimation
Chemical transport and emission modelling on regional, global and urban scales

Volume 8 Instrument Development for Atmospheric Research and Monitoring
Lidar profiling, DOAS and tunable diode laser spectroscopy

Volume 9 Exchange and Transport of Air Pollutants over Complex Terrain and the Sea
Field measurements and numerical modelling; aircraft, ship, ocean platform and laboratory measurements

Volume 10 Photo-oxidants, Acidification and Tools:
Policy Applications of EUROTRAC Results

Foreword by the Series Editors

EUROTRAC was the European co-ordinated research project, within the EUREKA initiative, studying the transport and chemical transformation of pollutants in the troposphere. That the project achieved a remarkable scientific success since its start in 1988, contributing substantially both to the scientific progress in this field and to the improvement of the scientific basis for environmental management in Europe, was indicated by the decision of the participating countries to set up, in 1996, EUROTRAC-2 to capitalise on the achievements. EUROTRAC, which at its peak comprised some 250 research groups organised into 14 subprojects, brought together international groups of scientists to work on problems directly related to the transport and chemical transformation of trace substances in the troposphere. In doing so, it helped to harness the resources of the participating countries to gain a better understanding of the trans-boundary, interdisciplinary environmental problems which beset us in Europe.

The scientific results of EUROTRAC are summarised in this report which consists of ten volumes.

Volume 1 provides a general overview of the scientific results, together with summaries of the work of the fourteen individual subprojects prepared by the respective subproject coordinators.

Volumes 2 to 9 comprise detailed overviews of the subproject achievements, each prepared by the respective subproject coordinator and steering group, together with summaries of the work of the participating research groups prepared by the principal investigators. Each volume also includes a full list of the scientific publications from the subproject.

Volume 10 is the complete report of the Application Project, which was set up in 1993 to assimilate the scientific results from EUROTRAC and present them in a condensed form so that they are suitable for use by those responsible for environmental planning and management in Europe. It illustrates how a scientific project such as EUROTRAC can contribute practically to providing the scientific consensus necessary for the development of a coherent atmospheric environmental policy for Europe.

A multi-volume work such as this has many contributors and we, as the series editors, would like to express our thanks to all of them: to the subproject coordinators who have borne the brunt of the scientific co-ordination and who have contributed so much to the success of the project and the quality of this report; to the principal investigators who have carried out so much high-quality scientific work; to the members of the International Executive Committee (IEC) and the SSC for their enthusiastic encouragement and support of EUROTRAC; to the participating governments and, in particular, the German Government (BMBF) for funding, not only the research, but also the ISS publication activities; and finally to Mr. Christian Witschel and his colleagues at Springer-Verlag for providing the opportunity to publish the results in a way which will bring them to the notice of a large audience.

Since the first volumes in this series were published in 1997 there have been many changes. In 1998 the International Scientific Secretariat moved from its old home at the Fraunhofer Institute in Garmisch-Partenkirchen to the GSF-Forschungszentrum für Umwelt und Gesundheit in München. Dr. Pauline Midgley was appointed as Executive Secretary for EUROTRAC-2 on the retirement in 1999 of Dr. Peter Borrell, who had also been the Scientific Secretary of EUROTRAC. With these came a change of the series editors. The present editors would like to thank their former colleagues, Dr. Tomislav Cvitaš and Ms. Kerry Kelly, for their contribution to the success of the final report and also, and in particular, Dr. Wolfgang Seiler, the former Director of the International Scientific Secretariat, without whose initial efforts EUROTRAC would never have got off the ground.

Peter Borrell
Patricia M. Borrell
P&PMB Consultants, Garmisch-Partenkirchen

Pauline Midgley
Executive Secretary
EUROTRAC-2 International Scientific Secretariat
GSF-Forschungszentrum für Umwelt und Gesundheit, München

Table of Contents

Part I: Air - Sea Exchange

Part II: The TRACT Experiment

List of Contributors

G. Adrian
Inst. für Meteorologie und Klimaforschung
Universität Karlsruhe / Forschungszentrum
Karlsruhe Postfach 6980
D-76021 Karlsruhe
Germany

P. Allerup
Danish Meteorological Institute (DMI)
Lyngbyvej 100
DK-2100 Copenhagen
Denmark

D. Amouroux
Laboratoire de Photophysique et
Photochimie Moléculaire
Université de Bordeaux I
351 Cours de la Libération
F-33405 Talence Cedex
France

M.O. Andreae
Max Planck Institute for Chemistry
Biogeochemistry Dept.
P.O.Box 3060
D-55020 Mainz
Germany

D. Anfossi
C.N.R. - Istituto di Cosmogeofisica
Corso Fiume 4
I-10133 Torino
Italy

W.A.H. Asman
National Environmental Research Institute
(NERI)
Frederiksborgvej 399
DK-4000 Roskilde
Denmark

W. Baeyens
Department of Analytical Chemistry
Free University of Brussels (VUB)
Pleinllan 2
B-1050 Brussels
Belgium

H. Bange
MPI für Chemie
D-55020 Mainz
Germany

R. Barthelmie
Meteorology and Wind Energy Department
Risø National Laboratory
DK-4000 Roskilde
Denmark

S. Belviso
Centre des Faibles Radioactivités
CNRS-CEA
Av. de la Terrasse
F-91198 Gif-sur-Yvette
France

G. Bergametti
Laboratoire Interuniversitaire des
Systèmes Atmosphériques
Universités Paris 7 et Paris 12
61 Av. du général de Gaulle
F-94010 Créteil cedex
France

R. Berkowicz
National Environmental Research Institute
(NERI)
Frederiksborgvej 399
DK-4000 Roskilde
Denmark

P. Berner
Institute for Energy Economics and the
Rational Use of Energy
University of Stuttgart
Heßbrühlstraße 49a
D-70565 Stuttgart
Germany

H.-J. Binder
Inst. für Meteorologie und Klimaforschung
Universität Karlsruhe /
Forschungszentrum Karlsruhe GmbH
Postfach 6980
D-76128 Karlsruhe
Germany

S. Bistacchi
CISE SpA
Environment Department
via R.Emilia 39
I-20090 Segrate, Milano
Italy

G. Bocchiola
CISE SpA
Environment Department
via R.Emilia 39
I-20090 Segrate, Milano
Italy

L. De Bock
Department of Chemistry,
University of Antwerp (U.I.A.),
B-2610 Antwerp-Wilrijk
Belgium

B. Bonsang
Centre des Faibles Radioactivités
Laboratoire mixte CNRS/CEA
F-91198 Gif-sur-Yvette
France

P. Borrell
P&PMB Consultants
Ehrwalder Straße 9
D-82467 Garmisch-Partenkirchen
Germany

W.J. Broadgate
School of Environmental Sciences
University of East Anglia
Norwich NR4 7TJ
UK

G. Brusasca
ENEL-CRAM
Via Rubattino 54
I-20134 Milano
Italy

P. Buat-Ménard
Dépt. de Géologie et Océanographie
Université Bordeaux 1
av. des Facultés
F-33405 Talence Cedex
France

H. Cachier
Centre des Faibles Radioactivités
Laboratoire mixte CNRS/CEA
F-91198 Gif-sur-Yvette
France

C. Cailliau
Lab. de Physique et Chimie Marines
Observatoire Océanologique de Villefranche
UPMC-CNRS-INSU, BP28
F-06230 Villefranche-sur-Mer
France

S. Cholbi
Laboratoire Interuniversitaire des
Systèmes Atmosphériques
Universités Paris 7 et Paris 12
Faculté des sciences
61 Av. du général de Gaulle
F-94010 Créteil cedex
France

U. Christaki
Centre des Faibles Radioactivités
CNRS-CEA
Av. de la Terrasse
F-91198 Gif-sur-Yvette
France

J. Christensen
National Environmental Research Institute
(NERI)
Frederiksborgvej 399
DK-4000 Roskilde
Denmark

J.L. Colin
Laboratoire Interuniversitaire des
Systèmes Atmosphériques
Universités Paris 7 et Paris 12
Faculté des sciences
61 Av. du général de Gaulle
F-94010 Créteil cedex
France

M. Corn
Centre des Faibles Radioactivités
CNRS-CEA
Av. de la Terrasse
F-91198 Gif-sur-Yvette
France

U. Corsmeier
Inst. für Meteorologie und Klimaforschung
Universität Karlsruhe / Forschungszentrum
Karlsruhe Postfach 3640
D-76133 Karlsruhe
Germany

W. Dannecker
Inst. für Anorganische und Angewandte Chemie
Abteilung Angewandte Analytik
Universität Hamburg
Martin-Luther-King-Platz 6
D-20146 Hamburg
Germany

O.F.X. Donard
Laboratoire de Photophysique et Photochimie Moléculaire
Université de Bordeaux I
351 Cours de la Libération
F-33405 Talence Cedex
France

J. Ducret
Centre des Faibles Radioactivités
Laboratoire mixte CNRS/CEA
F-91198 Gif-sr-Yvette
France.

J.C. Duincker
Institut für Meereskunde
D-24100 Kiel
Germany

A. Ebel
EURAD-Project, University of Cologne
Institute for Geophysics and Meteorology
Aachenerstraße 201-209
D-50931 Köln
Germany

R. Ebinghaus
GKSS Research Centre
D-21502 Geesthacht
Germany

J.B. Edson
Woods Hole Oceanographic Institution
Woods Hole
Maryland
USA

A.M.J. van Eijk
TNO Physics and Electronics Laboratory
P.O. Box 96864
NL-2509 JG The Hague
The Netherlands

N. Entstrasser
Institut für Physik der Atmosphäre
DLR
D-82234 Oberpfaffenhofen
Germany

H. Feldmann
EURAD-Project, University of Cologne
Institute for Geophysics and Meteorology
Aachenerstraße 201-209
D-50931 Köln
Germany

E. Ferrero
C.N.R. - Istituto di Cosmogeofisica
Corso Fiume 4
I-10133 Torino
Italy

F. Fiedler
Inst. für Meteorologie und Klimaforschung
Universität Karlsruhe / Forschungszentrum
Karlsruhe Postfach 6980
D-76128 Karlsruhe
Germany

S. Finardi
ENEL SpA
Environment and Material Research Centre
via Rubattino 54
I-20134 Milano
Italy

O.R. Flöck
Max Planck Institute for Chemistry
Biogeochemistry Dept.
P.O.Box 3060
D-55020 Mainz
Germany

F. François
Institute for Nuclear Sciences
Proeftuinstraat 86
B-9000 Gent
Belgium

R. Friedrich
Institute for Energy Economics
and the Rational Use of Energy
University of Stuttgart
Heßbrühlstraße 49a
D-70565 Stuttgart
Germany

L.L.S.Geernaert
Risø National Laboratory
DK-4000 Roskilde
Denmark

K. Granby
National Environmental Research Institute
(NERI)
Frederiksborgvej 399
DK-4000 Roskilde
Denmark

R. Van Grieken
Department of Chemistry,
University of Antwerp (U.I.A.)
B-2610 Antwerp-Wilrijk
Belgium

S.-E. Gryning
Risø National Laboratory (RISØ)
DK-4000 Roskilde
Denmark

P. Hakizimfura
Institut d'Astronomie et de Géophysique
Université catholique de Louvain
2 ch. du cyclotron
B-1348 Louvain-la-Neuve
Belgium

F.Aa. Hansen
Risø National Laboratory
DK-4000 Roskilde
Denmark

R.M. Harrison
Institute of Public & Environmental Health
School of Chemistry
University of Birmingham
Birmingham B15 2TT
UK

C.B. Hasager
Meteorology and Wind Energy Department
Risø National Laboratory
DK-4000 Roskilde
Denmark

H. Hass
EURAD-Project, University of Cologne
Institute for Geophysics and Meteorology
Aachenerstraße 201-209
D-50931 Köln
Germany

A.D. Hatton
School of Environmental Sciences
University of East Anglia
Norwich NR4 7TJ
UK

K. Hedegaard
Danish Meteorological Institute (DMI)
Lyngbyvej 100
DK-2100 Copenhagen
Denmark

O. Hertel
National Environmental Research Institute
(NERI)
Frederiksborgvej 399
DK-4000 Roskilde
Denmark

P. Hummelshøj
Meteorology and Wind Energy Department
Risø National Laboratory
DK-4000 Roskilde
Denmark

J. Injuk
Department of Chemistry
University of Antwerp (U.I.A.)
B-2610 Antwerp-Wilrijk
Belgium

H.J. Jakobs
EURAD-Project, University of Cologne
Institute for Geophysics and Meteorology
Aachenerstraße 201-209
D-50931 Köln
Germany

B. Jensen
National Environmental Research Institute
(NERI)
Frederiksborgvej 399
DK-4000 Roskilde
Denmark

N.O. Jensen
Meteorology and Wind Energy Department
Risø National Laboratory
DK-4000 Roskilde
Denmark

P. K. Jensen
National Environmental Research Institute (NERI)
Frederiksborgvej 399
DK-4000 Roskilde
Denmark

T.D. Jickells
School of Environmental Sciences
University of East Anglia
Norwich NR4 7TJ
UK

A.M. Jochum
Institut für Physik der Atmosphäre
DLR
D-82234 Oberpfaffenhofen
Germany

C. John
Institute for Energy Economics
and the Rational Use of Energy
University of Stuttgart
Heßbrühlstraße 49a
D-70565 Stuttgart
Germany

J. Jørgensen
Danish Meteorological Institute (DMI)
Lyngbyvej 100
DK-2100 Copenhagen
Denmark

N. Kalthoff
Inst. für Meteorologie und Klimaforschung
Universität Karlsruhe /
Forschungszentrum Karlsruhe Postfach 6980
D-76128 Karlsruhe
Germany

M.M. Kane
School of Environmental Sciences
University of East Anglia
Norwich NR4 7TJ
UK

H. -J. Kanter
Fraunhofer Institut für Atmosphärische
Umweltforschung
Kreuzbahnstr. 19
D-82467 Garmisch-Partenkirchen
Germany

N.A. Kilde
Risø National Laboratory (RISØ)
DK-4000 Roskilde
Denmark

A.-M. Kitto
Institute of Public & Environmental Health
School of Chemistry
University of Birmingham
Birmingham B15 2TT
UK

H.H. Kock
GKSS Research Centre
D-21502 Geesthacht
Germany

R. Koppmann
Institut für Atmosphärische Chemie
Forschungszentrum Jülich
D-52425 Jülich
Germany

M. Koßmann
Department of Geography
University of Canterbury
Private Box 4800
NZ-Christchurch
New Zealand

O. Krüger
GKSS Research Centre
D-21502 Geesthacht
Germany

J. Kuballa
GKSS Research Centre
D-21502 Geesthacht
Germany

K. Kuntze
Inst. für Meteorologie und Klimaforschung
Universität Karlsruhe / Forschungszentrum
Karlsruhe Postfach 6980
D-76128 Karlsruhe
Germany

J. Kuss
Institut für Meereskunde
D-24100 Kiel
Germany

S.E. Larsen
Department of Meteorology
Risø National Laboratory
DK-4000 Roskilde
Denmark

N. Le Bris
Laboratoire Interuniversitaire des Systèmes Atmosphériques
Universités Paris 7 et Paris 12
61 Av. du général de Gaulle
F-94010 Créteil cedex
France

M. Leermakers
Department of Analytical Chemistry
Free University of Brussels (VUB)
Pleinllan 2
B-1050 Brussels
Belgium

G. de Leeuw
TNO Physics and Electronics Laboratory
P.O. Box 96864
ML-2509 JG The Hague
The Netherlands

B. Lim
Centre des Faibles Radioactivités
Laboratoire mixte CNRS/CEA
F-91198 Gif-sur-Yvette
France.

C. Liousse
Centre des Faibles Radioactivités
Laboratoire mixte CNRS/CEA
F-91198 Gif-sur-Yvette
France.

P.S. Liss
School of Environmental Sciences
University of East Anglia
Norwich NR4 7TJ
UK

R.H. Little
School of Environmental Sciences
University of East Anglia
Norwich NR4 7TJ
UK

M. Löffler-Mang
Inst. für Meteorologie und Klimaforschung
Universität Karlsruhe /
Forschungszentrum
Karlsruhe Postfach 3640
D-76021 Karlsruhe
Germany

R. Losno
Laboratoire Interuniversitaire des Systèmes Atmosphériques
Universités Paris 7 et Paris 12
61 Av. du général de Gaulle
F-94010 Créteil cedex
France

S. Madec
Laboratoire Interuniversitaire des Systèmes Atmosphériques
Universités Paris 7 et Paris 12
61 Av. du général de Gaulle
F-94010 Créteil cedex
France

H. Madsen
Danish Meteorological Institute (DMI)
Lyngbyvej 100
DK-2100 Copenhagen
Denmark

W. Maenhaut
Institute for Nuclear Sciences
Proeftuinstraat 86
B-9000 Gent
Belgium

H. Van Malderen
Department of Chemistry
University of Antwerp (U.I.A.)
B-2610 Antwerp-Wilrijk
Belgium

G. Malin
School of Environmental Sciences
University of East Anglia
Norwich NR4 7TJ
UK

P. Marcacci
ENEL/SRI-CRAM
Thermal & Nuclear Research Centre
Via Rubattino 54
I-20134 Milano
Italy

J.C. Marty
Lab. de Physique et Chimie Marines
Observatoire Océanologique de Villefranche
UPMC-CNRS-INSU, BP28
F-06230 Villefranche-sur-Mer
France

A. Marzorati
ENEL/SRI-CRAM
Thermal & Nuclear Research Centre
Via Rubattino 54
I-20134 Milano
Italy

N.C. McArdle
School of Environmental Sciences
University of East Anglia
Norwich NR4 7TJ
UK

M. Memmesheimer
EURAD-Project, University of Cologne
Institute for Geophysics and Meteorology
Aachenerstraße 201-209
D-50931 Köln
Germany

P.G. Mestayer
Laboratoire de Mecanique des Fluides
Ecole Centrale de Nantes
B.P. 92101
F-44321 Nantes Cedex 3
France

C. Meuleman
Department of Analytical Chemistry
Free University of Brussels (VUB)
Pleinllan 2
1050 Brussels
Belgium

N. Mihalopoulos
Crete University
GK-71409 Heraklion, Crete
Greece

V. Mohnen
Fraunhofer Institut für Atmosphärische Umweltforschung
Kreuzbahnstr. 19
D-82467 Garmisch-Partenkirchen
Germany

G. Morselli
CISE SpA
Environment Department
via R.Emilia 39
I-20090 Segrate, Milano
Italy

M.I. Msibi
Institute of Public & Environmental Health
School of Chemistry
University of Birmingham
Birmingham B15 2TT
UK

K. Nester
Inst. für Meteorologie und Klimaforschung
Universität Karlsruhe /
Forschungszentrum Karlsruhe Postfach 3640
D-76021 Karlsruhe
Germany

B.C. Nguyen
Centre des Faibles Radioactivités
Laboratoire Mixte CNRS-CEA
Orme des Merisiers
F-91191 Gif-sur-Yvette
France

H. Nielsen
National Environmental Research Institute (NERI)
Frederiksborgvej 399
DK-4000 Roskilde
Denmark

P.D. Nightingale
School of Environmental Sciences
University of East Anglia
Norwich NR4 7TJ
UK

A. Obermeier
Institute for Energy Economics and the Rational Use of Energy
University of Stuttgart
Heßbrühlstraße 49a
D-70565 Stuttgart
Germany

C.J. Ottley
Institute of Public & Environmental Health
School of Chemistry
University of Birmingham
Birmingham B15 2TT
UK

S. Overgaard
Danish Meteorological Institute (DMI)
Lyngbyvej 100
DK-2100 Copenhagen
Denmark

C. Pecheyran
Laboratoire de Photophysique et
Photochimie Moléculaire
Université de Bordeaux I
351 Cours de la Libération
F-33405 Talence Cedex
France

U. Pellegrini
ENEL/SRI-CRAM
Thermal & Nuclear Research Centre
Via Rubattino 54
I-20134 Milano
Italy

M.H. Pertuisot,
Centre des Faibles Radioactivités
Laboratoire mixte CNRS/CEA
F-91198 Gif-sur-Yvette
France.

K. Pilegaard
Environmental Science and Technology
Department
Risø National Laboratory
DK-4000 Roskilde
Denmark

Ch. Plass-Dülmer
Meteorologisches Observatorium
Hohenpeissenberg
Albin-Schwaiger-Weg 10
D-82383 Hohenpeissenberg
Germany

J.-P. Putaud
Centre des Faibles Radioactivités
Laboratoire Mixte CNRS-CEA
Orme des Merisiers
F-91191 Gif-sur-Yvette
France

B. Quack
Institut für Meereskunde
D-21400 Kiel
Germany

S. Rapsomanikis
Max Planck Institute for Chemistry
Biogeochemistry Dept.
P.O.Box 3060
D-55122 Mainz
Germany

M. Ratte
Institut für Atmosphärische Chemie
Forschungszentrum Jülich
D-52425 Jülich
Germany

A. Rebers
Inst. für Anorganische und Angewandte Che
Abteilung Angewandte Analytik
Universität Hamburg
Martin-Luther-King-Platz 6
D-20146 Hamburg
Germany

A.R. Rendel
School of Environmental Sciences
University of East Anglia
Norwich NR4 7TJ
UK

J. Rudolph
Institut für Atmosphärische Chemie
Forschungszentrum Jülich
D-52425 Jülich
Germany

E.H. Runge
National Environmental Research Institute
(NERI)
Frederiksborgvej 399
DK-4000 Roskilde
Denmark

D. Sacchetti
C.N.R. - Istituto di Cosmogeofisica
Corso Fiume 4
I-10133 Torino
Italy

G. Schayes
Institut d'Astronomie et de Géophysique
Université catholique de Louvain
2, chemin du cyclotron
B-1348 Louvain-la-Neuve
Belgium

H. Schlager
Institut für Physik der Atmosphäre
DLR
D-82234 Oberpfaffenhofen
Germany

M. Schrader
Institut für Meereskunde
D-21400 Kiel
Germany

M. Schulz
Laboratiore des Sciences du Climat et de l'Evnironnement
CEA/CNRS - LSCE
L'Orme des Merisiers, Bat. 701
F-91191 Gif-sur-Yvette Cedex
France

J. Seier
Institute for Energy Economics and the Rational Use of Energy
University of Stuttgart
Heßbrühlstraße 49a
D-70565 Stuttgart
Germany

A.M. Sempreviva
Risø National Laboratory (RISØ)
DK-4000 Roskilde
Denmark

F. Slemr
Fraunhofer Institut für Atmosphärische Umweltforschung
Kreuzbahnstr. 19
D-82467 Garmisch-Partenkirchen
Germany

L.-L. Sørensen
National Environmental Research Institute (NERI)
Frederiksborgvej 399
DK-4000 Roskilde
Denmark

M.M. De Souza Sierra
Laboratoire de Photophysique et Photochimie Moléculaire
Université de Bordeaux I
351 Cours de la Libération
F-33405 Talence Cedex
France

L.J. Spokes
School of Environmental Sciences
University of East Anglia
Norwich NR4 7TJ
UK

T. Stahlschmidt
Inst. für Anorganische und Angewandte Chemie
Abteilung Angewandte Analytik
Universität Hamburg
Martin-Luther-King-Platz 6
D-20146 Hamburg
Germany

C. Strodl
Institut für Physik der Atmosphäre
DLR
D-82234 Oberpfaffenhofen
Germany

G.Tinarelli
ENEL s.p.A. - CRAM
via Rubattino, 54
I-20134 Milano
Italy

J. Tippke
EURAD-Project, University of Cologne
Institute for Geophysics and Meteorology
Aachenerstraße 201-209
D-50931 Köln
Germany

M. Tisserant
Laboratoire Interuniversitaire des
Systèmes Atmosphériques
Universités Paris 7 et Paris 12
61 Av. du général de Gaulle
F 94010 Créteil cedex
France

B. Tranchant
Laboratoire de Mécanique des Fluides
UMR CNRS 6598
Ecole Centrale de Nantes
F-44072 Nantes Cedex 03
France

S. Trini Castelli
C.N.R. - Istituto di Cosmogeofisica
Corso Fiume 4
I-10133 Torino
Italy

S.M. Turner
School of Environmental Sciences
University of East Anglia
Norwich NR4 7TJ
UK

G. Uher
University of Newcastle upon Tyne
Dept. Marine Sciences
and Coastal Management
Newcastle upon Tyne, NE1 7RU
UK

V.S. Ulshöfer
Max Planck Institute for Chemistry
Biogeochemistry Dept.
P.O.Box 3060
D-55020 Mainz
Germany

I. Vassali
Laboratoire Interuniversitaire des
Systèmes Atmosphériques
Universités Paris 7 et Paris 12
61 Av. du général de Gaulle
F-94010 Créteil cedex
France

F. Vejen
Danish Meteorological Institute (DMI)
Lyngbyvej 100
DK-2100 Copenhagen
Denmark

F. Vimeux
Laboratoire Interuniversitaire des
Systèmes Atmosphériques
Universités Paris 7 et Paris 12
61 Av. du général de Gaulle
F-94010 Créteil cedex
France

R. Vögtlin
Inst. für Meteorologie und Klimaforschung
Universität Karlsruhe / Forschungszentrum
Karlsruhe Postfach 6980
D-76128 Karlsruhe
Germany

H. Willeke
Institut für Physik der Atmosphäre
DLR
D-82234 Oberpfaffenhofen
Germany

Q. Xianren
Department of Analytical Chemistry
Free University of Brussels (VUB)
Pleinllan 2
B-1050 Brussels
Belgium

R.R. Yaaqub
School of Environmental Sciences
University of East Anglia
Norwich NR4 7TJ
UK

S. Yamulki
Institute of Public & Environmental Health
School of Chemistry
University of Birmingham
Birmingham B15 2TT
UK

H. Zimmermann
Institut für Meteorologie und Klimaforschung
Universität Karlsruhe / Forschungszentrum
Karlsruhe Postfach 6980
D-76128 Karlsruhe
Germany

Transport and Chemical Transformation of Pollutants in the Troposphere

Volume 9, Part I

Air–Sea Exchange

Søren Larsen (Roskilde)
Coordinator

ASE Steering Group

Willy Baeyens	Brussels
Sauveur Belviso	Gif-sur-Yvette
P. Buat Menard	Bordeaux
Olivier F.X. Donard	Talence
J.L. Colin	Creteuil
Roy M. Harrison	Birmingham
Gerrit de Leeuw	The Hague
Peter S. Liss	Norwich
Spyridon Rapsomanikis	Mainz
Michael Schulz	Hamburg

Chapter 1

Overview of Air-Sea Exchange

Søren E Larsen

Risø National Laboratory, Roskilde, Denmark

Substantial progress has been made in recent years in understanding the role of atmospheric transport of pollution for the marine ecosystems, as well as understanding the importance of the ocean as a source or sink for many of the important atmospheric trace gases influencing its composition and chemistry. The EUROTRAC subproject ASE has participated in and contributed significantly to this development. It has further contributed with results on a number of specific processes of importance for resolving the larger issues as well as with development of relevant measuring systems and computer programmes.

Introduction

Due to its huge surface and large biological activity, the global ocean is a significant and important source as well as sink for many substances of importance for atmospheric composition, chemistry and radiation balances. Although the coastal seas cover only a small fraction of the total area of the ocean, it is disproportionally more important for the balance of many atmospheric trace constituents, because the biological activity in the coastal ocean is very large, compared to the open sea. Simultaneously the coastal ocean matters relatively more to humanity, because much human activity takes place in the coastal region, and therefore the coastal seas act as the main depository for anthropogenic pollution. They also simultaneously act as the immediate upstream boundary for the atmospheric flow across the coastal lands.

Conforming to this dual role of the ocean, most of the activities within ASE have been focused either on estimating the role of atmospheric deposition on the pollution of the coastal ocean or on the role of the ocean as a source/sink for trace constituents of importance in atmospheric chemistry.

Aims of the subproject ASE

The aim of ASE has been to study the air-sea exchange of material relevant to atmospheric chemistry in general, and specifically to the transport and transformation of pollutants across sea surfaces. Substances of interest have included both natural and man-made compounds, with the exchange being considered between the gas phase, particulate particles or from rain. The programme has concentrated on estimating emissions, atmospheric transformations and deposition of sulfur and nitrogen species, low molecular weight natural organo-halogen and hydrocarbon gases in the marine atmosphere, and on quantification of atmospheric reactivity of particulate material as well as emission from and deposition to the sea surface. The research has involved direct field measurements, as well as laboratory and field studies of the mechanisms controlling emission, deposition and atmospheric transformation processes. Regions studied have included the inner Danish waters, the North Sea, the Mediterranean and the eastern North Atlantic.

Principal scientific activities and results

Structure of ASE and of this report

The ASE subproject was originally subdivided into five topic areas:

1 the contribution of biology and photochemistry of surface waters to sea-to-air trace gas fluxes;

2 trace gas emission measurements from surface sea water;

3 the role of atmospheric transformations of trace gases and aerosols in marine deposition processes;

4 the magnitude of the air-sea fluxes of natural and anthropogenic substances to European regional seas;

5 factors determining particle dynamics near air-sea surfaces.

However, as is apparent, there is an underlying unity to the five topics, since as their primary concern all have some aspects of mass transfer across the sea surface and its role in atmospheric chemistry and oceanic chemistry and biology. Therefore, in somewhat broader categories, the studies of the subproject centred around studies of processes of importance for deposition to the sea of atmospheric substances, and on studies of processes involved in exchange of biogases between the ocean and the atmosphere. This grouping was to some extent forced upon the subproject by the interests of the funding communities, and therefore on the form and focus of those ASE projects that were already organised when the project was originally formulated, and also on the joint and individual activities undertaken by the ASE groups during the lifetime of the project. With few exceptions these categories matched the working focus of the participating groups, in the sense that the first group involved studies of atmospheric processes, while the second group made most of their measurements in the liquid phase. This part therefore is organised into two further chapters according to these broad categories. Each chapter is introduced by several overview papers summarising the major results of the subproject, followed by contributions from individual groups. Some of the major joint activities conducted within ASE are also described.

Results and activities of aerosols and deposition to marine systems

An important aspect of the ASE activities has been the effort to establish the magnitude and importance of atmospheric deposition of pollutants to European coastal waters. Some ASE projects have focused on the load of heavy metals from the atmosphere to the marine ecosystems, and others on the deposition of nutrients in the form of nitrogen compounds which give rise to many of the eutrophication events that have occurred in the enclosed northern European seas.

The research within ASE has to a large extent been responsible for the recognition of the importance of the atmospheric deposition to the pollution of the marine ecology over the North Sea and the inner Danish waters. Atmospheric transport has been found to account for, typically, 30 % or more of the total load for the substances studied here. Since the atmospheric contribution is generally more bio-available and is delivered to the surface where the largest biological activity takes place, the ecological significance of the atmospheric contribution is larger than indicated by its mass ratio. Simultaneously with the research within ASE, researchers in other regions of the world have found similar results for the importance of the atmospheric deposition to coastal seas in the vicinity of large industrialised regions.

Aspects of these results are reported by many of the groups, and are placed in a general framework in the review by T.D. Jickells and L.J. Spokes (see Chapter 2.1).

Many detailed studies have been carried out within this subject, both to ensure that necessary data was obtained, and also to clarify relevant questions concerning the transport involved, the chemical transformation and wet and dry deposition processes. These activities and results are described in the contributions from the individual groups, together with a couple of larger joint ASE activities.

One group of have studied the production, dynamics and role of marine aerosols, this is summarised by De Leeuw *et al.* (see Chapter 2.3). The production of marine aerosols through various forms of droplet formation is important, both for the processes in the liquid phase controlling the exchange of gases across sea surfaces, and for the atmospheric processes involving multi-phase chemical transformations, as well as for deposition of aerosols and water-soluble gases to the ocean surface. Within ASE however, only atmospheric aspects of the formation of spray droplets were studied.

An important aspect of the research within ASE has been the joint experimental work over the North Sea, which has involved intercalibration activities between the different groups at Mace Head in Ireland, and the 1991 North Sea Experiment, involving two ships and an effort to conduct a pseudo-Lagrangian experiment between these two ships. A smaller measuring campaign on the German research platform, Nordsee, was conducted just before the platform was dismantled. The main activity here, the NOSE experiment is described in overview paper by Schulz (see Chapter 2.2) .

Results on biological processes and the air-sea exchange of biogases

The ASE subproject has aimed to increase our knowledge about the role of the open ocean, as well as coastal waters, as sources and sinks for a number of gases of biological origin and atmospheric importance, and those with a biological component in their production/destruction cycle. Also here ASE groups have participated in, as well as contributed significantly to, the generally increase of knowledge about the source/sink cycles of many of these gases.

DMS (dimethylsulfide) is probably one of the biogases that has been most extensively studied recently, both for its contribution to the global sulfur budget and for its possible significant contribution to the production of cloud condensation nuclei, through oxidation in the atmosphere. Here use of isotope signatures (Liss *et al.*, see Chapter 3.3) have indicated that about 30 % of the non-sea-salt sulfate in aerosols was derived from oxidation of DMS, but that this ratio must be considered highly variable. The pattern of DMS production in the surface waters has been mapped across the eastern Atlantic margin, showing that a production maximum in the spring follows the shelf break, where the biological production is highest. The marine cycle of DMS has been studied extensively, as has also the cycle of other marine trace constituents such as selenium. Another important contributor to the budget of atmospheric condensation nuclei is carbonyl sulfide, COS, for which the air-sea exchange budget has been determined in the Mediterranean and the eastern Atlantic.

Important natural atmospheric oxidising gases in the marine atmosphere are organo-halogens and non-methane hydrocarbons. The air-sea exchange for both, but especially the last species, have been extensively mapped and the ocean found to be a significant source for atmospheric constituents of these gases, as reported in Chapter 3.1. A review presentation on the current knowledge on the ocean's role in the source/sink cycles for a number of the more important biogases is described by Rapsomanikis.

Positive and negative experiences with ASE

Within its lifetime, the ASE project has had the privilege to participate in, and contribute to, the strong development of knowledge about the exchange processes occurring across the ocean-atmosphere interface, which has taken place during this period, both with respect to scientific insight into the exchange processes, and with respect to knowledge about the absolute and relative importance of these exchange processes for a number of environmental ocean and atmospheric issues on global as well as local scales. This development has occurred simultaneously and to a large extent fuelled by the growing awareness in society about such environmental issues. The general awareness that we do not know enough about the global environment has led to questions on air-water pollution for the coastal countries of Europe.

ASE has provided a framework for multi-disciplinary research groups to study the environmental issues as they became apparent. In this context, the main handicap of ASE has been that it was never able to secure any long-term funding aiming specifically at the ASE objectives and it proved difficult to maintain a well defined ASE profile. In spite of this, groups participating in the subproject, found a very effective forum for the formulation of new concepts.

Acknowledgements

Support of the individual ASE contributions is acknowledged in the reports of the principal investigators. Here we wish to acknowledge the repeated support of EC-Environment, through the IGAC office, to the annual ASE workshops, hereby contributing significantly to the scientific productivity and continuing development of ideas and concepts within the subproject.

Chapter 2

Aerosols and Deposition to Marine Systems

2.1 A Review of the Magnitude and Significance of Atmospheric Inputs of Metals and Nutrients to Marine Ecosystems

T.D. Jickells and L.J. Spokes

School of Environmental Sciences University of East Anglia,
Norwich NR4 7TJ, UK.

One of the overall aims of the EUROTRAC ASE project has been to study the transport and transformation of material across the air-sea interface. In this short contribution we try to briefly review the state of knowledge of the scale and effects of atmospheric deposition to the marine environment to provide a context for the individual EUROTRAC reports.

Review

Over the time of the EUROTRAC project, the critical importance of atmospheric transport of metals and fixed nitrogen (*i.e.* nitrogen other than N_2 gas) to the marine environment has come to be recognised much more widely than before. Two developments in particular have contributed to this. The first such development was the work of the GESAMP Studies of the State of the Marine Environment, particularly the review of atmospheric inputs (published in an abbreviated form by Duce *et al.* [1]). This provided very well documented evidence of the magnitude of atmospheric inputs in comparison to riverine fluxes. These results are summarised in Table 2.1.1 and emphasise that for many metals and for fixed nitrogen the atmosphere contributes as much material to the oceans as rivers, or in some cases substantially more. Furthermore, while the riverine inputs occur at the coast, the atmospheric input is much more widely dispersed. Hence away from the coastal margins, the atmospheric input is dominant. The metals travel predominantly in the aerosol phase (apart from Hg) while for fixed nitrogen both gas phase species (nitrogen oxides and ammonia for example) and aerosol phases are important. In prehistoric times, the main source of metal transport to the oceans was via crustal dust transport. The magnitude of this dust flux has changed over time as a result of human activity, via changing farming practices, and through changes in climate, since most dust arises from the desert. For some metals (*e.g.* Al and Fe) this dust source still dominates the input to the oceans and therefore the flux of metals associated with the dust will also have varied over time. However, as discussed by Jickells [2], the atmospheric fixed nitrogen fluxes and those of many trace metals (*e.g.* Pb, Cd and Zn) have been massively increased as a result of human activity, particularly via combustion processes. Thus the current picture implied by Table 2.1.1 is markedly different to that in prehistoric times. Furthermore, the release of metals via combustion sources results in metals associated with aerosols of generally smaller size and different chemical composition to those associated with crustal material, which means that both the deposition and solubility of this material is markedly altered. The scale of atmospheric perturbation of these metal inputs is eloquently recorded in the lead concentration of annual growth rings of calcium carbonate laid down by corals. These are believed to faithfully record the surface water lead concentrations, which are themselves derived almost entirely from the atmosphere. These records show a rapid increase in lead from 1945 to the late 1970s as a result of increased emissions from cars, followed by a decline in recent years with the introduction of unleaded petrol in Europe and North America [3]. The scale of chemical perturbation of the oceans arising from atmospheric inputs has been clearly documented (see Jickells [2] for review). Recently it has become clear that even in coastal areas where riverine inputs exert their maximum effect, atmospheric inputs still contribute substantial amounts of material (*e.g.* North Sea [4] and the Mediterranean [5]). Since such coastal waters are used for waste disposal, mining, fishing and recreation, it is vital that their environmental status be fully evaluated and pollution threats, including atmospheric inputs, be managed carefully.

Table 2.1.1: Comparison of present day fluxes of fixed nitrogen and various metals to the global oceans (10^9 g yr^{-1}) From Duce *et al.* [1].

Element	Riverine Flux*	Atmospheric Flux
N (excluding N_2	(21–50) × 10^3	30 × 10^3
Pb	2	90
Cd	0.3	2.3–4
Zn	6	44–230
Fe	1.1× 10^3	32 × 10^3

*Note only dissolved river flux is considered here since most riverine particulate material is deposited in estuaries.

The second major development which forced increasing attention on the atmospheric inputs of metals to the ocean was the "Iron Hypothesis" of the late John Martin [6]. Martin argued that, in certain areas of the ocean, iron supply limits primary productivity. This iron supply is overwhelmingly atmospheric in origin for central ocean regions. Thus the ocean areas most susceptible to iron limitation are those furthest from land, and in particular those furthest from arid land areas. Martin went on to suggest that during the last ice age atmospheric dust inputs to the oceans were higher, due to stronger winds and increased aridity, and hence productivity would also have been higher. Such an increase in productivity could then contribute to the lower atmospheric CO_2 levels characteristic of the ice ages. Thus iron supply to the oceans could be an important factor in the regulation of climate. Within the last year some experimental evidence that iron addition can increase primary productivity has become available but this is still controversial [7]. It has also been suggested that other metals including Co, Zn, Cu and Mn may influence marine productivity (see Jickells [2] for review). In addition Zhuang *et al.* [8] have suggested an alternative mechanism by which atmospheric inputs of metals can influence climate. They suggested that atmospheric inputs of metals could enhance the biological production of dimethyl sulfide (DMS) from sea water, resulting in increased emissions to the atmosphere. Once in the atmosphere, DMS is photolysed in a series of reactions with the resultant formation of sulfuric acid which can go on to produce cloud condensation nuclei, and thus regulate climate by influencing albedo. This linkage of sulfur and metal cycles may yet be more complex. The formation of sulfuric acid from DMS may in part be catalysed by transition metals while the acidity in the clouds affects the solubility of the metals and hence their subsequent bioavailablity [9–10]. Bioavailablity is also a complex function of processes occurring at or below the air-sea interface [2] which are outside the scope of this short review. Thus it is clear that atmospheric transport is a major component of global biogeochemical cycles and dominates the input of many species to the oceans. Furthermore these inputs can have a direct effect on oceanic productivity and on climate. They also provide the most rapid mechanism for the long range transport of contaminants. However, despite the growing awareness of the importance of atmospheric transport to marine systems, the accurate quantification of this input is still difficult, and the ultimate goal of adequately modelling the transport and deposition of this material, still illusive. There are a number of impediments to progress. The first is simply one of insufficient data. The dynamic nature of the atmosphere means that long sampling campaigns are needed to adequately estimate average fluxes. In this context, the coastal areas of particular concern within EUROTRAC represent a particularly challenging environment because of their complex meteorology and steep concentration gradients, perhaps best exemplified by the Mediterranean. Here atmospheric inputs associated with both desert dust storms from the south and industrial contamination from the north, meet and interact (*e.g.* Remoudaki *et al.* [11]). Due to its high cost and short duration, ship board sampling often cannot provide sufficient data for average flux calculations. The alternative of shore based sampling requires an extrapolation which may or may not justified. For instance Kane *et al.* [12] have shown that for most species, extrapolation from land based coastal sites is appropriate but not for nitrogen species, principally because of the interactions of nitric acid with sea salt ([13]; Harrison *et al.* this volume; Jickells *et al.* this volume). The work of the EUROTRAC ASE group has contributed significantly to improving our knowledge of the complexity of atmospheric transport and deposition in the marine environment. For many areas we do now have good flux estimates [2, 4, 5] and in some cases we can predict the bioavailablity of this flux [9]. However, much remains to be done before the net effects of these inputs on the oceans are properly understood, leave alone the challenges of predicting the interactions in the past, for instance during the last glaciation, or in the future as the climate changes due to human activity.

Acknowledgements

This work was supported under NERC grant number GR3/8404A.

References

1. R.A. Duce *et al.*; The atmospheric input of trace species to the world ocean *Global Biogeochem. Cycles* **5** (1991) 193–259.
2. T. Jickells; Atmospheric inputs of metals and nutrients to the oceans: their magnitude and effects. *Marine Chem.* **48** (1995) 199–214.
3. G.T. Shen, E.A. Boyle; Lead in corals: reconstruction of historical industrial fluxes to the surface ocean. *Earth Planet. Sci. Lett.* **82** (1987) 289–304.
4. R. Chester *et al.*; The atmospheric distributions of trace metals, trace organics and nitrogen species over the North Sea. *Phil. Trans. Roy. Soc. London* **A 343** (1994) 543–556.
5. J.M. Martin, F. Elbaz-Paulichet, C. Guieu, M-D. Loye-Pilot, G. Han; River versus atmopheric iputs of material to the Mediterranean Sea: an overview. *Marine Chem.* **28** (1989) 159–182.
6. J.H. Martin; Glacial-interglacial CO_2 change: the iron hypothesis. *Palaeoceanography* **5** (1990) 1–13.
7. J.H. Martin *et al.*; Testing the iron hypothesis in ecosystems of the equatorial Pacific Ocean. *Nature* **371** (1994) 123–129.
8. G. Zhuang, Z. Yi, R.A. Duce, P.R. Brown; Link between iron and sulfur suggested by tehdetection of Fe(II) in mrine aerosols. *Nature* **355** (1992) 537–539.
9. B. Lim, T.D. Jickells, J.L. Colin and R. Losno Solubilities of Al, Pb, Cu and Zn in rain sampled in the marine environment over the North Atlantic Ocean and Mediterranean Sea. *Global Biogeochem. Cycles* **8** (1994) 349–362.
10. L.J. Spokes, T.D. Jickells, B. Lim; Solubilisation of aerosol trace metals by cloud processing: a laboratory study. *Geochim. Cosmochim. Acta* **58** (1994) 3281–3287.
11. E. Remoudaki, G. Bergaetti, P. Buat-Menard; Temporal variability of atmospheric lead concentrations and fluxes over the northwestern Mediterranean. *J. Geophys. Res.* **96** (1991) 1043–1055.
12. M.M. Kane, A.R. Rendell and T.D. Jickells. Atmospheric scavenging processes over the North 5ea. *Atmos. Environ.* **28** (1994) 2523–2530.
13. C.J. Ottley, R.M. Harrison, The spacial distribution and particle size of some organic nitrogen, sulfur and chlorine species over the North Sea, *Atmos. Environ.* **26** (1992) 1687–1699.

2.2 The North Sea Experiment 1991 (NOSE): A Lagrangian-type experiment

M. Schulz

Institut für Anorganische und Angewandte Chemie, Universität Hamburg,
Martin-Luther-King-Platz 6, D-20146 Hamburg, Germany

Introduction

The direct measurement of air-sea-exchange fluxes of atmospheric pollutants is very difficult. Especially dry deposition can only be assessed in an indirect way. Highest fluxes are associated with rough sea conditions. These bring about considerable experimental difficulties. A mass budget approach may help in this respect in that it offers a realistic test of our general understanding of transport and deposition. To arrive at mass budgets for a transported air mass was the primary motivation for the Lagrangian-type NOSE experiment which is described in this chapter. The ASE - North Sea Experiment 1991 combined atmospheric measurements at two ships with a multi-station network at the coast around the North Sea. The two research vessels were during two weeks repeatedly realigned in wind direction and positioned 200 km apart from each other in the middle of the North Sea. By this approach we attempted to sample the same air mass twice during its transport over sea. Since absolute changes in atmospheric concentration might also result from dilution, emphasis was put on detecting relative changes of the composition of the air mass. A change in size distribution *e.g.* could tell whether air-sea-exchange processes would affect all size classes of the aerosol in the same way.

The Lagrangian experimental set-up is not useful to study the removal of atmospheric pollutants by wet deposition. Frontal passages and shower weather extend characteristically to three dimensions, bringing along rapid turns in wind direction. These allow no useful interpretation of ground level concentration differences from two stations. However, it was felt that the experiment provided a good chance to explore basin wide single wet deposition events. The only requirement was a synchronous timing of the sampling and harmonised analytical schemes. Such data sets did not exist and hence would improve our knowledge on deposition gradients in the North Sea region.

Prior to the NOSE experiment an intercalibration experiment was performed at Mace Head, Ireland. The timing between this and the NOSE experiment was, due to limited financial resources, very tight. Not all groups could participate and thus not all instruments were checked beforehand. Problems detected in Mace Head with regard to sampling and analytical precision of the measurements could not be solved to the extent, which would have been desirable for a perfect mass budget experiment. The intercomparisons achieved in Mace Head demonstrated that to measure the aerosol and its constituents, as well as rain water concentrations (especially trace metals) is still no trivial task, especially in the marine environment. Problems can be due to contamination problems but very often are equally due to sampling artefacts in windy conditions.

Since altogether 13 research groups from 6 countries have participated in the experiment only a broad overview and references to published work can be provided here. This overview is organised along the division in different atmospheric parameter classes. Entrainment at the top of boundary layer as a source of dilution appeared to be a major point of concern to the original concept. Entrainment has been suggested to be responsible for the significant decrease in concentration found between the upwind and the downwind ship positions for many of the atmospheric parameters. Recommendations for future Lagrangian experiments are derived.

Scientific findings and results

To arrive at a budget for the atmospheric constituents the difference in ambient air concentration between the upwind and the downwind ship was used as a basis for all further calculations. Table 2.2.1 provides an overview of concentrations. Since this experiment was essentially new in concept, many findings show up only qualitatively. Due to differing sampling frequencies as a result of difficult operating conditions during the two storms not all parameters can be easily compared. The number of 6-hourly intervals covered by measurements for any parameter is therefore attached in Table 2.2.1.

The 6h-intervals formed the basic time unit, where ships stayed at one position, heading in the wind and thus allowed contamination free sampling. Table 2.2.1 also reports on the removal rates and first order loss rates as inferred using missing layer height as estimated from radio soundings. Note that removal rates can not be interpreted as deposition velocities due to the important entrainment process. For discussion see the individual sections.

a) Trace element aerosol concentration and aerosol size distribution

M. Schulz, T. Stahlschmidt, University of Hamburg, Germany
W. Maenhaut, F. Francois, University of Gent, Belgium
J. Injuk, H. van Maldern, R. van Grieken, University of Antwerp, Belgium

Three sets of identical impactors, stacked filter units and bulk samplers have been utilised on both ships to measure trace metal size distributions and concentrations. Trace elements in the aerosol were analysed by a variety of methods such as TXRF, NAA, PIXE and EDXRS.

Measurements showed varying but largely corresponding losses of trace metals from the upwind to the downwind ship. The comparison for the most complete data set of University of Gent and the University of Hamburg showed for the downwind/upwind concentration ratios values from 0.4–0.9 for the different elements.

Table 2.2.1: Mean ambient air concentrations during the NOSE experiment at the upwind (Alkor) and downwind (Belgica) ship position 16–26.9.91. First order loss was assumed in a boundary layer of 1000 m height during 4.9 h air mass travel time to calculate the removal rates.

	Alkor	Belgica	Concentration decrease	Removal rate	Intervals covered
	Concentrations		[%]	$cm\ s^{-1}$	#
Harrison *et al.* [1]	LMA-3 and denuder/filter packs				
NO_2 ppb	4.64	2.01	56.6	4.70	17
*NO_{2day}	4.28	1.90	55.5	1.98	7
*NO_{2night}	4.69	2.13	54.5	1.86	6
SO_4^{2-} $\mu g\ m^{-3}$	3.13	1.98	36.7	2.57	12
$\Sigma(N0_3^-,HNO_3)$	~5.0	~3.0	40.0	2.70	12
*$\Sigma(NH_4^+,NH_3)$	~0.9	~0.3	64.7	2.10	11
* corrected for entrainment as derived from sulfate removal					
Schulz *et al.* [2]	Berner impactor				
Cl^- $\mu g\ m^{-3}$	4.27	4.47	–4.6	–0.26	24
NO_3^-	3.31	3.28	1.1	0.07	24
SO_4^{2-}	3.88	3.23	16.6	1.02	24
Baeyens *et al.* [3], Ebinghaus *et al.* [4]			bulk sampler		
Hg ng m^{-3}	1.5	1.10	26.6	1.74	24
Francois *et al.* [5]		stacked filter unit			
As ng m^{-3}	0.25	0.16	36.0	2.51	24
Fe	68.8	40.50	41.1	2.98	24
Mn	3.29	1.96	40.4	2.91	24
Pb	8.18	5.77	29.4	1.96	24
Se	0.74	0.51	31.0	2.09	24
S	940.9	917.40	2.5	0.14	24
Quack [13]	bulk absorption sampler				
Cl_3C_2H ng m^{-3}	76.6	33.40	56.4	4.66	25
Br_2ClCH	2.2	1.00	54.5	4.43	25
1,2 Br_2CH_2	1.5	0.80	46.6	3.53	25
Br_3CH	25.9	16.30	37.0	2.60	25
1,2 $Cl_2C_2H_4$	56.8	38.60	32.0	2.17	25
Cl_4C_2	98.7	75.20	23.8	1.53	25
Br_2CH_2	9.3	11.00	–18.2	–0.94	25
$BrCl_2CH$	1.7	1.90	–11.7	–0.62	25
$Cl_3C_2H_3$	2.1	3.10	–47.6	–2.19	25

University of Hamburg values showed slightly higher removals, which was attributed to the different inlet characteristics. When comparing these ratios from individual intervals, a large scatter is found. The data set was used to investigate possible reasons for this scatter [6]:

1) A displacement of the ships from the ideal trajectory could have resulted in concentration changes due to plumes entering the North Sea. This effect could be ruled out because of the good positioning of the ships in combination with dispersion assumptions of a common anthropogenic trace metal, Pb.

2) Also the actual timing of the intervals was investigated, since the integration over time would not be correct at both ships, if concentration fluctuations would dominate the integral value. On the basis of high time resolved particle concentration data this could be excluded.

3) Sedimentation of large particles would favour the removal of varying fractions of trace metal mass. Computation revealed that indeed 7 % of the concentration decrease can be explained by gravitational settling.

However, the removal rates for five selected elements (As, Fe, Mn, Pb, Se), even after accounting for sedimentation, still showed some dependency on particle size. For Fe and Mn this was most obvious. Removal rates for the fraction corresponding to a diameter of 0.35 μm where only 0.2 cm/s while those at 5.5 μm were removed with a velocity of 1.7 cm/s. For Pb, only the very small particles were removed slower then the larger ones. This smaller removal rate of the very fine particles challenges our understanding of the aerosol transformation and deposition processes. It might indicate that entrainment was not as important as suggested by other parameters.

b) Sulfur compounds

M. Schulz, University of Hamburg, Germany
W. Maenhaut, F. Francois, University of Gent, Belgium
R.M. Harrison, M.I. Msibi, A.-M. Kitto, S. Yamulki, University of Birmingham, UK;
L.-L. Sørensen, W. Asman, National Environmental Research Centre, Roskilde, Denmark

Sulfate, respectively sulfur was measured with two types of impactor (Berner and PIXE), stacked filter units and filter packs. The impactor samples span the whole period [2, 5]. No significant decrease in concentration was observed for both independent samplings. The size distributions measured on both ships were very similar for both impactors, varying on the other hand significantly from sample to sample. These findings led to the assumption that removal and new production of sulfate was balanced during transport over sea during NOSE. Another hypothesis suggested that the entrained air carried already some long-range transported aged sulfate and could thus not dilute the boundary layer air to the extent suggested for other substances.

Harrison *et al.* [1] during 10 sampling intervals measured the mean non-sea salt sulfate concentration to decrease by 37 %. Assuming a first order loss in a boundary layer of 1000 m for a mean travel time of 4.9 h an interfacial transfer velocity for the entrainment process of 2.6 cm/s was calculated.

c) Nitrogen compounds

R.M. Harrison, M.I. Msibi, A.-M. Kitto, S. Yamulki, University of Birmingham, UK
L.-L. Sørensen, W. Asman, O. Hertel, National Environmental Research Centre, Roskilde, Denmark
M. Schulz, Univ. of Hamburg, Germany
H. Bange, S. Rapsomanikis, MPI für Chemie, Mainz, Germany
O. Krueger, GKSS, Geesthacht, Germany

Measurements comprised both the measurement of gaseous compounds NO_x, NH_3, N_2O and the particulate constituents NH_4^+ and NO_3^-. The measurements at sea were supported by coastal station measurements notably at the Danish coast at Husbyklit, being most of the time downwind of the two ships.

Nitrogen dioxide was measured almost continuously on both ships. On the average the concentration decreased by 67 % on the downwind ship[1]. Since dry deposition and precipitation scavenging is negligible, such substantial changes in concentration must be due to photochemical decay or entrainment. After considering a first order loss from sulfate (see above) the residual was considered to be due to decay by atmospheric chemistry processes. Three major decay pathways have been discussed by Harrison and co-workers: During daytime the reaction of NO_x with OH radicals can well explain the mean loss. Equally during night time a decay by reaction with available ozone has been shown to be sufficient to explain the only slightly slower decay during night. It was thus not necessary to invoke the occurrence of the reaction of NaCl with NO_2 to explain the concentration changes.

Bulk particulate nitrate and nitric acid have been measured by a combination of annular denuders and filter packs. Because of the ready interconversion of nitric acid and particulate nitrate it is not practicable to consider each independently in terms of changes occurring between the two ships. Loss rates are rather similar to that of sulfate, and entrainment should be responsible for the concentration decrease observed. This implies that dry deposition removal and nitrate formation from NOx were equal in magnitude [1].

Size distributions of particulate nitrate have been measured with a Berner impactor [2]. Though integrated over several 6h-intervals the samples cover 24 intervals and are thus not easy to compare with the above mentioned data set. Compared to sulfate, the size distributions differ slightly more between the two ships. However, these shifts are small. Slight increases in coarse nitrate were observed. As for the trace metals the size distributions do however change from sample to sample. This confirms that the same air mass was measured. It further implies that the shape of the size distribution evolved closer to the sources.

Absolute changes in concentration measured were small. The size segregated anions measured by the Berner impactor thus belong to the few parameters not showing a large decrease in concentration. This may not be in contradiction to the bulk data since different intervals were integrated. Compared to the trace metal measurements, which span the same period, it seems that formation of nitrate outweighed the depletion by entrainment and deposition.

Two models were used to predict the evolution of the nitrogen compounds during the NOSE period [7, 8]. For this purpose both at the ship and at the coastal station Husbyklit NH_3 could be measured with a new continuous scrubber technique in addition to the denuders [9]. Both models could well reproduce the pollution episode around the 21st September and reproduced the concentrations with random fluctuations being most of the time in agreement with the measurements within a factor of 2. However, the steady westerly winds certainly facilitated the simulation. Most remarkable is the overestimation of the HNO_3 and NH_3 concentrations at the downwind coastal station at the coast of Denmark [7]. Possibly the Lagrangian model missed either the entrainment process and/or large removal of these two reactive components.

Measurements of N_2O could unfortunately not be performed on both ships. However, the measurements on the Alkor served for the establishment of a data base on the oceanic emissions of this greenhouse gas [10].

d) Mercury compounds

W. Baeyens, M. Leermakers, Free University of Brussels, Belgium
R.. Ebinghaus, H.H. Kock, J. Kuballa, GKSS Geesthacht, Germany

Total mercury was measured on both ships. The concentrations were found to be very low, indicating a substantial contribution of clean background air. A slight decrease was also found for mercury between the Alkor and the Belgica [3, 4]. This was a little surprise, because entrainment should only decrease the concentration, if the entrained air has substantially lower concentrations. Since the concentrations were already low, this decrease is a surprising indication of very low air concentrations in the free troposphere. However, it could also indicate deposition of a fraction. Emission from the sea is not very likely to be of importance in the cases studied.

e) Individual particle composition

J.H. Injuk, L. De Bock , H. van Maldern, R. van Grieken, Univ. of Antwerp, Belgium

Variations in the composition of individual aerosol particles was studied by performing electron-probe X-ray microanalysis (EPXMA) on particles > 1μm. Samples were taken size segregated in an isokinetically operated 6-stage May impactor. Altogether 69 aerosol samples were analysed [11, 12]. Multivariate techniques were used for the reduction of the data set. Based on the relative abundance's of different particle types, found by hierarchical cluster analysis, three to eight different aerosol types were distinguished. (freshly produced and aged NaCl; Ca and S-rich; sea water recrystallisation; aluminosilicates as fly ash and soil dust; Si-rich particles, Fe-rich particles and organic particles). After crossing the North Sea most notably large aluminosilicate particles decreased in abundance. This would be in accordance with significant sedimentation of the large particles as discussed above. NaCl and sea water recrystallisation products on the other hand increased. Aged sea salt disappeared and was replaced by pure NaCl particles. Sea salt particles in general dominated the particle spectrum. Half of the samples did not contain organic particles.

f) Light halogenated organic compounds

B. Quack, J. Kuss, J.C. Duincker, Institut für Meereskunde, Kiel, Germany

Air samples have also been analysed on both ships for a set of light halogenated hydrocarbons [13]. Since these substances may also be re-emitted from the sea surface, measurements of sea water concentrations were performed at the same time [14]. A data set of nine substances has been identified to be most reliable in terms of analytical quality. Samples were taken for about 1–3 hours and were paired for a mass budget analysis. Because of the shorter measuring intervals short term fluctuations probably contributed to the concentration differences. However, for six of the substances a significant decrease in concentration was measured. According to calculations of the equilibrium across the air-sea interface, the sea acted for all substances except 1,2-Br_2CH_2, $BrCl_2CH$ and 1,2-$Cl_2C_2H_4$ as a source. Mass fluxes from this air sea exchange contribution were taken into account when evaluating the overall change in concentration during transport over sea. Highest sea-to-air fluxes are reported for Br_2CH_2, $Cl_3C_2H_3$, and Br_2ClCH contributing about 10 % to the total mass of these constituents in the boundary layer. After having estimated sea-to-air fluxes the overall decrease in concentration can be expressed as removal rates, being of the order of 2.5–3.9 cm/s for 1,2-Br_2CH_2, Br_3CH, Cl_3C_2H and Br_2ClCH. These high removal rates may serve as a maximum entrainment velocity. It may also be possible that photochemical destruction contributed to the loss observed. Less removal was observed for 1,2-$Cl_2C_2H_4$ and Cl_4C_2 being in the order of 1.7–1.2 cm/s which is in accordance with their smaller susceptibility to photochemical attack. The remaining substances showed slight increases in concentration (Br_2CH_2, $Cl_3C_2H_3$, and $BrCl_2CH$) in accordance with a marine source hypothesis.

g) Wet deposition of major ions and trace metals

T. Jickells, L. Spokes, University of East England, Norwich, UK
M. Schulz, A. Rebers University of Hamburg, Germany
M. Leermakers, W. Baeyens, Free University of Brussels, Belgium

Atmospheric inputs via wet deposition are highly variable in space and time. It was thus not expected that measurements at the two ships would easily correspond to each other. Spatial information on the spread of atmospheric deposition are very scarce. The coastal network of 8 stations around the North Sea together with the two ships in open sea served ideally for the purpose of investigating this spatial variability. Although it rained several times during NOSE, only during one period widespread rain was observed and could be linked to a common frontal passage. A case study has thus been performed for this event on the 21st/22nd September [15, 16]. Sodium concentrations increased in rains at sea and were higher on downwind coasts. Total acid soluble trace metals (*e.g.* Pb and Fe) from both anthropogenic and crustal sources showed comparatively small variation particularly in the context of the relatively large analytical/sampling variability observed during the intercomparisons. In contrast the acid salts (ammonium nitrate and sulfate) showed large and relatively systematic variations with > 10 or more times higher concentrations over the whole eastern North Sea. It is remarkable that ammonium correlates with this deposition pattern also over sea. Meteorological conditions preceding this rain event had invoked a small episode with advection of polluted air from southerly directions.

To explain the large deposition gradient from east to west, it was suggested that higher ammonia emissions in the countries bordering the North Sea to the southeast were responsible. With these emissions affecting the North Sea region during southerly winds, the in-cloud oxidation of the acid precursors SO_2/NO_x is suggested to be increased.

The results also demonstrated that large scale gradients exist, though not as simple as thought on beforehand. The gradients in the particular case were rather from East to West, then from the coast to open sea regions. The single event also showed to be highly efficient in supplying within a few hours 1 % of the yearly N-input to the North Sea.

h) Evaluation of meteorological conditions during NOSE

M. Schulz, Univ. of Hamburg, Germany
M. Schrader, Institut für Meereskunde Kiel, Germany
S.E. Larsen, F.A. Hansen, Risø National Laboratory, Denmark
O. Krüger, GKSS Geesthacht, Germany

Standard meteorology was recorded on both ships and radiosondes were launched from the upwind ship every 6 hours. A sonic anemometer was deployed on the front mast of the Alkor to measure turbulent fluxes. During the experiment forecasted trajectories were faxed every 6-12 hours to the Alkor by the UK Meteorological office to help in the positioning of the ships. Finally back-trajectories have been computed by using the LAM50 model of the Norwegian Weather Service.

The whole period was characterised by two stormy periods interrupted by rather calm southerly winds. Rain appeared to be present in 40 % of the measuring intervals. Due to the autumn conditions a rather unstable boundary layer was present with mixing heights of 500–1500 m according to EMEP model calculations [8]. The radiosondes confirm this, except that for the period of southerly winds a low mixing height of 200 m was observed [17]

The wind analysis and the trajectory calculations confirmed that the ships had been positioned along trajectory for most of the time. Standard meteorological parameters at both ships were highly correlated. This confirmed that it is possible to conduct such an experiment at all.

Turbulent exchange and heat fluxes determine the evolution of the boundary layer. This necessitates measurements of bulk parameters, vertical structure and turbulent properties. However, moving ships distort the cospectra of wind and energy. With dissipation theory certain features of the turbulent spectrum may be retrieved even on a moving ship. First comparisons with bulk formulae indicate that heat fluxes derived for selected periods fall within 50 % of the mean value of both methods. This indicated that both methods operated satisfactorily [18].

With regard to entrainment velocities no substantial progress was made due to the lack of data and funded modelling groups. The chemical measurements provided more evidence of a substantial entrainment during the period. However, a large chemical and meteorological data set is available for the period.

Conclusions

While this experiment may not have served all expectations, it clearly brought progress in that the experimental strategy has been tested in many aspects and proved to be feasible. The hypothesis, that the mass budget approach would allow to quantify air-sea-exchange for a suite of components was difficult to test, given the logistical constraints and very little financial resources. This might be the main reason, that the real scientific novelties and findings of NOSE are somewhat hidden.

The experiment pointed clearly to the fact that the vertical distribution of atmospheric pollutants has been very little considered by atmospheric scientists, when these attempt to understand the abundance of the pollutants. A considerable entrainment at the top of the boundary layer has been suggested to be the main cause for the observed decrease in concentration during NOSE. Although, a dry deposition process larger then expected could not be ruled out. This has several consequences: First, any similar experiment needs to considerably improve the information about the vertical distribution of pollutants and the boundary layer and their dynamics. This might include the use of tracers, vertical profile measurements of all kinds and improved modelling of the evolution of the boundary layer and its exchange properties with the free troposphere. Second, since dilution could have been only effective if much cleaner air was entrained, then we have to assume a large well mixed reservoir of clean free tropospheric air in the case of NOSE. This has been probably North-Atlantic background air, since westerly winds prevailed. Third, removal of atmospheric pollutants from the boundary layer by entrainment might be more effective then dry deposition. This has implications for the extent to which dry deposition may compete with wet deposition with regard to the removal of atmospheric constituents. If pollutants are removed by entrainment from the boundary layer, they are more likely to be washed out by rain further on in their atmospheric residence time.

References

1. Harrison R. M., M.I. Msibi, A.-M.N. Kitto, S. Yamulki; Atmospheric chemical transformations of nitrogen compounds measured in the North Sea experiment, September 1991. *Atmos. Environ.* **28** (1994) 1593-1599.
2. De Bock L.A., H. van Madern, R.E. van Grieken; Individual aerosol particle composition variations in air masses crossing the North Sea. *Environ. Sci. Technology* **28** (1994) 1513–1520.
2. Schulz M, T. Stahlschmidt, A. Rebers; Internal report, University of Hamburg 1992.
3. Baeyens W., M. Leermakers, C. Meulemann, Q. Xianren, P. Lansens; Air-Sea exchange fluxes of mercury and atmospheric inputs to the sea. *EUROTRAC Ann. Rep. 1992, part 3 ASE*, EUROTRAC ISS, Garmisch-Partenkirchen 1993.
4. Ebinghaus R.., H.-H. Kock, J. Kuballa; Measurements of atmospheric mercury during the North sea Experiment 1991 - first results from the FS Alkor, internal report 1992.
5. Francois F., J. Cafmeyer, Gilot C., W. Maenhaut; Chemical composition of size fractionated atmospheric aerosols at some coastal stations and over the North sea. in: P.M. Borrell, P. Borrell, T. Cvitaš, W. Seiler (eds), *Proc. EUROTRAC Symp. '92,* SPB Academic Publishing bv, The Hague 1993, pp. 788–791.
6. Schulz M., T. Stahlschmidt, F. Francois, W. Maenhaut and S.E. Larsen (1994) The change of aerosol size distribution measured ina Lagrangian-type experiment to study deposition and transport processes in the marine atmosphere. in: P.M. Borrell, P. Borrell, T. Cvitaš, W. Seiler (eds), *Proc. EUROTRAC Symp. '94,* SPB Academic Publishing bv, The Hague 1994, pp. 702–706.
7. Sørensen, L.L., Hertel, O., Christensen, J., Schulz, M., Harrison, R.M.; Fluxes of nitrogen gases to the North Sea. in: P.M. Borrell, P. Borrell, T. Cvitaš, W. Seiler (eds), *Proc. EUROTRAC Symp. '94,* SPB Academic Publishing bv, The Hague 1994, pp. 626–631.
8. Krueger O.; Europaweite Ablagerung von Schwefel- und Stickstoffverbindungen: Neue Ansaetze bei Trocken- und Nassdeposition, Ph.D. Thesis, University of Hamburg, Hamburg, Germany, 1992.
9. Sørensen, L.L., Granby, H. Nielsen, K., Asman, W.A.H.; Diffusion Scrubber. A Technique used for measurements of atmospheric ammonia. *Atmos. Environ.* **28** (1994) 3637–3645.
10. Bange H., and S. Rapsomanikis; Measurements of atmospheric and dissolved Nitrous Oxide (N_2O) F.S. Alkor (16.–26.9.1991), internal report 1992.
11. Injuk J.H., H. van Maldern, R. van Grieken, E. Swietlicki, J. M. Knox, R. Schofield; EDXRS study of aerosol composition variations in air masses crossing the North Sea. *X-Ray Spectrometry* **22** (1993) 220–228.
13. Quack B.; Dissertation, Institut fuer Meereskunde, Universität Kiel, 1993.
14. Kuss J.; Dissertation, Institut fuer Meereskunde, Universität Kiel, 1993.
15. Spokes L., T. Jickells, A. Rendell, M. Schulz, A. Rebers, W. Dannecker, O. Krueger, M. Leermakers, W. Baeyens; High atmospheric nitrogen deposition events over the North Sea. *Marine Poll. Bull.* **26** (1993) 698–703.

16. Rebers A., M. Schulz, L. Spokes, T. Jickells, M. Leermakers; A regional case study of wet deposition in the North Sea area. in: P.M. Borrell, P. Borrell, T. Cvitaš, W. Seiler (eds), *Proc. EUROTRAC Symp. '94,* SPB Academic Publishing bv, The Hague 1994, pp. 621–625.

17. Schulz M, S.E. Larsen; Overview of the meteorological conditions during the North Sea Experiment 1991. internal report 1992.

18. Larsen S.E., F. Aa. Hansen; Comparison between heat fluxes from the drag formulations and the eddy correlations during the NOSE North Sea experiment, Sept. 1991, internal report 1995.

2.3 Factors Determining Particle Dynamics over the Air-Sea Interface

Gerrit de Leeuw[1], Søren E. Larsen[2] and Patrice G. Mestayer[3]

[1]TNO Physics and Electronics Laboratory, P.O. Box 96864, 2509 JG The Hague, The Netherlands

[2]Department of Meteorology, Risø National Laboratory, Roskilde, Denmark

[3]Laboratoire de Mecanique des Fluides, Ecole Centrale de Nantes, B.P. 92101, F-44321 Nantes Cedex 3, France

Work done in the framework of the ASE subproject, topic 5, on factors determining particle dynamics over the air-sea interface, is briefly reviewed. Emphasis is on the cooperative efforts between the authors, covering a period of roughly 8 years, from 1988 until 1996 [1–16], which in part were carried out in cooperation with other institutes. The work has focused on the description of the aerosol life cycle in the marine atmosphere, including production of sea spray aerosol, transport of sea spray aerosol from the surface into the surface layer and the mixed layer, the dispersal of aerosol throughout the marine atmospheric boundary layer (MABL), and the removal at the surface and at the top of the MABL. Continental aerosol, advected over the oceans in off-shore winds, is an important factor in the MABL, and processes involving aerosols of non-marine origin were included in our studies, in particular as regards their deposition on water surfaces.

The emphasis of the studies was on:

* the assessment of the effects of sea spray aerosol on the transfer of humidity and latent heat between the ocean and the atmosphere, and

* the effect of bubbles on the deposition velocity of sub-micron aerosol particles.

In addition, studies were performed on the source function of sea spray aerosol. Results from theses were applied to a study of heterogeneous processes involving sea-salt aerosol, clouds and nitrogen compounds.

The work was based on both a strong modelling component and a strong laboratory experimental effort. In total, six models were developed, with different degrees of sophistication and complexity, and two major laboratory experiments were conducted supported by several smaller experiments. Field experimental data were used for comparison and interpretation of the results. In the framework of EUROTRAC, a unique data set was collected during an experiment at the Forschungsplatform Nordsee (FPN).

Aims of the research

The aim of the research is to yield a description of the factors determining the life cycle of the aerosol in the marine atmospheric surface layer, including production, transport from the production zone into the surface layer and the mixed layer, dispersal throughout the marine atmospheric boundary layer and removal at the air-sea interface and at the top of the boundary layer, through field experiments and laboratory investigations in combination with numerical and analytical modelling.

The aerosol in the marine atmospheric boundary layer consists of a complex mixture of aerosols of maritime and continental origin. The continental aerosol, man made or natural, is advected over the sea in off-shore winds and gradually disappears from the boundary layer due to removal by dry or wet deposition at the air-water interface or by entrainment at the top of the boundary layer, or simply by the detachment of the aerosol layer from the surface due to the formation of an internal boundary layer. The latter mechanisms obviously do not remove the aerosol, and eventually the particles will be re-entrained in the boundary layer, likely after significant changes due to physical and chemical processes.

The maritime aerosol consists of sea spray produced from bubbles bursting at the sea surface, and, at high wind speeds ($u > 9$ m/s), droplets produced by direct tearing from the wave crests. In addition, several physical and chemical processes contribute to the formation of aerosol. The latter are usually found in the fine aerosol fraction, whereas sea spray droplets have sizes larger than 1 μm.

Our studies in EUROTRAC were focused on two topics:

* the life cycle of sea spray aerosol and their effects on the transfer of water vapour and latent heat between the sea surface and the atmosphere; later heterogeneous processes involving sea spray droplets were included; and

* the effect of bursting bubbles on the deposition of aerosol particles on water surfaces.

Both these topics were studied through intensive laboratory experiments in which representative situations were simulated, to provide the data for the development of comprehensive models describing the life cycle of the marine aerosol. Most of the laboratory experiments were made on fresh water droplets, and thus the models were initially developed for this situation. Later, the fresh water droplets were replaced by sea spray, *i.e.* salt water droplets. This complicated the numerical code because in contrast to fresh water droplets, sea spray will only partly evaporate until an equilibrium is reached between the water vapour pressure at the droplet surface and the relative humidity of the surrounding air, which depends on the salt content of the droplet. In addition, for application over sea the effects of waves need to be taken into account. This requires that wave-induced air flow perturbations are considered.

The work described in this contribution was carried out by the authors' institutes, and in close co-operation with other institutes. Work on aerosol production and source function estimates, together with the applcation of the SeaCluse model to open conditions is described in [4]. For the use of the results in applications on heterogeneous processes involving sea salt aerosol, see [9, 12].

Principal scientific results

Laboratory experiments

Two main experiments were conducted in the large air-sea interaction simulation tunnel of the Laboratoire Interactions Océan-Atmosphère, Institut de Recherche sur les Phénomènes Hors Equilibre (IRPHE, formerly IMST), Laboratoire de Luminy, Marseilles, France. These experiments were supported by several smaller experiments in both Luminy and Groton, Connecticut, USA, and some tests at TNO-FEL, The Hague, and at the Naval Postgraduate School in Monterey, California, USA. The experiments in Groton allowed for the use of salt water in the tank, thus providing comparisons between fresh and salt water droplets in the absence of waves. In the large air-sea interaction tunnel in Luminy bubbles are produced by breaking wind waves (note that also mechanical waves were created with a paddle). To increase the bubble fluxes, up to an equivalent whitecap cover of about 20 %, the tunnel was equipped with arrays of aerated frits. For a more detailed description of the aerosol production and estimates of the source function, see De Leeuw and Van Eijk [4].

Influence of sea spray on transfer of latent heat and water vapour

The first of the experiments in Marseilles, Grand-Cluse in 1988, was aimed at the determination of the influence of aerosols on the transfer of latent heat and water vapour. Aerosol size distributions were measured at a range of levels with a suite of optical particle counters and also with an impaction device; at the same time, turbulent fields of temperature, humidity and air flow were measured. Also some efforts were made to determine the influence of waves on the air flow, the humidity and the aerosol, to test earlier hypotheses [17].

Two models were developed based on the CLUSE experiments, *i.e.* a Lagrangian model [18–20] and a Eulerian model [21–23] describing the life cycle of fresh water droplets over a flat water surface and their effects on the temperature and humidity profiles.

SeaCluse: application to open-ocean conditions

The SeaCluse model was developed from Rouault's Eulerian model for extension to oceanic conditions. The influence of waves on the air flow had to be taken into account [24] and fresh water droplets had to be replaced by sea-spray droplets.

In contrast to fresh water droplets, that can evaporate completely, sea water droplets can evaporate only to a certain extent determined by the salt concentration and the ambient relative humidity [25]. Therefore, the model not only separates the droplets according to their size but also according to their salt concentration. Condensation and evaporation constitute sources and sinks through which droplets can be exchanged between different size categories.

The module describing the air flow over waves calculates the horizontal and vertical wind components over the wave surface, based on a simple assumption of constant stress and a 'local' logarithmic wind profile. An 8th order Stokes wave with capillary waves superimposed is introduced to describe the surface. The wind field over this wave is used to calculate droplet trajectories that yield flight times, final fall velocities and reference profiles for droplet concentrations. This information, together with the ejection velocities, is stored in look-up tables for use during the actual model run.

The SeaCluse model applies to droplets in the 5–150 μm radius range. Droplet profiles calculated with SeaCluse have been compared by Tranchant and Van Eijk [26] with field data from the North Sea [27]. The comparison shows that the trends are similar, although the absolute concentrations over the North Sea, as predicted by SeaCluse, are rather low. This is not surprising, considering that advection is not taken into account in the model, only local production.

For smaller particles, 0.05–5 μm in radius, the CLUSA model was developed to describe the evolution of particles of continental origin over water surfaces [15, 16]. This requires a description of the air flow structure in a coastal atmospheric boundary layer. After the introduction of the profiles of wind speed and turbulent viscosity in the coastal zone, as well as the upper boundary condition [6], the lower boundary condition at the air-sea interface is analysed. In particular the usual condition that the sea surface provides a 'perfect sink' for particles implies that particle concentrations are zero in the immediate vicinity of the surface. This requires a very fine grid to obtain a correct representation of the aerosol concentration profiles in the viscous sub-layer, which depends on particle size, and a gradually coarser grid in the transition layer and the turbulent mixing layer. The results indicate the necessity to use a physically unreasonably fine grid that indicates that the hypothesis that the concentration must be zero at the surface only applies for very fine aerosols which behave like gases. For larger particles the interaction with the surface depends more on the gravitational fall velocity than on Brownian motion and the zero surface condition assumption must be reconsidered.

In the next step, the interaction between the size categories is introduced, creating local concentration gradients and generating turbulent fluxes to restore the concentration profiles. A model has been developed for the local source functions describing the interaction between different aerosol categories, which is based on the equations of Pruppacher and Klett [28] and follows the work of Mestayer [25].

Particle deposition on water surfaces

To study the effects of breaking waves and bubbles on the deposition of aerosol on water surfaces, the TWO-PIE experiments were conducted in the Large Air-

Sea Interaction Simulation Tunnel in Marseilles, in May/June 1990 [29]. Tracer aerosol particles were released in the tunnel and their removal was determined from the decrease of the concentrations with time [6]: the removal is not only determined by the deposition at the water surface, but also by the walls and the surfaces in the return flow channel), and from profiles of particle concentrations. The influence of the breaking waves and bubbles was determined by repeating the measurements at four simulated whitecap coverages, *i.e.* by changing the bubble concentrations through doubling and tripling the amount of aerated frits, while keeping all other parameters (wind speed, relative humidity, air and water temperature) constant.

The data were analysed using the deposition model developed by Hummelshøj (see [13, 14] and references cited therein). As discussed by Larsen *et al.* [13, 14] the absence of a systematic variation of the deposition velocity with the bubble flux leads to the conclusion that bursting bubbles and associated spray do not influence the deposition of the test particles, at least not significantly. This conclusion applies to tracer aerosol particles in the 0.1–1 μm radius range, for a wide range of relative humidities and a whitecap cover of up to about 20 %, equivalent to conditions of wind speeds of up to 25 m/s in the natural marine boundary layer.

As an *ad hoc* argument for this conclusion, which is contrary to expectations, one may argue that although the resistance of the laminar sub-layer to particle deposition is locally eliminated during the bubble bursting process, the particles are at the same time pushed away from the surface by the air escaping from the bursting bubble. Therefore, an enhancement of the deposition will have to depend on the secondary circulation, during the short time after the first eruption. Another argument could be found in 'surface renewal theory' for the transfer through the laminar sub-layer. According to this theory, most of the transfer takes place in bursts, and hence the bursting bubbles will add another process that may replace already on-going processes. Under all circumstances, the experimental result is that the bursting bubbles and spray have no significant influence on the deposition rate of small atmospheric particles.

Conclusions

The work described in this contribution was carried out in projects from which the results contributed to the goals of EUROTRAC [1, 9, 12–14]. Details are thus described in many reports, symposium proceedings and refereed articles. The work has resulted in three PhD theses [18, 21, 31].

In total, six models were developed, with different degrees of sophistication and complexity. These include the CLUSE Eulerian [21, 23] and Lagrangian [18, 19] models describing fresh water aerosols, the SeaCluse model [26, 31, 32] that evolved from the Eulerian model to describe sea salt aerosol in the presence of surface waves, the CLUSA model describing deposition of small particles over sea in off-shore winds [15], a deposition model [14], and a simple model to calculate sea spray source functions from bubble size distributions [33].

The TWO-PIE experiments were designed to test the hypothesis that bursting bubbles would increase the deposition velocity of small particles. However, the absence of a systematic variation of the deposition velocity with the bubble flux leads to the conclusion that bursting bubbles and associated spray do not significantly influence the deposition of the test particles [12–14]. This conclusion applies to tracer aerosol particles in the 0.1–1 μm radius range, for a wide range of relative humidities and a whitecap cover of up to about 20 %, equivalent to conditions of wind speeds of up to 25 m/s in the natural marine boundary layer.

Acknowledgements

The work described in this contribution was supported by the Commission of the European Communities (EG DG XII contracts STEP-CT90-0047, MAS2CT930056, MAS2CT930069 and EV5V-CT93-0313). TNO-FEL efforts were further supported by the Netherlands Ministry of Defence (Contracts A86KM106, A92KM615, A92KM635, A92KM767, A93KM699) and the US Office of Naval Research (ONR, Grants N00014-87-J-1212, N00014-91-J-1948). Part of the numerical modelling was made while Alexander M.J. van Eijk held a position as maître de conférence invité at the Ecole Centrale de Nantes, and while at other times Benoit Tranchant made an extended visit to TNO Physics and Electronics Laboratory. ECN received additional support from the Contract MRT 89W0081 of the P.I.R. Environnement, French joint CNRS/Ministry of Environment committee for EUROTRAC (grant Environnement 92N82/0019). The work performed by Risø was further supported in the framework of the Marine Research Programme 90 in Denmark. Part of the TWO-PIE efforts by Soren Larsen was made while he held a position as Professeur Invité at the Ecole Centrale de Nantes. The EUROTRAC projects are a cooperation between TNO Physics and Electronics Laboratory (The Netherlands), Risø National Laboratory (Denmark) and Ecole Centrale de Nantes (France).

References

1. De Leeuw, G.; The physics of marine aerosols and their role in atmospheric chemistry. in: P.M. Borrell, P. Borrell, K. Kelly, T. Cvitaš, W. Seiler (eds), *Proc. EUROTRAC Symp. '96, Vol. 1,* Computational Mechanics Publications, Southampton 1996, pp. 67–71.
2. De Leeuw, G., and A.M.J. van Eijk; Factors determining particle dynamics near the air-sea interface. Study of the production and dynamics of marine aerosols. *EUROTRAC Ann. Rep. 1989, part 3 ASE*, EUROTRAC ISS, Garmisch-Partenkirchen 1990, pp. 36–37.
3. De Leeuw, G., and A.M.J. van Eijk; Generation, transport and deposition of marine aerosols. *EUROTRAC Ann. Rep. 1991, part 3 ASE*, EUROTRAC ISS, Garmisch-Partenkirchen 1992, pp. 56–62.
4. De Leeuw, G., and A.M.J. van Eijk; Dynamics of aerosols in the marine atmospheric surface layer. this volume.

5. De Leeuw, G., S.E. Larsen, P.G. Mestayer, A.M.J. van Eijk, H. Dekker, A. Zoubiri, P. Hummelshoj and N.O. Jensen; Aerosols in the marine atmospheric surface layer: production, transport and deposition. *EUROTRAC Ann. Rep. 1991, part 3 ASE*, EUROTRAC ISS, Garmisch-Partenkirchen 1992, pp. 61–67.
6. De Leeuw, G., S.E. Larsen and P.G. Mestayer; Factors determining particle dynamics over the air-sea interface. *EUROTRAC Ann. Rep. 1992, part 3 ASE*, EUROTRAC ISS, Garmisch-Partenkirchen 1993, pp. 32–38.
7. De Leeuw, G., A.M.J. van Eijk and H. Dekker; Studies on aerosols in the marine atmospheric surface layer. *EUROTRAC Ann. Rep. 1992, part 3 ASE*, EUROTRAC ISS, Garmisch-Partenkirchen 1993, pp. 94–102.
8. De Leeuw, G., A.M.J. van Eijk, P.G. Mestayer and S.E. Larsen; Aerosol dynamics near the air-sea interface. in: P.M. Borrell, P. Borrell, T. Cvitaš, W. Seiler (eds), *Proc. EUROTRAC Symp. '94,* SPB Academic Publishing bv, The Hague 1994, pp. 707–710.
9. De Leeuw, G., A.M.J. van Eijk, A.I. Flossmann, W. Wobrock, P.G. Mestayer, B. Tranchant, E. Ljungstrom, R. Karlsson, S.E. Larsen, M. Roemer, P.J.H. Builtjes; Transformation and vertical transport of N-compounds in the marine atmosphere. in: P.M. Borrell, P. Borrell, K. Kelly, T. Cvitaš, W. Seiler (eds), *Proc. EUROTRAC Symp. '96 Vol. 1,* Computational Mechanics Publications, Southampton 1996, pp. 195–198.
10. Larsen, S.E., J.B. Edson, P.G. Mestayer, C.W. Fairall, G. de Leeuw; Sea spray and particle deposition, air/water tunnel experiment and its relation to over-ocean conditions. in: P. Borrell, P.M. Borrell, W. Seiler (eds), *Proc. EUROTRAC Symp. '90,* SPB Academic Publishing bv, The Hague 1991, pp. 71–75.
11. Larsen, S.E., P. Hummelshøj, M.P. Rouault; Aerosol dynamics, transport and deposition near the sea surface. *EUROTRAC Ann. Rep. 1991, part 3 ASE*, EUROTRAC ISS, Garmisch-Partenkirchen 1992, pp. 50–55.
12. Larsen, S.E., G. de Leeuw, A. Flossmann, E. Ljungstrom, P.G. Mestayer; Transformation and removal of N-compounds in the marine atmosphere, *Final report on the STEP project CT90–0047,* Risø National Laboratory, Roskilde, Denmark 1994.
13. Larsen, S.E., J.B. Edson, P. Hummelshoj, N.O. Jensen, G. de Leeuw, P.G. Mestayer; Laboratory study of the particle dry deposition velocity over water. in: P.M. Borrell, P. Borrell, T. Cvitaš, W. Seiler (eds), *Proc. EUROTRAC Symp. '94,* SPB Academic Publishing bv, The Hague 1994, pp. 698–701.
14. Larsen, S.E., J.B. Edson, P. Hummelshoj, N.O. Jensen, G. de Leeuw, P.G. Mestayer; Dry deposition of particles to the ocean surfaces. *Ophelia* **42** (1995) 193–204.
15. Mestayer, P.G.; Numerical simulation of marine aerosol dynamics. *EUROTRAC Ann. Rep. 1992, part 3 ASE*, EUROTRAC ISS, Garmisch-Partenkirchen 1993, pp. 112–116.
16. Mestayer, P.G., A. Zoubiri and J.B. Edson; Experimental and numerical study of aerosol dynamics and deposition at the sea surface; 2PIE, GWAIHIR & CLUSA. *EUROTRAC Ann. Rep. 1990, part 3 ASE*, EUROTRAC ISS, Garmisch-Partenkirchen 1991, pp. 66–74.
17. de Leeuw, G., Vertical profiles of giant particles close above the sea surface, *Tellus* **38B** (1986) 51–61.
18. Edson, J.B.; Lagrangian model simulation of the turbulent transport of evaporating jet droplets. Ph.D. thesis, Dept. Meteorol., Penn. State University, 1989.

19. Edson, J.B., C.W. Fairall; Spray droplet modeling I: Lagrangian model simulation of the turbulent transport of evaporating spray droplets. *J. Geophys. Res.* **99** (1994) 25295–25311.

20. Edson, J.B., S. Anquetin, P.G. Mestayer and J.F. Sini; Sea spray modelling Part II, A combined Eulero-Lagrangian simulation, *J. Geophys. Res.* **101** (1996) 1279–1293.

21. Rouault, M.P.; Modélisation numérique d'une couche limite uni-dimensionelle stationnaire d'embruns. Thèse de Doctorat, Univ. d'Aix-Marseille II, France 1989.

22. Rouault M.P., S.E. Larsen; Spray droplets under turbulent conditions. *Rep. Risø-M-2847*, Risø Natl. Lab., Denmark 1990.

23. Rouault, M.P., P.G. Mestayer, R. Schiestel; A model of evaporating spray droplet dispersion. *J. Geophys. Res.* **C96** (1991) 7181–7200.

24. Van Eijk, A.M.J., P.G. Mestayer, G. de Leeuw; Conversion of the CLUSE model for application over open ocean; progress report. TNO Physics and Electronics Laboratory, *report FEL-93-A035*, 1993.

25. Mestayer, P.G.; Sea water droplet evaporation in CLUSE model. in: P.G. Mestayer, E.C. Monahan, P.A. Beetham (eds), *Modeling the fate and influence of marine spray. Proc. Workshop in Luminy,* Marseille, 6–8 June 1990, pp. 65–76.

26. Van Eijk, A.M.J. and B. Tranchant (1996). Numerical simulations of sea spray aerosols. *J. Aerosol Sci.* **27,** Suppl. 1 (1996) 61–62.

27. De Leeuw, G.; Profiling of aerosol concentrations, particle size distributions and relative humidity in the atmospheric surface layer over the North Sea. *Tellus* **42B** (1990) 342–354.

28. Pruppacher, H.R., J.D. Klett; *Microphysics of clouds and precipitation.* D. Reidel, Dordrecht, The Netherlands 1978.

29. De Leeuw, G., S.E. Larsen, P.G. Mestayer, J.B. Edson, D.E. Spiel; Deposition, dynamics and influence of particles in the marine atmospheric boundary layer, laboratory study. TNO Physics and Electronics Laboratory, October 1991.

30. De Leeuw, G.; Aerosols near the air-sea interface. *Trends in Geophys. Res.* **2** (1993) 55–70.

31. Tranchant, B.; Simulations numeriques des embruns marins. Thèse de Doctorat, Ecole Centrale de Nantes et Université de Nantes, France 1997.

32. Mestayer, P.G., A.M.J. van Eijk, G. de Leeuw, B. Tranchant; Numerical simulation of the dynamics of sea spray over the waves. *J. Geophys. Res.* **101** (1996) 20771–20797.

33. De Leeuw, G. (1990a). Spray droplet source function: from laboratory to open ocean. in: P.G. Mestayer, E.C. Monahan, P.A. Beetham (eds), *Modelling the fate and influence of marine spray. Proc.Workshop in Luminy*, Marseille, 6–8 June 1990, pp. 17–28.

2.4 Aerosol Dynamics Modelling in the Marine Atmospheric Surface Layer

Patrice G. Mestayer and Benoit Tranchant

Laboratoire de Mécanique des Fluides, UMR CNRS 6598
Ecole Centrale de Nantes, F-44072 NANTES Cedex 03, France

The research is a contribution to the ASE topic, 'Factors determining particle dynamics over the air-sea interface', a co-operation between Risø National Laboratory, TNO Physics and Electronics Laboratory, and ECN. It included two parts, an experimental study of sub-micronic particle deposition at the water surface in a wind-water tunnel, and a theoretical study of the dynamics of aerosols in the atmospheric surface layer over the sea surface. This second part consisted in developing two numerical models of the behaviour of (a) the sea spray droplets generated at the wave surface and dispersed by the turbulent motions of air while they partially evaporate, (b) the transport of sub-micronic aerosols and their deposition at the sea surface.

Aims of the research

A large part of the air-sea mass exchanges takes place under the form of particles. The aerosols that are present in the marine atmospheric surface layer (MASL) are crudely of two types: those that are light enough to have been transported from far away and those that have been locally generated. The particles of the first type are smaller than a micron and they are of various origins, but in the coastal seas they are essentially dry particles transported from the continent: when reaching the more humid air of the marine atmosphere they eventually grow in size and weight, by condensing water vapour, and depose at the surface. This is for instance one of the pathways for the land-sea fluxes of nitrogen compounds responsible of eutrophication. The spray generated by breaking waves form most of the locally-produced particles: they are either torn off from the wave crests by the very strong winds or ejected from the surface by the bursting bubbles, as shown by the Fig. 2.4.1. Many of these droplets fall back to the sea surface, after having exchanged sensible heat and water vapour with the air, and some of them are eventually diffused high enough by turbulence to be transported higher and further. This net input of aerosols from the sea to the atmosphere, as well as the net deposition of particulate matter from the atmosphere to the sea surface, are partially controlled by similar processes: the coupling of particle transformations due to thermodynamic and mass exchanges with the air layers through evaporation-condensation, with their dispersion by the turbulent motions of air. In addition, the air motion induced by the wavy surface is expected to largely influence the initial dynamics of sea spray and their behaviour, while the presence of a thin mm-thick “diffusive layer” or “viscous sub-layer" just over the interface is expected to be the main resistance to sub-micron particle deposition mainly driven by turbulent mixing. Therefore, to assess these net upward and downward fluxes as a function of meteorological and sea state parameters, it appears necessary to model correctly the interaction of particle dynamics and thermodynamics, which is the aim of this research.

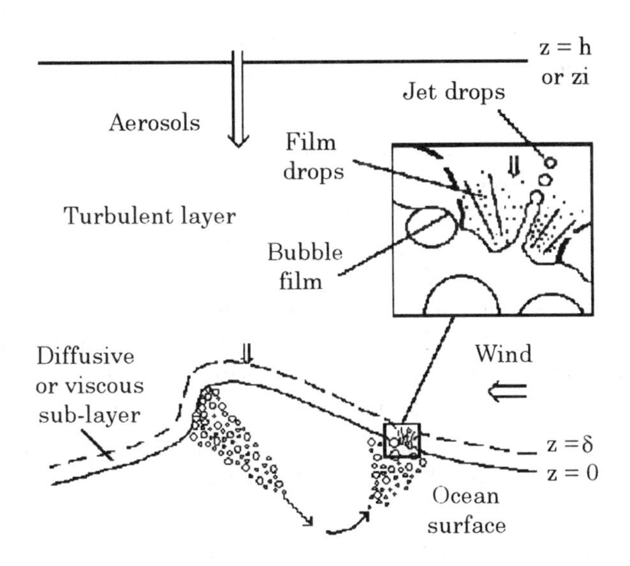

Fig. 2.4.1: Model of the marine atmospheric surface layer over breaking waves including spray generation and aerosol deposition at the surface

The programme includes the development of two numerical codes in parallel. SeaCluse concerns the generation of marine aerosols by the evaporating sea spray droplets, *i.e.* the dynamics of the spray droplets whose diameters are between 10 and 500 μm, where the emphasis is on the ejection process from the sea surface and on the coupling between partial evaporation and vertical dispersion by turbulence. CLUSA concerns the evolution of continental aerosols, from 0.05 to 5 μm in diameter, when they are transported over the marine atmosphere of the coastal seas: the emphasis is on the aerosol growth due to the condensation of the atmospheric water vapour on the nuclei, and on the deposition process to the water surface. The final aim of these latter numerical simulations is to estimate the deposition of particulate compounds of anthropogenic origin to the coastal seas. In addition, we co-operated with the scientists from Risø and FEL-TNO (and also J.B. Edson from Woods Hole Oceanographic Institution and D. Spiel from Naval Postgraduate School, USA) in the experimental assessment of the role of bubbles bursting at the surface on the deposition of sub-micron particles through the diffusive layer [1, 2].

Principal scientific results

The major result of the experiment, driven in the Air-Sea Interaction Simulation Tunnel at IMST (now Institut de Recherche sur les Phénomènes Hors Equilibre), Marseilles, France, is to prove that the rupture of the viscous sub-layer by the bursting bubbles has a negligible mechanical effect on the deposition of non-hygroscopic sub-micronic particles. This experiment and the analysis of results is reported more extensively elsewhere [3] and in this volume (see G. De Leeuw and S.E. Larsen).

The two numerical codes are developed from the same basic structure CLUSE created by Rouault, Mestayer and Schiestel [4]. They are stationary, one dimensional on the vertical and simulate the distribution of aerosol particles by a series of interacting scalar fields: one field for each size bin or category, plus the water vapour concentration and the sensible heat. At each level of the vertical grid, the model solves simultaneously the mass concentration budget equations of the interacting fields, with a closure on each flux term, including the particle inertial slip. The turbulent fluxes are modelled with a K-type closure, *i.e.* in the MASL, a turbulent diffusivity proportional to z ($\nu_t = k\ u_* \ z$). The particle transformations due to droplet evaporation and vapour condensation are modelled by source-sink terms that couple the budget equations of the particle and water vapour mass concentrations.

The development of the CLUSA model includes the generation and implementation of a model for the vertical profile of $U(z)$ and $K(z)$ as functions of U_{10} and X, the fetch from the coast [5]. This model produces profiles for all wind speeds from the top of the surface layer, say 100 m, down into the viscous sub-layer in contact with the water surface, averaging the flow structure in the wave layer. The model is based on Deardorff's assumption [6] that the aerodynamic momentum transfer is partitioned into a "form drag" responsible for the development of the wave field and a "viscous drag" that transfers momentum to the water surface. The wind structure over the wave crests and the form drag are deduced from Geernaert *et al.* [7] and Geernaert [8]. The viscous drag is then deduced by subtraction and introduced into a model of the viscous/diffusive sub-layer (either the classical model for a rigid surface or the model derived by Coantic [9] for gas transfers through a mobile surface).

A second part of this work concerns the conditions applied at the lower boundary of the calculation domain (*i.e.* at the surface) for the sub-micron particles. The surface is considered as a perfect sink for particles (no rebounds); this assumption can be translated in the numerical code in two ways:

* either (1) a free-surface Neumann condition, that expresses that the particles cross the surface "without feeling it",

* or (2) a condition of zero concentration, that expresses that immediately close to the surface the Brownian movement is blocked by the water surface and the particles loose agitation freedom. This is the usual condition for gas transfers.

The first condition is easily accepted by the code and produces constant vertical profiles of concentrations. The second one ($\rho_n(z = 0_+) = 0$) can be accepted only when the grid mesh close to the surface is extremely thin, inversely proportional to the particle size and as small as 10^{-6} µm for 10^{-2} µm particles, and even 10^{-9} µm for 1 µm particles! This demonstrates that this surface condition is not physically sound for the larger particles that behave more like drops (for which the gravity fall is too fast for really experiencing the viscous sub-layer) than gases. The separation between gas-like particles and drop-like particles seems to be around 10^{-2} µm.

A model for the aerosol transformations due to condensation of water vapour has been derived from the cloud nucleation model of Pruppacher and Klett [10], for aerosols composed of an NaCl solution and an insoluble nucleus. This model is to be implemented in CLUSA in the form of source-sink functions allowing for transfers between aerosol categories. This is delicate due to the extremely small time constants of the air-particle thermodynamic and mass exchanges.

The SeaCluse development is lead in co-operation with the FEL-TNO group. It included the creation of a pre-processor for computing spray droplet statistics (flight times, final velocities, trajectories, concentration distributions) in the flow over the waves in the absence of turbulence and evaporation. This pre-processor includes three successive computations: the 2-D mean air flow over waves; the droplet trajectories; the average droplet statistics. The wave field is represented by a 2-D 5th-order Stokes wave, whose height H_o and phase ω are computed from the reference wind speed U_{10} using parameterisations similar to Geernaert *et al.* [7]. The mean wind field over this 2-D wave is assumed to exhibit the flow structure of a turbulent boundary layer over a flat surface, locally in the wave layer ($-1/2H_o < z \leq H_o$) (*i.e.* for each position along the wave profile) and on the average in the surface layer $z > H_o$ (where the z origin is the mean sea level). The pre-processor computes the trajectories of particles of all sizes ejected from the surface (along the wave profile) in the absence of turbulence and of evaporation, in this 2-D mean wind field. An averaging process along the wave profile allows to obtain the reference mean vertical profiles of droplet concentrations $\rho_{n_0}(z)$ needed to model the flux due to their mean movement in the 1-D budget equation model.

The main part of the SeaCluse model solves the budget equations for the various categories of droplets generated at the interface, sorted by their size, and of droplets generated by the partial evaporation of these "parent" droplets, their "offspring" sorted by size and salt concentration, since the droplet evaporation rate depends on their environment, their size and their salt concentration which increases during the evaporation process. These budget equations are coupled by the source-sink functions that model the mass transfers between droplet bins due to this partial evaporation, while the vertical transfers through the air layers are modelled by flux terms for the mean movement (ejection + gravity fall) and the turbulent diffusion. This last term includes an explicit modelling of the reduction of diffusion due to droplet inertia.

The Figs. 2.4.2 and 2.4.3 [11] illustrate the behaviour of the spray droplets generated at the interface. They present the concentrations of droplets as a function of height, for two sizes of droplets, 10 and 100 μm, as computed without turbulent diffusion (dots in frame b) and with three different models of turbulent diffusion reduction by droplet inertia. The scale on the right is normalised by the mean wave height. The droplet concentrations are normalised by their surface flux, giving results independent of the size distribution and strength parameterisation of the surface source function which is still a matter of debate. The droplets are generated uniformly at all points of the sea surface. The mean motion of air, induced by the motion of the waves, a rotor-like flow structure, is shown to be an efficient transport mechanism for the droplets when compared to the profiles obtained over a flat surface: 70 to 80 % of them are transported higher than - $H_o/4$ and about 25 % reach the wave crest height.

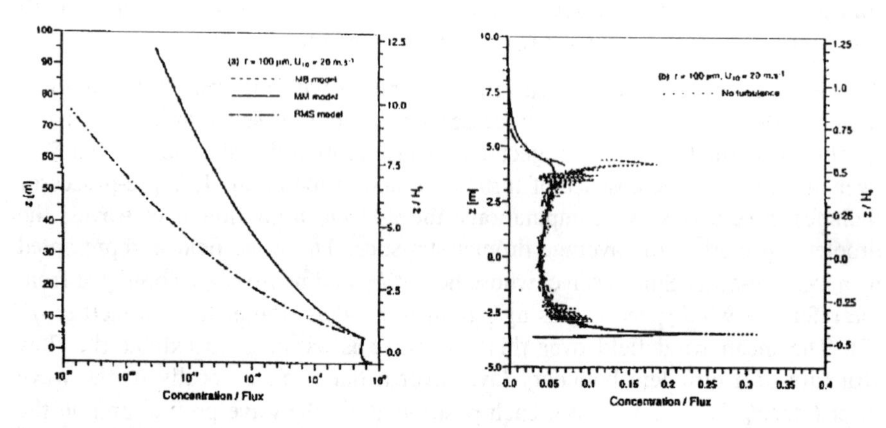

Fig. 2.4.2: Normalised profiles of 100 μm sea spray droplet concentrations without turbulence (dots) and with various models of turbulence, (left) in the whole atmospheric surface layer, (right) in the wave region.

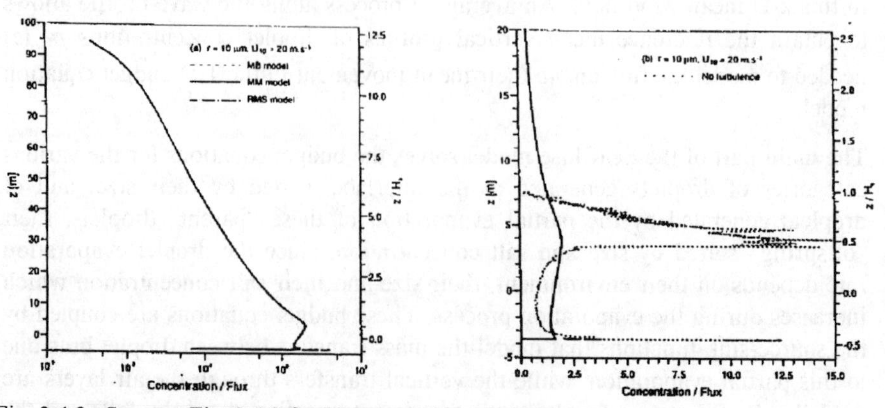

Fig. 2.4.3: Same as Fig. 2.4.2 for 10 μm sea spray droplets.

As regards the turbulent diffusivity, the two models MM (Mostafa and Mongia [12]), and MB (Melville and Bray [13]) appear to produce unexpectedly close results considering their functional differences, but Rouault *et al.*'s diffusivity reduction [4] due to inertia appears to increase much more rapidly with increasing droplet size, reducing the diffusion of the 250 μm droplets by turbulence to practically zero. For the 10 μm droplets, the effect of inertia is so small that the lines corresponding to the three different models cannot be separated.

The primary effect of turbulent diffusion, *mixing*, is to reduce the gradients and to smooth the peaks. This is clearly illustrated in the (b) frames where the low level profiles are well smoothed, although the very weak diffusivity in the wave trough is unable to erase the lowest peaks. Since the concentrations are zero at the top of the calculation domain, the secondary effect of turbulent diffusion is to *transfer* droplets upwards, from high concentration layers to low concentration layers. It appears that only about 10 % of the droplets are transported by turbulence higher than the highest level reached without turbulence, except for the 10 μm droplets that are well transported upwards and for which this number is on the order of 50 %. For those droplets that have non-negligible inertia and weight, turbulence does not appear to be a very efficient transport process. Except for the smallest category, no more than 12-18 % of the droplets reach the wave crest level or go higher, and no more than 3–5% at 20 m/s (1–2 % at 10 m/s) reach a level 1 m above the wave crest. This last comment addresses the experimental attempts to deduce the surface source function from the droplet concentrations measured in the atmosphere. This method to deduce the surface flux function from fits to experimental data appears very delicate and can easily produce erroneous estimates: the errors can be extremely large if the number of measurement levels is small and the model used to analyse the data is too crude or poorly assessed; especially since droplet partial evaporation is expected to highly complicate the transfer functions.

Conclusions

The result obtained with CLUSA, concerning the boundary condition at the surface for modelling sub-micronic aerosol deposition is quite important since the assumption of zero concentration at the surface is a key of the resistance layer models derived from Slinn and Slinn [1]: this casts a serious doubt on his results for those most unknown particles between 10^{-2} and 1 μm radius. More important it is one basic assumption for deducing the particle fluxes from the measurements of concentrations at one reference elevation, $F_{surf} = V_d \cdot C_{ref}$ where V_d is the gravitational or turbulent deposition velocity. Unfortunately, no better assumption has been implemented up to now [14].

The SeaCluse code has now capabilities to simulate most of the dynamics of the evaporating sea spray droplets, their behaviour, and their influence on the structure of the lower marine atmosphere. It is able to describe and quantify the non-linear interactions between the droplets and the scalar fields of water vapour concentration and temperature. After validation versus experimental data from the North Sea, HEXMAX, MAPTIP, it will be used to assess the spray surface source function and to predict the marine aerosol content of the atmospheric surface layer as a function of the basic micro-meteorological parameters. The main results obtained to date are the following.

The largest droplet concentrations exist on the lee side of the waves, based on the extended droplet suspension times at that location and the observation of droplet trajectories.

The mean air flow induced by the motion of the wavy surface results into a large rotor-like structure whose position and intensity depends on the wind speed and wave amplitude. This mean flow appears as a powerful transport process for the droplets of all sizes, typically propelling a large proportion of them about the wave crest height $H_o/2$ or even higher. Due to the displacement of the centre of the rotor-like structure, the efficiency of this transport process decreases relatively with increasing wind speed. This relative decrease is more than compensated for by the increase in wave height which is proportional to the square of the wind speed. Yet, this transport does not appear to change dramatically the trajectories of most droplets and the average flight times, as well as the maximum flight times, of the medium and large droplets appear to depend rather weakly on the wind speed and to increase much less rapidly than the wave height, in contradiction with the model of residence times equal to $H_{1/3}/V_g$, increasing as U_{10}^2 [15].

Over the waves, the role of turbulence appears much more to *diffuse* than to *transport* the droplets. Diffusion by turbulence does generate an important transfer of the passive scalars like gases and small droplets that have a negligible weight, from the regions where the concentrations are high towards those where they are low; but for those larger droplets that have non-negligible weights, the downward gravitational flux opposes this turbulent transfer, while this transfer is also reduced because these droplets have non-negligible inertia. Turbulence therefore appears to have a relatively weak influence on the vertical profiles of most droplet concentrations: these profiles are mainly determined by the structure of the mean flow induced by the wave motion, and only extremely small parts of the droplet concentrations are diffused higher than the wave region. Therefore a very fine tuning of the droplet diffusion model is necessary to assess the vertical profiles of droplet concentrations and the estimation of the atmospheric aerosol content close to, and away from, the sea surface.

Acknowledgements

This research has been supported by the French Ministry of Environment (Comité EUROTRAC) and the Centre National de la Recherche Scientifique (Programme PACB)(92N82/0019, 93N82/0115, 94N82/0079). It also benefited from the programs of the European Union STEP (CT90-0047) and Environment (EV5V-CT93-0313) and the US Office of Naval Research (N00014-94-1-0498)

References

1. Slinn, S.A., W.G.N. Slinn, Predictions for particle deposition on natural waters. *Atmos. Environ.* **14** (1980) 1013–1016.
2. Williams, R.M., A model for the dry deposition of particles to natural water surfaces. *Atmos. Environ.* **16** (1982) 1933–1038.
3. DeLeeuw, G., S.E. Larsen, P.G. Mestayer, J.B. Edson, D.E. Spiel, Deposition, dynamics and influence of particles in the marine atmospheric boundary layer, laboratory study. *TNO Report,* TNO Physics and Electronics Laboratory, The Hague 1991.
4. Rouault, M.P., P.G. Mestayer, R. Schiestel, A model of evaporating spray droplet dispersion, *J. Geophys. Res.* **96** (1991) 7181–7200.
5. Mestayer, P.G., Modèle synthétique de profils de vent et de diffusivité turbulente au dessus de la surface marine côtière, *Note interne L.M.T.T.D.-H.91.01*, 1991.
7. Geernaert, G.L., K.B. Katsaros, K. Richter, Variation of the drag coefficient and its dependence on sea state, *J. Geophys. Res.* **91** (1986) 7667–7679.
8. Geernaert, G.L., Drag coefficient modeling for the coastal zone, *Dyn. of Atmos. and Oceans* **11** (1988) 307–322.
9. Coantic, M., A model of gas transfer across air-water interfaces with capillary waves, *J. Geophys. Res.* **91** (1986) 3925–3943.
10. Deardorf, J.W., Aerodynamic theory of wave growth with constant wave steepness. *J. Ocean. Soc. of Japan* **23** (1967) 278-–297.
10. Pruppacher, H.R., J.D. Klett, *Microphysics of Clouds and Precipitation*, D.Reidel Publ. Co., Dordrecht 1978, 714 p.
11. Mestayer, P.G., 1994, Dry deposition of particles to ocean surfaces, *2nd Int. Conf. of Air-Sea Interaction and Meteorology and Oceanography of the Coastal Zone*, A.M.S., Lisbon 1994, pp. 243–244.
12. Mostafa, A.A., H.C. Mongia, On the modeling of turbulent evaporating sprays: Eulerian versus Lagrangian approach, *J. Heat Mass Transfer* **12** (1987) 2583-2593.
13. Melville, W.K., K.N.C. Bray, A model of the two-phase turbulent jet, *Int. J. Heat Mass Transfer* **22** (1979) 647–656.
14. Mestayer, P.G., A.M. J. VanEijk, G. DeLeeuw, Simulation of spray droplet dynamics over waves, *2nd Int. Conf. of Air-Sea Interaction and Meteorology and Oceanography of the Coastal Zone*, A.M.S., Lisbon 1994, pp. 228–229.
15. Andreas, E.L., Sea spray and turbulent air-sea fluxes, *J. Geophys. Res* **C97** (1992) 11429–11441.

2.5 Dynamics of Aerosols in the Marine Atmospheric Surface Layer

Gerrit de Leeuw and Alexander M.J. van Eijk

TNO Physics and Electronics Laboratory, P.O. Box 96864, 2509 JG The Hague, The Netherlands

This article gives a brief summary of the work performed by the TNO Physics and Electronics Laboratory (TNO-FEL) on dynamics of aerosols in the marine atmospheric surface layer. The work was mainly carried out as a cooperative effort between TNO-FEL, Ecole Centrale de Nantes (France) and Risø National Laboratory (Roskilde, Denmark), as described in De Leeuw *et al.* [1] over a period of 8 years [2–16]. The work has focused on a description of the aerosol life cycle in the marine atmosphere.

A task for TNO-FEL was to determine the source function of sea spray aerosol, and to provide data from field experimental studies. TNO-FEL also participated in a EUROTRAC experiment from the Forschungsplatform Nordsee (FPN), in 1992. Results from the various studies have been applied to heterogeneous processes involving sea-salt aerosol, clouds and nitrogen compounds, which has been coordinated by TNO-FEL during the last three years.

Aims of the research

The aim of the research has been to describe the factors determining the life cycle of the aerosol in the marine atmospheric surface layer.

Detailed studies have been made of the production processes of sea spray aerosol, the subsequent transport from the production zone into the surface layer and the dispersal throughout the atmospheric boundary layer, including transformation due to evaporation and condensation, and the removal of both sea spray aerosol and particles of continental origin. Bubble-mediated droplet production has been studied in detail in laboratory experiments, together with theoretical analyses of the bubble-breaking mechanism. Adequate sea-spray source functions are described for oceanic conditions as a function of meteorological conditions. To this end, a bubble measuring system has been developed and built for deployment in the open sea. The contribution to the oceanic source function of spume droplets, that are generated by the direct action of the wind on the wave crests, is also taken into account.

The life cycle of the aerosol and the effect of sea spray on the surface-layer temperature and humidity profiles have been investigated through models based on laboratory experiments, and field data were used for the extension to oceanic conditions. The oceanic model requires adequate descriptions of the flow over the surface wave field.

The deposition velocity of sub-micron particles is not well known: uncertainties of one order of magnitude are reported. To understand the causes for this large uncertainty, a laboratory study has been carried out to determine the effect of breaking waves and bubbles on the deposition velocity. Bubbles protruding from the surface disrupt the laminar sublayer, which offers the largest resistance against deposition of sub-micron particles. Thus the deposition velocity should be enhanced as more bubbles reach the surface and disrupt the laminar sub-layer. This study has now been completed and our analyses lead to the conclusion that the deposition velocity of sub-micron aerosol particles is not significantly influenced by bubbles [10, 12–14].

The effects of sea-spray droplets, including their role in atmospheric chemistry, are studied in the framework of several international and multi-disciplinary cooperation.

In our studies we use data from laboratory experiments in which specific processes are simulated in wind-wave tunnels, and data from field experiments. The latter include particle size distributions measured over a wide range of meteorological and oceanographic conditions, mostly at fixed height (ship deck level of about 10 m), as well as a comprehensive data set of particle size distribution profiles [17]. In addition, bubble size distributions are now measured by TNO-FEL to gain an insight into the effect of bubbles on air-sea exchange. This includes studies on gas exchange, in particular CO_2 fluxes, and the bubble-mediated production of sea-spray aerosol.

Our studies in EUROTRAC were focused on two topics, see [1]. As part of these studies, in close co-operation with ECN and Risø, TNO-FEL undertook the task to include sea spray, *i.e.* salt water droplets, in the Eulerian CLUSE model, and further to include the effects of surface waves on the air flow and thus also on the aerosol transport near the sea surface. This complicated the numerical code because in contrast to fresh water droplets, sea spray will only partly evaporate until an equilibrium is reached between the water vapour pressure at the surface of the droplet and the relative humidity of the surrounding air, which depends on the salt content.

For the use of the results in applications on heterogeneous processes involving sea-salt aerosol, see Larsen *et al.* [12] and de Leeuw *et al.* [2].

Principal scientific results

Aerosol production and source function estimates

In the large air-sea interaction tunnel in Luminy, bubbles were produced by breaking wind waves (note that also mechanical waves were created with a paddle). To increase the bubble fluxes, up to an equivalent whitecap cover of about 20 %, the tunnel was equipped with arrays of aerated frits. When the bubbles burst at the water surface they produce two types of droplets. Film droplets are formed following breaking of the film cap. A detailed study of this mechanism and the number of film droplets produced has been made by Spiel [20]. Earlier estimates of the number of film droplets produced from a bursting bubble are available from, *e.g.* the work of Blanchard [21] and Resch and Afeti [22, 23]. Jet droplets are produced after the film has broken and a jet shoots up and breaks into a small number of droplets (typically on the order of 6). The number of jet droplets produced from a bubble with a known size has been extensively studied by Blanchard [21, 24] and recently new evidence was presented by Spiel [25]. Based on the work published by Kientzler *et al.* [26], an analytical model describing the bubble bursting process and the subsequent formation of jet droplets was developed by Dekker and De Leeuw [27]. All these studies were (are) made on single bubbles.

Source functions can be derived in different ways. Aerosol size distributions can be used to derive an effective source function by application of a model assuming steady state and a balance between production and removal of the sea spray aerosol [28, 29]. Alternatively, bubble size distributions can be combined with information on bubble rise velocities and bubble-mediated droplet production to derive the source function (see [30] and cited references). This latter method ignores the production of spume droplets, because at present no quantitative information is available on this process (see the discussions in [31, 32] and Katsaros and de Leeuw [33]). Also, in a plume the bubbles influence each other and phenomena observed in the plume are different from those derived from studies on single bubbles [34]. These effects cannot be taken into account because all current information on bubble-mediated droplet production applies to single (isolated) bubbles. A third method to derive the sea spray source function is based on laboratory experiments in combination with whitecap cover estimates [35].

In spite of these efforts, current estimates of the sea spray source functions vary by several orders of magnitude [36], even when similar methods are used to derive this important parameter. Efforts are underway to provide better estimates based on simultaneously bubble size distributions and aerosol size distributions [37, 38].

Direct measurements were made to estimate the source function in the laboratory [39]. To this end, glass slides coated with MgO were used to measure the droplet flux at a range of levels above the water surface, at different positions with respect to the bubbler. These measurements resulted in a series of profiles of fluxes of freshly produced droplets which were extrapolated to yield the surface flux. In a second effort, an optical particle counter was used to measure only the upward travelling particles. This was achieved by application of an optical array probe (OAP, Particle Measuring Systems, Boulder, CO, USA) and covering the area above the laser beam to catch the sampled droplets, and to prevent any other droplets to enter the sample volume from above. These measurements were made in still air, so the freshly produced jet droplets travelled in principle only in the vertical. The OAP was moved over the tunnel width to cover a large area and thus provide a mean source function that is representative for the whole water surface.

SeaCluse: application to open-ocean conditions

Most of the laboratory experiments were made on fresh water droplets, and thus the models were initially developed for this situation [18, 19]. In a close cooperation with ECN and Risø, TNO-FEL undertook the task to include sea spray, *i.e.* salt water droplets, in the Eulerian CLUSE model, and further to include the effects of surface waves on the air flow and thus also on the aerosol transport near the sea surface. This complicated the numerical code because in contrast to fresh water droplets, sea spray will only partly evaporate until an equilibrium is reached between the water vapour pressure at the surface of the droplet and the relative humidity of the surrounding air, which depends on the salt content [44]. The development of the SeaCluse model appeared to be a major task, as described in de Leeuw *et al.* [1]. The first comparisons of the results from SeaCluse compare reasonably with experimental data. The variation of the concentrations with wind speed are similar to observed behaviour, and the gradient of the particle concentrations is well predicted for wind speeds exceeding 10 m/s. However, the absolute concentrations are smaller than observed. This is not surprising since only local production is taken into account by SeaCluse. Details are presented in Van Eijk *et al.* [44], Mestayer *et al.* [456], Van Eijk and Tranchant [46] and Tranchant [47].

Field experiments

Field experimental data were used to compare the models with field conditions and to interpret the model predictions in terms of open-ocean conditions. Field experimental data are available form the numerous experiments by TNO-FEL at a variety of locations, providing a large range of meteorological and oceanographic conditions (see [7, 17] for overviews). In the framework of EUROTRAC, a unique data set was collected from the Forschungsplatform Nordsee (FPN) combining for the first time optical particle counters and Rotorod rotating impaction samplers, together spanning diameters between 0.1 and 100 μm, at heights from about 2 m up to 30 m above the sea surface.

The EUROTRAC ASE experiment was conducted from the Forschungsplatform Nordsee (FPN), 1-22 September, 1992, in a cooperation between seven institutes from Germany (Univ. Hamburg, MPI, Mainz), Belgium (Univ. Antwerp, Univ. Brussels, Univ. Gent), Denmark (Risø) and The Netherlands (TNO-FEL). The main objective of TNO-FEL participation in this experiment was the extension of our data base on the height dependence of particle size distributions in the marine atmospheric surface layer, for a wide variety of conditions. This data base is used for studies of the factors that are important for the dynamics of the aerosol in the marine boundary layer. Results from the statistical analyses will be used to improve the physical descriptions of the aerosol properties, including production, dispersal and removal (deposition). The final goal is the development of a comprehensive physical model that is generally applicable.

The emphasis was on the smaller particles which were measured with optical particle counters, in addition to the giant particles for which a relatively large data base is now available [17]. The FPN offered the opportunity to combine both techniques for the first time. To this end, optical particle counters from TNO (a CSASP-200 from Particle measuring Systems, Boulder, CO, USA) and from the University of Hamburg (size range 0.2–20 μm in diameter) were mounted in a cage which was hooked to the 20-meter crane at the SW corner of the platform [48]. The height dependence of the particle size distributions was measured at a distance of about 15 m upwind from the platform, at levels between 2 and 30 m above mean sea level. These aerosol spectra were extended to larger diameters (100 μm maximum) by simultaneous Rotorod [40] measurements from the cage at the same heights. Additional measurements were made with the Rotorods mounted on a wave-following buoy system at levels between 0.35 m and 3 m and at fixed heights between 1 m and 5 m. Relative humidity and air temperature were also recorded from the cage. During these profiles, Risø recorded particle size distributions at a fixed height of about 27 m, as a reference to detect changes in the air mass that could also influence the profile during the course of its measurement. Prior to and during the experiment, particle counter comparison experiments were conducted.

In between the profile measurements, the cage was put at a fixed height of 4–6 m above the sea surface, where 1-min. averaged particle size distributions were continuously recorded. These short intervals were chosen to study the response of the aerosol concentrations to changes in meteorological conditions.

The profile measurements were made between 9th and 20th September. In the afternoon of 16th September the TNO optical particle counter failed, probably due to a wave washing over the cage when the wind rapidly picked up to a storm. Repair was not possible, but fortunately data was available from the optical particle counters from the University of Hamburg.

Conclusions

The work described in this contribution was carried out in projects from which the results contributed to the goals of EUROTRAC [2, 9, 13, 14]. Details are thus described in many reports, symposium proceedings and refereed articles. One PhD thesis [47] resulted from work on the SeaCluse model that was partly carried out at TNO-FEL, and that was co-supervised by A.M.J. van Eijk from TNO-FEL who spent two months per year at Ecole Centrale de Nantes.

TNO-FEL had a major role in the development of the SeaCluse model (see [45–47] for conclusions) that evolved from the Eulerian CLUSE model [19, 42, 43], to provide a description of the sea salt aerosol life cycle and its effects in the presence of surface waves. Also a simple model was presented to calculate sea spray source functions from bubble size distributions [30]. Further efforts in the field of the production of sea spray aerosol are the development and application of a bubble measuring system [37, 49] and the use of the bubble spectra in conjunction with information on bubble-mediated droplet production and simultaneously measured aerosol size distributions to evaluate a sea spray source function [38].

The TNO-FEL data base on particle size distribution gradients has been extended with valuable information on particles smaller than the minimum size of 13 μm in diameter that we usually measure with our Rotorod impaction samplers [40], *i.e.* down to 0.2 μm.

Acknowledgements

The work described in this contribution was supported by the Commission of the European Communities (EG DG XII contracts STEP-CT90-0047, MAS2CT930056, MAS2CT930069 and EV5V-CT93-0313). TNO-FEL efforts were further supported by the Netherlands Ministry of Defence (Contracts A86KM106, A92KM615, A92KM635, A92KM767, A93KM699) and the US Office of Naval Research (ONR, Grants N00014-87-J-1212, N00014-91-J-1948). Part of the numerical modeling was made while Alexander M.J. van Eijk held a position as Enseignant Invité at the Ecole Centrale de Nantes, and while at other times Benoit Tranchant made an extended visit to TNO Physics and Electronics Laboratory.

References

1. De Leeuw, G., S.E. Larsen, P.G. Mestayer; Factors determining particle dynamics over the air-sea interface. this volume.
2. De Leeuw, G.; The physics of marine aerosols and their role in atmospheric chemistry. in: P.M. Borrell, P. Borrell, K. Kelly, T. Cvitaš, W. Seiler (eds), *Proc. EUROTRAC Symp. '96 Vol 1,* Computational Mechanics Publications, Southampton 1996, pp. 67–71.
3. De Leeuw, G., A.M.J. van Eijk; Factors determining particle dynamics near the air-sea interface. Study of the production and dynamics of marine aerosols. In: P.S. Liss *et al.* (Eds.) *EUROTRAC Ann. Rep. 1989, part 3 ASE*, EUROTRAC ISS, Garmisch-Partenkirchen 1990, pp. 36–37.
4. De Leeuw, G., A.M.J. van Eijk; Generation, transport and deposition of marine aerosols. *EUROTRAC Ann. Rep. 1990, part 3 ASE*, EUROTRAC ISS, Garmisch-Partenkirchen 1991, pp. 56–62.

5. De Leeuw, G., S.E. Larsen, P.G. Mestayer, A.M.J. van Eijk, H. Dekker, A. Zoubiri, P. Hummelshøj, N.O. Jensen; Aerosols in the marine atmospheric surface layer: production, transport and deposition. *EUROTRAC Ann. Rep. 1991, part 3 ASE*, EUROTRAC ISS, Garmisch-Partenkirchen 1992, pp. 61–67.
6. De Leeuw, G., S.E. Larsen, P.G. Mestayer; Factors determining particle dynamics over the air-sea interface. *EUROTRAC Ann. Rep. 1992, part 3 ASE*, EUROTRAC ISS, Garmisch-Partenkirchen 1993, pp. 32–38.
7. De Leeuw, G., A.M.J. van Eijk, H. Dekker; Studies on aerosols in the marine atmospheric surface layer. *EUROTRAC Ann. Rep. 1992, part 3 ASE*, EUROTRAC ISS, Garmisch-Partenkirchen 1993, pp. 94–102.
8. De Leeuw, G., A.M.J. van Eijk, P.G. Mestayer, S.E. Larsen; Aerosol dynamics near the air-sea interface. in: P.M. Borrell, P. Borrell, T. Cvitaš, W. Seiler (eds), *Proc. EUROTRAC Symp. '94,* SPB Academic Publishing bv, The Hague 1994, pp. 707–710.
9. De Leeuw, G., A.M.J. van Eijk, A.I. Flossmann, W. Wobrock, P.G. Mestayer, B. Tranchant, E. Ljungstrom, R. Karlsson, S.E. Larsen, M. Roemer, P.J.H. Builtjes; Transformation and vertical transport of N-compounds in the marine atmosphere. in: P.M. Borrell, P. Borrell, K. Kelly, T. Cvitaš, W. Seiler (eds), *Proc. EUROTRAC Symp. '96 Vol. 1,* Computational Mechanics Publications, Southampton 1996, pp. 195–198.
10. Larsen, S.E., J.B. Edson, P.G. Mestayer, C.W. Fairall, G. de Leeuw (1990). Sea spray and particle deposition, air/water tunnel experiment and its relation to over-ocean conditions. in: P. Borrell, P.M. Borrell, W. Seiler (eds), *Proc. EUROTRAC Symp. '90,* SPB Academic Publishing bv, The Hague 1991, pp. 71–75.
11. Larsen, S.E., P. Hummelshøj, M.P. Rouault; Aerosol dynamics, transport and deposition near the sea surface. *EUROTRAC Ann. Rep. 1990, part 3 ASE*, EUROTRAC ISS, Garmisch-Partenkirchen 1991, pp. 50–55.
12. Larsen, S.E., G. de Leeuw, A. Flossmann, E. Ljungstrom, P.G. Mestayer; Transformation and removal of N-compounds in the marine atmosphere. *Final report on the STEP project CT90-0047*, Risø National Laboratory, Roskilde, Denmark 1994.
13. Larsen, S.E., J.B. Edson, P. Hummelshøj, N.O. Jensen, G. de Leeuw, P.G. Mestayer (1994b). Laboratory study of the particle dry deposition velocity over water. in: P.M. Borrell, P. Borrell, T. Cvitaš, W. Seiler (eds), *Proc. EUROTRAC Symp. '94,* SPB Academic Publishing bv, The Hague 1994, pp. 698–701.
14. Larsen, S.E., J.B. Edson, P. Hummelshøj, N.O. Jensen, G. de Leeuw, P.G. Mestayer; Dry deposition of particles to the ocean surfaces. *Ophelia* **42** (1995) 193–204.
15. Mestayer, P.G.; Numerical simulation of marine aerosol dynamics. *EUROTRAC Ann. Rep. 1992, part 3 ASE*, EUROTRAC ISS, Garmisch-Partenkirchen 1993, pp. 112–116.
16. Mestayer, P.G., A. Zoubiri, J.B. Edson; Experimental and numerical study of aerosol dynamics and deposition at the sea surface; 2PIE, GWAIHIR & CLUSA. *EUROTRAC Ann. Rep. 1990, part 3 ASE*, EUROTRAC ISS, Garmisch-Partenkirchen 1991, pp. 66–74.
17. De Leeuw, G.; Aerosols near the air-sea interface. *Trends in Geophys. Res.* **2** (1993) 55–70.
18. Edson, J.B.; Lagrangian model simulation of the turbulent transport of evaporating jet droplets. Ph.D. thesis, Dep. of Meteorol., Penn. State Univ., University park 1989.

19. Rouault, M.P.; Modélisation numérique d'une couche limite uni-dimensionelle stationnaire d'embruns. Thèse de Doctorat, Univ. d'Aix-Marseille II, France. 1989

20. Spiel, D.E.; A hypothesis concerning the peak in film drop production as a function of bubble sizes. *J. Geophys. Res.* **102** (1997) 1153–1161.

21. Blanchard, D.C.; The electrification of the atmosphere by particles from bubbles in the sea. *Prog. Oceanography* **1** (1963) 171–202.

22. Resch, F., G.M. Afeti; Film drop distributions from bubbles bursting in seawater. *J. Geophys. Res.* **96** (1991) 10681–10688.

23. Resch, F., G.M. Afeti; Submicron film drop production by bubbles in seawater. *J. Geophys. Res.* **97** (1992) 3679–3683.

24. Blanchard, D.C.; The size and height to which jet droplets are ejected from bursting bubbles in sea water. *J. Geophys. Res.* **94** (1989) 10999– 11002.

25. Spiel, D.E.; The number and size of jet drops produced by air bubbles bursting on a fresh water surface. *J. Geophys. Res.* **99** (1994) 10289–10296.

26. Kientzler, A.B., A.B. Arons, D.C. Blanchard, A.H. Woodcock; Photographic investigation of the projection of droplets by bubbles bursting at a water surface. *Tellus* **6** (1954) 1–7.

27. Dekker, H.J., G. de Leeuw; Bubble excitation of surface waves and aerosol droplet production: a simple dynamical model. *J. Geophys. Res.* **98(C6)** (1993) 10223–10232.

28. Smith, M.H., P.M. Park, I.E. Consterdine; Marine aerosol concentrations and estimated fluxes over the sea. *Quart. J. Roy. Meteorol. Soc.* **119** (1993) 809–824.

29. Fairall, C.W., K.L. Davidson, G. Schacher; An analysis of surface production of sea-salt aerosol. *Tellus* **35B** (1983) 31–39.

30. De Leeuw, G.; Spray droplet source function: from laboratory to open ocean. In: P.G. Mestayer, E.C. Monahan and P.A. Beetham (eds), *Modelling the fate and influence of marine spray*. Proc. of workshop in Luminy, Marseille 1990, pp. 17–28.

31. Andreas, E.L.; Sea spray and turbulent air-sea fluxes. *J. Geophys. Res.* **97** (1992) 11429–11441.

32. Andreas, E.L.; Reply. *J. Geophys. Res.* **99** (1994) 14345–14350.

33. Katsaros, K.B., G. de Leeuw; Sea spray and the turbulent air-sea heat fluxes - Comments. *J. Geophys. Res.* - Oceans **99C7** (1994) 14339–14343.

34. Mestayer, P.G., C. Lefauconnier; Spray droplet generation, transport and evaporation in tunnel during the Humidity Exchange Over the Sea experiments in simulation tunnel. *J. Geophys. Res.* **93** (1989) 572–586.

35. Monahan, E.C., D.E. Spiel, K.L. Davidson; A model of marine aerosol generation via whitecaps and wave disruption. in: E.C. Monahan and G. MacNiocaill (eds), *Oceanic whitecaps and their role in air-sea interaction exchange processes*. Reidel, Dordrecht 1986, pp. 167–174.

36. Andreas, E.L., J.B. Edson, E.C. Monahan, M.P. Rouault, S.D. Smith; The spray contribution to net evaporation from the sea: A review of recent progress. *Boundary Layer Meteorol.* **72** (1995) 3–52.

37. De Leeuw, G., L.H. Cohen; Bubble size distributions in coastal seas. in: B. Jähne, E.C. Monahan (eds), *Air-Water Gas Transfer*, AEON Verlag & Studio, Hanau 1995, pp. 325–336.

38. Spiel, D.E., G. de Leeuw. Formation and production of sea spray aerosols. *J. Aerosol Sci.* **27** Suppl. 1 (1996) 65–66.

39. De Leeuw, G., S.E. Larsen, P.G. Mestayer, J.B. Edson, D.E. Spiel; Deposition, dynamics and influence of particles in the marine atmospheric boundary layer, laboratory study. TNO Physics and Electronics Laboratory, October 1991.

40. De Leeuw, G.; Vertical profiles of giant particles close above the sea surface. *Tellus* **38B** (1986) 51–61.

41. Edson, J.B., C.W. Fairall; Spray droplet modeling I: Lagrangian model simulation of the turbulent transport of evaporating spray droplets. *J. Geophys. Res.* **99** (1994) 25295–25311.

42. Rouault M.P., S.E. Larsen; Spray droplets under turbulent conditions. *Rep. Risø-M-2847*, Risø Natl. Lab., Denmark 1990.

43. Rouault, M.P., P.G. Mestayer, R. Schiestel; A model of evaporating spray droplet dispersion. *J. Geophys. Res.* **C96** (1991) 7181–7200.

44. Van Eijk, A.M.J., P.G. Mestayer, G. de Leeuw; Conversion of the CLUSE model for application over open ocean; progress report. TNO Physics and Electronics Laboratory, *report FEL-93-A035*, 1993.

45. Mestayer, P.G., A.M.J. van Eijk, G. de Leeuw, B. Tranchant; Numerical simulation of the dynamics of sea spray over the waves. *J. Geophys. Res.* **101** (1996) 20771–20797.

46. Van Eijk, A.M.J., B. Tranchant; Numerical simulations of sea spray aerosols. *J. Aerosol Sci.* **27 Suppl. 1** (1996) 61–62.

47. Tranchant, B.; Simulations numeriques des embruns marins. Thèse de Doctorat, Ecole Centrale de Nantes et Université de Nantes, France, 6 Mars 1997.

48. Schulz, M., W. Dannecker; *EUROTRAC Ann. Rep. 1990, part 3 ASE*, EUROTRAC ISS, Garmisch-Partenkirchen 1991, pp. 80–83.

49. De Leeuw, G., L.H. Cohen; Measurements of oceanic bubble size distributions. OCEANS94, *Proc. Vol. II*, Brest, France 1994, pp. 694–699.

2.6 Fluxes in the Marine Atmospheric Boundary Layer

S.E. Larsen[1], L. L. S. Geernaert[1], F.Aa. Hansen[1], P. Hummelshøj[1], N.O. Jensen[1] and J.B. Edson[2]

[1]Risø National Laboratory, Roskilde, Denmark
[2]Woods Hole Oceanographic Institution, Woods Hole, Maryland, USA

The work performed within the framework of ASE subproject has been aimed mainly on the following subjects:

1 Deposition of submicron particles were studied in the TWO-PIE experiment in the IMST wind-wave tunnel in Marseilles. The TWO-PIE experimental data focused on deposition of tracer aerosols to the water surfaces in the presence of a simulated whitecap and associated sea spray aerosol. The results show that presence of bubble bursting and sea spray does not influence the deposition velocity of submicron particles significantly.

2 The budget for aerosols and other trace constituents in the marine atmospheric boundary layer (MABL) were studied both theoretically and experimentally. The role of entrainment for ventilating the MABL were found from general budget consideration within the EC-NTRANS projects and experimentally from the data obtained during the two-ship pseudo Lagrangian EUROTRAC NOSE (North Sea) 1991 experiment. The importance of the exchange between marine aerosol and soluble gases as HNO_3 and NH_3 for the deposition of such gases was clarified in a number of experiments conducted at sites within the inner Danish waters, in the Anholt and the Vindeby experiments. The data showed that particle- gas exchange was responsible for up to about half of deposition rates, estimated at standard heights.

In connection with the NOSE experiment, the Risø group had a sonic anemometer operating semi-automatically on one of the participating ships. Here was the robustness of the dissipation method that was once again demonstrated, and the low-frequency co-spectral method, employing the universality of the co-spectra corresponding to vertical fluxes, was tested here for the first time for stress and heat flux, comparing to standard drag formulations. These results were employed in:

3 The air-sea exchange of CO_2 was studied in the ASGASEX experiment in 1993 in the EC-MAST2 targeted project OMEX (Ocean Margin EXchange), where fluxes stress, heat, water vapour and CO_2 was successfully measured, from cruising ships in 1995, using a mixture of dissipation technique, co-spectral technique and the exchange coefficient formulations.

Aims of the research

The overall aim of the research was to improve the understanding of air-sea fluxes and their role for atmospheric processes in the marine boundary layer.

The work on the TWO-PIE experiment and data analysis has been part topic 5 within the ASE subproject: factors determining particle dynamics and deposition at the air-sea interface. This topic aims at describing the production of bubbles and sea spray aerosols at the water surface and the effect of these processes on the trace constituents in the atmosphere, their transformation, and deposition. The TWO-PIE 1990 experiment is described in earlier EUROTRAC ASE annual reports. This experiment was part of series of measuring campaigns conducted in the wind/water tunnel at Institut Mecanique Statistique de la Turbulence (IMST), Marseilles, to study the dynamics of sea spray aerosols as well as the importance of the bubble/spray production for the deposition of submicron aerosols.

The work on budgets and fluxes of trace constituents in the marine atmospheric boundary layer (MABL) is part of the ASE topic 3: the role of atmospheric transformations of trace gases and aerosols in the marine deposition process. The overall aim of the EC-NTRANS projects have been to summarise aspects of the budgets for nitrogen compounds in the MABL, their transformation and exchanges with the boundaries, through entrainment at the top and deposition at the sea surface.

The aim and the results of the NOSE 1991 campaign is reported by M. Schulz in this volume. It was centred on establishing *pseudo* Lagrangian data from simultaneous measurements from two ships, that in conjunction with a mass-balance model, could be used to describe fluxes and other meteorological and chemical processes of importance for the load of atmospheric trace constituent to the North Sea. The Risø group participated with model developments and experimentally by measurements of stress and heat flux from one of the ships in the experiment. The specific aim of this research was primarily to support the NOSE experiment with meteorological data. The secondary aim was to test our ability to perform such unmanned micrometeorological measurements in connection with ship-based measuring campaigns.

The aim of the Vindeby experiments in 1993 and 1994 was to establish simultaneous measurements of gradient of water soluble nitrogen gases (NH_3 and HNO_3) and aerosols within the marine atmospheric surface layer, thus allowing modelling of fluxes and chemical exchange of these species near the surface.

Also, the joint FPN92-EUROTRAC air-sea exchange experiment is described in this report by M. Schulz. The overall aim of the measurements was to study the air-sea exchange of gaseous and particulate matter over the North Sea. The specific aim of the Risø group was to study the relation between meteorological conditions and the concentrations of submicron aerosols over the North Sea.

The measurement of CO_2 exchange within the ASGASEX and the OMEX projects have been part of topic 4 within ASE: the magnitude of the air-sea flux of anthropogenic and natural substances in European regional seas. The aim of the Risø group, that also in this project formed a joined group with the TNO partner within ASE, was to test and use micrometeorological instrumentation to measure the air-sea fluxes of CO_2, and therefore also of momentum, heat and water vapour (these meteorological parameters are all essential for determining the interfacial fluxes of trace constituents as CO_2) over the European Marginal Sea. Within the ASGASEX experiment the measurements were conducted at the platform Meet-Poost Nordweijk, 10 km off the Dutch coast, while the OMEX project focused on the shelf break between Ireland and Brittany. The micrometeorological measurements was compared with simultaneous results based on exchange coefficient expressions to study the fluxes under inhomogeneous and non-stationary situations, where the latter have been considered in doubt.

Research programme

As is apparent from the above, the research conducted by the Risø group within the framework of ASE has been composed of several sub-activities. These have all been carried out in cooperation with other partners within ASE as well as with partners from outside the EUROTRAC project.

Deposition of submicron particles

In the deposition part of the TWO-PIE experiment test particles (here 0.2–1 μm diameter) were injected into the IMST tunnel under conditions of different bubbling rates at the water surface. The bubbling was produced and controlled through a subsurface array of aquarium frits, through which the airflow could be varied. All deposition models for such particles indicate that the main resistance against deposition is posed by the laminar layer close to the water surface. Since bursting bubbles disrupt this layer, the working hypothesis for the experiment was that bubble bursting would increase the deposition velocity. These considerations were confirmed and quantified by model developments, presented by Hummelshøj *et al.* [1]. However, the results obtained from the analysis show that within the uncertainty of the experiment no such increase of deposition can be found in the data. Indeed, the analysis indicates that the deposition velocity is unchanged within 30 % for change of whitecap cover between 0 and 20 %, corresponding to a velocity range between 0 and 25 m/s over natural waters [2–5]. An illustration of the data is presented in Fig. 2.6.1. A possible explanation of this result was found in the so called "surface renewal model" for the transport across the laminar surface layer. In this model the transport mainly takes place through a number of surges penetrating down to the water surface from the air above the laminar surface layer. Since this would be exactly the process induced by bursting bubbles, the bubbles do not contribute with a new process to the transfer. Rather they add one more bursting process that contributes with a strength corresponding to the relative area of bursting bubbles being roughly equal to the relative whitecap area of the water surface (typically a few per cent).

It may even be less since the bubble bursting to some extent might replace other bursting processes.

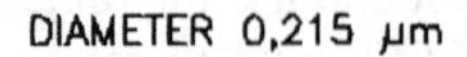

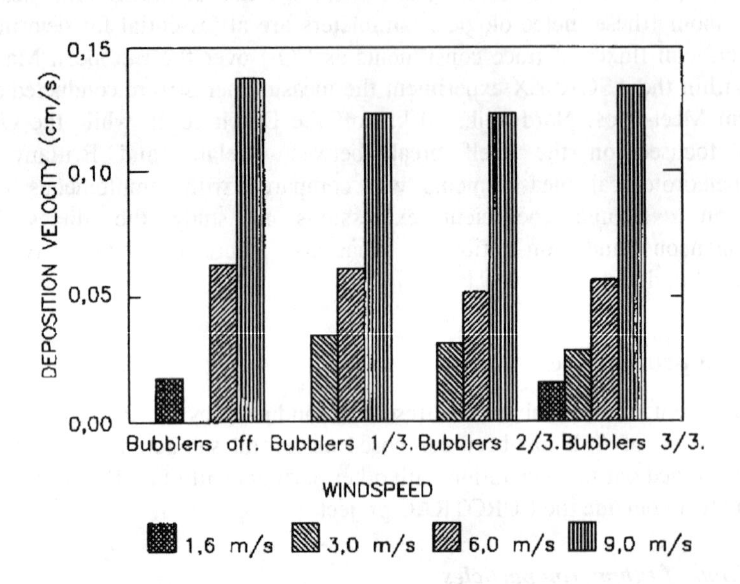

Fig. 2.6.1: Example of the measured deposition velocities for different velocities and different bubbling intensities from the TWO-PIE 1990 experiment in the wind-wave tunnel at IMST in Marseilles. The example shown pertains to 0.2 micrometer particles. The bubbling intensities used correspond approximately to the following white cap covers over the ocean (%): 1/3 (6.6 %), 2/3 (12.8 %) and 3/3 (19.2 %), [5].

Heterogeneous processes and mass balances

In connection the EC-NTRANS projects and the analysis of the data from the NOSE 1990 experiment, the Risø group has focused on studying the mass balance for air masses in the MABL. For a homogeneous boundary layer with only small lateral or vertical concentration variations, we consider the mass balance for a volume of air limited by the height of the atmospheric boundary layer, h. We will follow the air mass as it is advected by the mean wind, and we assume that the boundary layer is well mixed throughout its depth, meaning that all concentrations are constant with height. We denote the concentration of the species considered by $C = C(t)$, where t is the flight time in a Lagrangian sense. Similarly the other parameters in (1) can be considered functions of t.

$$\frac{d}{dt}C = -\frac{v_d}{h}C - \frac{w_e}{h}(C - C_t) - \lambda C + E_C + P \qquad 1$$

I II III IV V VI

In equation (1) term II is the decrease in C due to deposition at the surface. From physical considerations are found, that v_d can be described by:

$$v_d \leq u_* \frac{u_*}{U} + V_g , \qquad 2$$

where u_* is the friction velocity, U the mean air velocity and V_g the gravitational settling velocity, nonzero only for aerosols. A typical value coming out of equation (2) is v_d being about or less than 1 cm/sec. Species as ammonia and nitric acid show apparent (see below) deposition velocities close to the upper limit, given by (2), NO, CO_2 and NO_2 show considerably smaller values, because they are slowly dissolving gases, for which the transfer across the laminar interfacial water layer is strongly inhibited.

Particles will have larger deposition velocities for diameters larger than, say, 10 μm . The submicron particles on the other hand show deposition velocities considerably smaller than (2), typically one to two orders of magnitude less, depending on actual size. The TWO-PIE experiment described above, was aimed to clarify some of the uncertainties associated with the understanding of the particle deposition velocity, and confirmed that the deposition rate stayed small even with an active sea with white caps and sea spray.

Term III describes the depletion of C due to entrainment of air from the free troposphere, across the boundary layer top. It is controlled by the entrainment velocity, w_e, and the difference in concentration C in the boundary layer and the concentration C_t in the free troposphere. The entrainment rate or the entrainment velocity is related to the rate of change in the boundary layer height and is typically of the order of 0.1 to 2 cm/sec, but can increase to more than 5 cm/sec during frontal passages, and further up to more than 10 cm/sec at cloud bases. Since the nitrogen species considered in this project are produces in the boundary layer, C_t is often considerably less than C, and with respect to the boundary layer and species C, term III will mostly signify detrainment. The fairly large value of w_e, relative to v_d, therefore indicate that the entrainment processes often will dominate the deposition processes as the reason for depletion of the boundary layer.

The entrainment velocity is difficult to determine in general. A method often used employs knowledge about the structure of the entrainment zone:

$$w_e = A w_* / Ri_* \qquad 3$$

where R_i is a Richardson number based on the boundary layer height, h, the temperature gradient across the inversion zone and the convective velocity scale w_* and A is coefficient of order one [6]. Other methods employ the concept of top-down/bottom-up diffusion [7, 8].

In (1) a boundary layer is envisaged with a simple interface at height h. For the clear sky entrainment processes, one must in more realistic models account for the existence of an entrainment zone, Δh, where the transition from boundary layer to the free atmosphere occurs, see [6]. This transition zone is often so deep, of the order of 0.2–0.6 times h, that with typical entrainment velocities, the transport through the layer takes many hours.

For the wet marine aerosol, reaching this height, such a transition time means that the particles passing through the entrainment zone will have time to adapt to the usually much lower humidity here, and reach the free atmosphere, as relatively dry particles.

Term IV describes the loss from the boundary layer through precipitation, with λ being the wash out coefficient for the species considered in the equation. The term is of course zero, when there is no precipitation. The wash out coefficient λ increases with rain intensity and depend on the size distribution of the rain drops *etc.* as well as the characteristics of the species considered. For dissolvable gases, *i.e.* HNO_3, H_2O_2 and NH_3 as well as small particles λ varies typically between 5 and $20 \cdot 10^{-5}$ sec^{-1} for a rain intensity between 1 and 10 mm/hr. For other particles and slowly dissolvable gases it, λ is smaller even zero.

Term V corresponds to term IV, in the sense that it indicates re-injection of C into the boundary layer, by evaporating drops on the way to the surface in rain situations, or by rising evaporating spray aerosols, that received C by reaction between C and spray droplets close to the surface. The precipitation part is of course zero, when there is no precipitation. It may even be close to zero during precipitation, if the evaporation is minimal. However, since about 80 % of the wet deposition at the surface is due to within clouds scavenging and nucleation, even a small evaporation may counterbalance the wash out loss in term IV. The role of the term for dry conditions as a redistributor of C, that was washed out by spray drops is part of an ongoing project on the effect of spray particle–gas interaction on deposition rates of highly soluble gases such as NH_3 and HNO_3, see below.

Term VI reflects the production/removal of C in the boundary layer for the species considered by one of the many microphysical or chemical processes occurring. For species on aerosol form, the term might reflect both changes in aerosol size, due to feed back between the humidity field and the particle size (Mestayer *et al*, and De Leeuw *et al* in this volume), and chemical changes due to reaction with the gaseous species around.

For the studies on the importance of interaction between turbulent fluxes and gas-particle interactions for fluxes of solvable gases within the marine surface layer, it is more convenient to write (1) on a differential and more basic form, retaining only in the most important terms:

$$\frac{dC}{dt} = -\frac{\partial}{\partial z}(\langle cw \rangle - V_g C) + \alpha C , \qquad 4$$

where we have introduced the Reynolds convention for the species concentration, separating it into a mean value, C, and a fluctuating part, c.

Further, we have allowed for a fall velocity, V_g, for particles, and modelled all the concentration changes, due to chemical processes and/or particulate changes, as a first order process, which of course just leave us with the job to describe α including the relevant processes for the individual species.

In (4),the flux term, <cw>, must be related to C and the reaction rates. Within the EUROTRAC ASE project the equation was used by Geernaert *et al.* [9, 10] in analysing the profile data for marine aerosols, HNO_3 and NH_3 obtained at the Anholt and the Vindeby experiments, assuming stationarity and realistic values for α for the reactions between the aerosols and the two species, including the assumption that the rates were proportional to the total surface area of the sea salt aerosols available during the individual experiments. Examples of a profile analysis is presented in Fig. 2.6.2 taken from Geernaert *et al.* [10].

These experiments were initiated with measurements in 1991 and 1992 at shore of the island of Anholt in Kattegat during the effort to determine realistic deposition rates for nitrogen compounds to be used in the models being developed to describe the atmospheric contribution to the eutrophication of the inner Danish waters, see Asman *et al.* in this volume and [9, 11]. They were continued at an off shore site, Vindeby, in the southern part of the Great Belt in 1994 and 1995 [10].

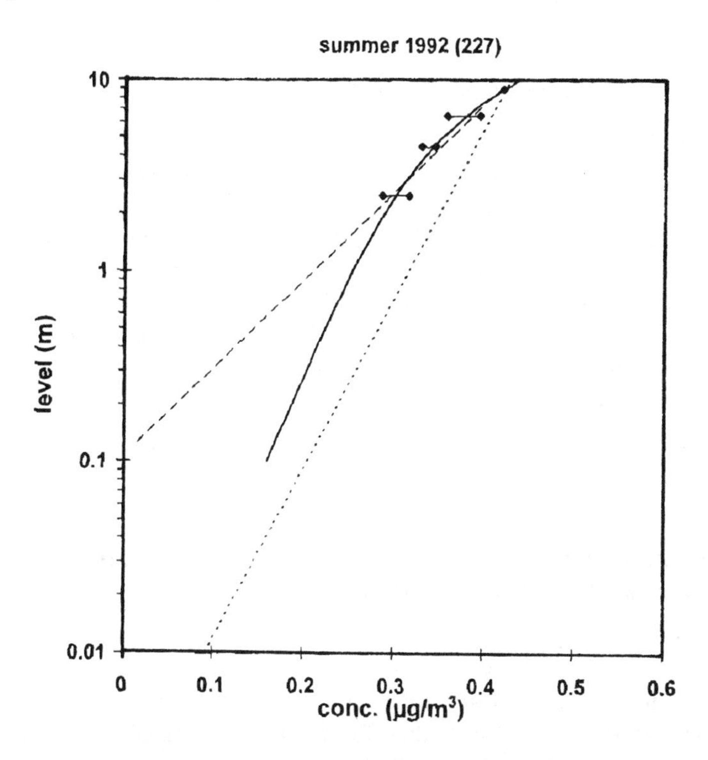

Fig. 2.6.2: Examples of measured and calculated HNO_3 gradient, taken from [10].
- - - -: gradient from extrapolation of the measured concentrations.
¨: measured concentrations of HNO_3
..........: gradient calculated from resistance theory,
------------ : gradient calculated taking the sink term in (4) into account.

As shown in Fig. 2.6.2 the experiments and the subsequent analysis indicate clearly the need for inclusion of the reactivity term within (4), when trying to infer the surface flux from measurements at "standard heights". Also the need for continued work is implied, because part of the spray aerosol that have absorbed the gases under consideration, will continue to fall back into the ocean, and hence constitute a new deposition pathway, while other parts of the aerosols will diffuse up into the boundary layer proper and here, as they evaporate, release the gases absorbed at the surface. Hence they will present an other process of the term IV type in equation (1).

Considerations of equation (1) for the situations during the NOSE experiment showed that a 30 % reduction of the concentration for most species should be expected between the two ships, by either one or more combinations of the many terms listed above. Since different terms were important for different species, however, it was still possible to derive the importance of the individual processes for most species by a multiple regression approach, see Schulz in this volume and [12].

Micrometeorological estimation of air-sea fluxes

The analysis of the turbulence data from the NOSE experiment has concentrated on flux estimation [5, 13]. It was found that the different methods of estimating the surface fluxes of momentum and heat agreed with each other with an standard deviation of about 15 %. The different methods applied were the exchange (or the drag) coefficient method, the dissipation method and the co-spectral method. Specifically, it was noted that both the inertial dissipation method and the co-spectral method agreed well with each other and seemed remarkably robust in the face of the severe flow distortion and platform motion induced by the moving ship.

For the treatment of the indirect flux estimation techniques and their validity, it is convenient to carry out the discussion using co-spectra and power-spectra based on the Fourier transforms of the signals:

The power spectrum is the expansion of the fluctuations into harmonics with frequency f [Hz] and integrates to the variance.

$$\langle \gamma'^2 \rangle = \int_o^\infty S_\gamma(f)df \qquad 5$$

The co-spectrum is the corresponding frequency expansion between the correlation between two signals. Note that if one correlate a signal with itself the corresponding co-spectrum becomes a power spectrum as the corresponding co-variance becomes a variance.

$$\langle \zeta'\gamma' \rangle = \int_o^\infty Co_{\zeta\gamma}(f)df, \qquad 6$$

where ζ and γ stands for any of the turbulence variables we have considered above. Note that if one of the variables in the co-variance and the co-spectrum is w the covariance becomes the vertical turbulence flux, and the co-spectrum is then a frequency expansion of this flux.

Spectra and co-spectra of atmospheric turbulence are known to show a high degree of universality, when properly scaled. This universality of the forms are especially valid in the high frequency end of the power spectra, where they follow the so called inertial range behaviour of a power law as function of frequency:

For the spectra of interest these inertial forms look:

$$f\, S_u(f) = \alpha \varepsilon^{2/3} (\langle u \rangle /2\pi)^{2/3} f^{-2/3} \; . \qquad (7)$$

$$f\, S_\gamma(f) = \beta N_\gamma \varepsilon^{-1/3} (\langle u \rangle /2\pi)^{2/3} f^{-2/3} \; . \qquad (8)$$

In equations (7) and (8) we have introduced a number of parameters: ε is the dissipation of velocity fluctuations and is the rate by which viscous forces dissipate the turbulent velocity fluctuations. Correspondingly we have dissipation for γ, where γ as above indicate any passive scalar, *e.g.* T, q or q_{CO_2} fluctuations, they are denoted by N with the appropriate subscript. The letter q denotes the mixing ration, without a subscript it refers to water vapour, while any other gas are indicated with a subscript. α and β are the Kolmogorov constants for inertial range spectra for the longitudinal velocity and scalars respectively. They are considered universal constants with $\alpha = 0.55$ and $\beta = 0.8$.

The dissipation terms ε and N_γ can be described in terms of Monin-Obuchkov similarity functions. The definitions are:

$$\phi_\varepsilon\left(\frac{z}{L}\right) = \frac{\kappa z \varepsilon}{u_*^3} \; , \quad \phi_{N_\gamma}\left(\frac{z}{L}\right) = \frac{\kappa z N_\gamma}{u_* \gamma_*^2} \; , \qquad (9)$$

where γ_* follows the definition : $\langle w'\gamma' \rangle = -\, u_* \gamma_*$, and L is the Monin-Obuchkov stability scale.

The so called inertial dissipation method for determination of the surface fluxes consist in reversing (7,8) to determine the scale parameters, and thereby the fluxes, using the definition in (9) for the dissipation functions.

$$u_* = \left[\frac{f\, S_u(f)}{\alpha}\right]^{1/2} \left[\frac{2\pi}{\langle u \rangle} \frac{\kappa z}{\phi_\varepsilon} f\right]^{1/3} , \qquad (10)$$

$$|\langle w'\gamma' \rangle| = \left[\frac{\alpha}{\beta} \frac{\phi_\varepsilon}{\phi_{N_\gamma}} \frac{f\, S_\gamma(f)}{f\, S_u(f)}\right]^{1/2} u_*^2 \qquad (11)$$

where it is understood that the spectral values have to be taken in within the inertial subrange, where (7) and (8) apply. As indicated in (11) the dissipation method can not tell about the sign of the fluxes, these have to be deduced from other information available, for example the co-spectra or the air-water difference. The stability dependency can be included iteratively, using (9) .

The advantage of the dissipation method becomes clear when one considers how data obtained from ship borne measurements unavoidably must be distorted by flow perturbation by the ship itself and also by the motion of the ship. The distortion is illustrated in Fig 2.6.3, showing spectra of the three velocity components obtained during an NOSE cruise. It is seen how the ships motion enters as additional spectral energy at 0.1–1.0 Hz, while the inertial ranges of the spectra show relatively low distortion, and the dissipation method can be applied here.

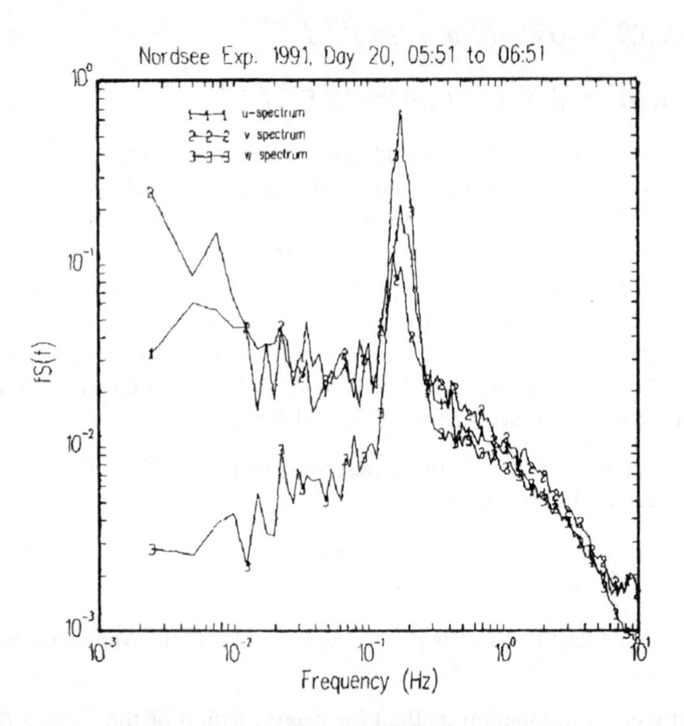

Fig, 2.6.3: Power spectra of the three velocity components *u*, *v* and *w* obtained during the Poseidon cruise in connection with the NOSE 1991 experiment.

Some uncertainty will however always exist as to the importance of more subtle flow distortion, not showing up as dramatically as in Fig. 2.6.3. These aspects of flow distortion are very complicated bordering to impossible to take fully into account [14, 15]. Therefore it becomes important to derive the fluxes from as many methods as possible, and preferably from methods with different sensitivities to the flow distortion and ship motion. Of such methods we have already described the exchange coefficient method, that is certainly different from the dissipation method, in that it is based on mean values, while the dissipation estimates are based on high frequency turbulence measurements.

The co-variance method, as defined by (6), in a way is the basic definition of the flux. We have further extended the method by use of the co-spectrum. Review of the literature [16–22] on the co-spectra reveals that for unstable to neutral conditions the co-spectra have a fairly broad maximum in a normalised frequency range between 0.01 and 0.1. The maximum of the spectra are found to be:

$$f\,Co_{uw}(f) \approx 0.22 < u'w' > \text{ in the interval } 0.008 < f < 0.12 \text{ Hz},$$

$$f\,Co_{w\gamma}(f) \approx 0.20 < w'g' > \text{ in the interval } 0.007 < f < 0.12 \text{ Hz},$$

With the main influence of the ships motion to be found around 0.1–1.0 Hz, the universal and simple form of the co-spectra for unstable to neutral conditions suggests that the amplitude around 0.01 Hz offers the best conditions for determining the corresponding flux. For stable conditions the conditions are more difficult, but since stable data have not been important in the context of the present report, this aspect shall not be further discussed here.

The flow distortion depends on the scale of the turbulence and the scale of the distorting platform, and generally it must be expected to be quite different for small scale and large scale turbulence. These considerations have also been carried into quantitative estimates for some distorting bodies and platforms, see discussions in [14, 15, 23–25]. Therefore, although it is virtually impossible to provide an accurate description of the flow distortion effects on the turbulence around a moving ship, the inertial dissipation method and the co-spectral method, provides two estimates of the flux, that depends on frequency ranges of the signal, that is outside the main distortion from the ships motion, and furthermore the residual distortional effects on the two estimates are know to be different in magnitude and character, owing to the scale differences of the turbulence involved in the two estimates. Conversely, if the two methods yield similar results, the flow distortion can not have a serious influence on the estimate.

Air-sea fluxes of CO_2

The overall objectives of the Risø group within topic 4 of EUROTRAC ASE has been to develop and use a micrometeorological instrumentation to derive estimates of air-sea fluxes of momentum, heat, water vapour, with emphasis on CO_2, to test the exchange coefficient expressions, especially for conditions of instationarity and inhomogeneity.

Measurement of fluxes of CO_2 by micrometeorological methods is associated with the same flow distortion problems as discussed above. Furthermore, an additional problem for most slowly dissolving gases with very small exchange rates is that even the very small and not measurable mean vertical velocity existing close to the surface contributes to the flux. We have followed the usual method in applying the 'Webb' correction, which in turn demands that determination of the CO_2 flux necessitates determination of the heat and the water vapour flux as well [26].

Since the temperature measured by the sonic, is derived from the speed of sound the so called sonic virtual temperature, that is a mix between temperature and humidity, the Webb correction takes a slightly different form from the usual one [27]:

$$F_{qCO2} = \langle w' q_{CO2} \rangle + 1.1 \langle q_{CO2} \rangle \langle w'q' \rangle + \langle q_{CO2} \rangle \frac{\langle w' T_{sv} \rangle}{\langle T \rangle} \qquad 12$$

where we have neglected $\langle q \rangle$ compared to 1, and also used $T \approx T_{sv}$.

For all the data considered here the Webb correction was of the same magnitude as the turbulence CO_2 flux alone, making the flux corrections a very critical exercise.

Furthermore the sensors presently available for the CO_2 measurements are only marginally well suited for measurement of the low signals levels over the ocean. In the present project this problem was reduced by that the two partners TNO and Risø each bought an instrument allowing for increasing the signal/noise ratio by cross-correlation between their outputs. The sensors used were of the type Advanet, an infra red device measuring both CO_2 and water vapour concentrations [28].

The ASGASEX experiment took place in September 1993. The measurements were conducted on board the Dutch platform Meet-Poost Nordweijk, and a comprehensive description of the experiment is given in Oost [29]. The overall purpose of the experiment was to establish a comparison between CO_2 flux measurements established by the different groups working in the field, and each applying different sensors. Unfortunately most of the CO_2 sensors failed as a proof of the present early stage of development for these sensors. However, one of our sensors survived and provided quite satisfactory data, that largely supported the exchange coefficient proposed by Wanninkhof [30], but with monumental scatter [3, 31, 32]. Further work was conducted at the Vindeby site, already mentioned above, refining and testing the sensor set-up and the methods of analysis.

During the OMEX project measurements were conducted successfully in connection with three cruises in 1995, must be mentioned a relatively successful comparison between the micrometeorological fluxes and the ΔpCO_2 method, reported by Keir *et al.* [33], from the Poseidon 211 cruise from Reykjavik to Lisbon. In the first part of that cruise a fairly constant under saturation of water CO_2 was present, allowing a relatively simple comparison between the two methods, without being concerned about inhomogeneous conditions. The wind speed however changed from virtually calm to slightly less than 20 m/s in connection with the cruise, hence it was possible to test for the influence of non-stationarity on the exchange rate. The analysis of these data showed that fluxes estimated by the dissipation method and the co-spectral method agreed well with each other and also reasonably well with fluxes derived by the exchange coefficient methods [27]. In Fig. 2.6.4 we present and compare the CO_2 fluxes estimated by the dissipation and by the co-spectral technique from the Poseidon cruise. As for the other parameters measured the methods show good comparison, a bit worse that the other parameters, but given the technological difficulties quite acceptable. Spectacular are the high values at 6 a.m. the 5th of September and after 6 p.m. on the 6th, both corresponding to rapidly changing conditions.

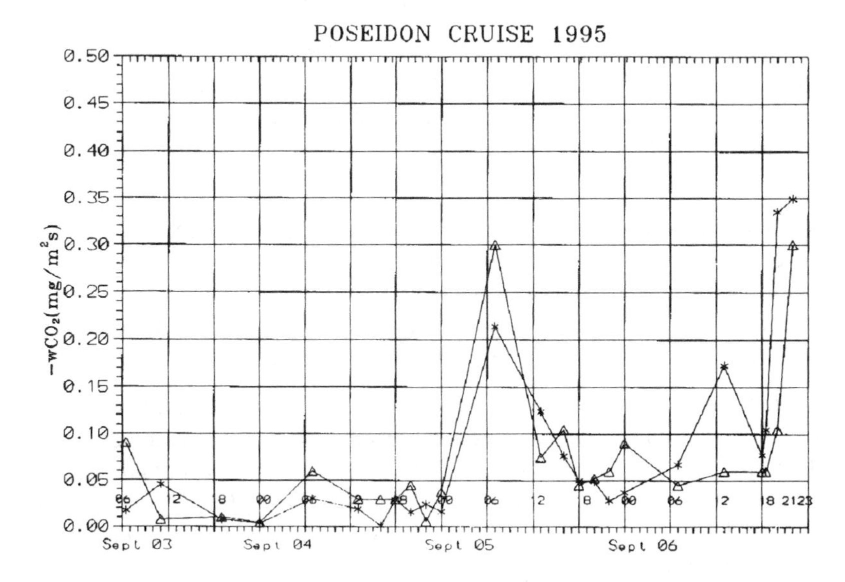

Fig. 2.6.4: Time trace of the turbulent CO_2 flux estimated by the co-spectral method, _, and the dissipation method, *. It must be pointed out that the fluxes presented here are the pure turbulence fluxes. The Webb corrections to be applied are the same for the two estimation methods, but are not applied here [27].

Principal results

The main result of the TWO-PIE data analysis is that the deposition rate of submicron aerosols seems unaffected by bubble and spray processes at the water surface [34]. This result was unexpected.

The analysis of the NOSE data has illustrated the robustness of the dissipation method when used to estimate fluxes from moving flow-distorting platforms like a ship. The results showed the possibility of operating micrometeorological sensor packets semi automatically during ship cruises, and provided the first comparison between heat and momentum fluxes measured both by dissipation method and the co-spectral method.

Furthermore, simple analysis of the boundary layer mass-balance for the simple Lagrangian experiment with only two ships as in the NOSE experiment was shown to predict an overall reduction in air concentration of about 30 % between two ships being 500 km apart. The objectives of the experiment could be fulfilled, though, provided that a sufficient number of different species was measured allowing for determination of the individual processes from a multiple regression type of analysis.

The mass balance analysis shows as well that entrainment processes are likely to be more important for the depletion of the marine atmospheric boundary layer for most trace constituents than are the deposition processes.

Detailed studies of the deposition processes for water soluble gases as NH_3 and HNO_3 showed the necessity of including the exchange between the gases and the spray aerosols in the marine surface layer to properly understanding the data, and therefore also in properly parametrising the deposition of such species to the water surface.

The experience with flux measurements within the NOSE experiment resulted in development of a proto flux packet for ship measurements to determine fluxes of momentum, heat, water vapour and CO_2 at the air water interface by means of a combination of dissipation and co-spectral estimation techniques. In connection with the OMEX project shown to worked very satisfactorily.

The turbulent measurements of CO_2 fluxes were during the ASGASEX and the OMEX campaigns found to agree reasonably well with the exchange coefficient method, following the parameter choices suggested by Wanninkhof [30], provided conditions were steady. Some data obtained during OMEX suggest that the turbulent fluxes exceeded the exchange coefficient estimates considerably during unsteady conditions.

Acknowledgements

The work described above has been supported by EC DGXII, contracts No. STEP-CT90-0047 and ENVIRONMENT - CT93-0313 and MAST2 (OMEX - project), the Danish Marine Research Programme, contract No. 4-01, an by the EUROTRAC/EUREKA grant by the Danish Ministry of Industry. The Vindeby experiments has been conducted as a PhD study, supported by the Danish Research Academy, and further supported by US Office of Naval Research and the Nordic Counsil of Ministers. The senior author performed much of the TWO-PIE analysis as a visiting professor at Ecole Centrale de Nantes.

References

1. Hummelshøj, P., N.O. Jensen, S.E. Larsen; Particle dry deposition to the seasurface. in: S.E. Schwartz, W.G.N. Slinn (eds), *Precipitation Scavenging and Atmosphere-Surface Exchange, Vol. 2.* Hemisphere Publishing Corporation, Washington 1992, pp. 829–840.
2. DeLeeuw, G., S.E. Larsen, P.G. Mestayer (1993). Factors determining particle dynamics over the air-sea interface. *EUROTRAC Annual Report 1992, Part 3 ASE*, EUROTRAC ISS, Garmisch-Partenkirchen 1993, pp. 32–38.
3. De Leeuw, G., G.J. Kunz, S.E. Larsen, F. Aa Hansen; Measurements of CO_2 fluxes and bubbles from a tower during ASGASEX. *Proc. 2nd Int. Conf. Air-Sea Interaction and on Meteorology and Oceanography of the Coastal Zone, Amer. Meteorol. Soc.* 1994, pp. 267–268.

4. Larsen, S.E., G. de Leeuw, A. Flossmann, E. Ljungström, P.G. Mestayer (1994b). *Final report on the STEP project CT90-0047; transformation and removal of N compounds in the marine atmosphere*. Risø National Laboratory, Roskilde 1994, 75 pp.

5. Larsen, S.E., F.Aa. Hansen; Comparison between heat fluxes from the drag formulation and the eddy correllation during the NOSE: The North Sea Experiment, Sept.1991. *Risø-I-report* 1995.

6. Gryning, S.E., E. Batcharova; Parameterization of depth of entrainment zone above the daytime mixed layer. *Quart. J. Roy. Met. Soc.* **120** (1994) 47–58.

7. Moeng, C.H., J.C. Wyngaard; Statistics of conservative scalars in the convective boundary layer. *J. Atmos. Sci.* **41** (1984) 3161–3169.

8. Wyngaard, J.C., R.A. Brost; Top-down and Bottom Diffusion of a scalar in a convective boundary layer. *J. Atmos. Sci.* **41** (1983) 102–112.

9. Geernaert, L.L.S., G.L. Geernaert, K. Grandby, W.A. Asman; Fluxes of solluble gases in the marine atmospheric surface layer. *Tellus* **50B** (1998) 111–127.

10. Geernaert, L.L.S., G. deLeeuw, J. Højstrup; Flux divergence for nitric acid in the marine atmospheric boundary layer. *Atmos. Environ.* submitted.

11. Asman, W.A., O. Hertel, R. Bercowicz, J. Christensen, E.H. Runge, L.L. Sørensen, K. Granby, H. Nielsen, B. Jensen, S.E. Gryning, A.M. Sempreviva, S.E. Larsen, P. Hummelshøj, N.O. Jensen, P. Allerup, J. Jorgensen, H. Madsen, S. Overgaard, F.Vejern; Atmospheric Nitrogen to the Kattegat Strait. *OPHELIA* **42** (1995) 5–29.

12. Schulz, M. T. Stahlschmidt, F. Francois, W. Maenhaut, S.E. Larsen; The change of aerosol size distribution measured in a Lagrangian-type experiment to study deposition and transport processes in the marine atmosphere. in: P.M. Borrell, P. Borrell, T. Cvitaš, W. Seiler (eds), *Proc. EUROTRAC Symp. '94,* SPB Academic Publishing bv, The Hague 1994, pp. 702–706.

13. Larsen, S.E., F.Aa. Hansen, M. Schulz, 1992. Initial analysis of wind data from a sonic anemometer on FS Alkor in the NOSE 1991 campaign. in: T.D. Jickells, L.J. Spokes (eds), EUROTRAC Air-Sea Exchange Experiment North Sea, 14–27 September 1991, School of Environmental Sciences, University of East Anglia, Norwich 1992, pp. 1–12.

14. Oost, W.A., C.W. Fairall, J.B. Edson, S.D. Smith, R.J. Anderson, R.J. Wills, K.B. Katsaros, J. DeCosmo; Flow distortion calculations and their application in HEXMAX, *J. Atmos. Ocean. Technol.* **11** (1994) 366–386.

15. Yelland, M.J., B.I. Moat, P.K. Taylor, R.W. Pascal, J. Hutchings, V.C. Cornell; Wind Stress Measurements from the open ocean corrected for air flow distortion by the ship. *J. Phys. Oceanography* **28** (1998) 1511–1526.

16. Kaimal,J.C., J.C. Wyngaard, Y. Izumi, O.R. Coté; Spectral characteristics of surface layer turbulence. *Quart. J. Roy. Met. Soc.* **98** (1972) 563–589.

17. McBean G.A., M. Miyake; Turbulent transfer mechanisms in the atmospheric surface layer, *Quart. J. Roy. Met. Soc.* **98** (1972) 383–398.

18. Pond, S., G.T. Phelps, J.E. Paquin, G. McBean, R.W. Steward; Measurements of the turbulent fluxes of momentum,moisture and sensible heat over the ocean. *J. Atmos. Sci.* **28** (1971) 109–917.

19. Schmitt, G E., C.A. Friehe, C.H. Gibson; Structure of the marine surface layer. *J. Atmos. Sci.* **36** (1979) 602–618.

20. Larsen, S.E., Hot-wire measurements of atmospheric turbulence. Risø-R-232, 1986.
21. Smith, S.D., R.J. Anderson; Spectra of humidity, temperature and wind over the sea at Sable Island, Nova Scotia. *J. Geophys. Res.* **89(C2)** (1984) 2029–2040.
22. Wyngaard, J.C., O.R. Coté; Cospectral similarity in the atmospheric surface layer. *Quart. J. Roy. Met. Soc.* 98 (1972) 590–603.
23. Wyngaard, J.C., J.A. Businger, J.A. Kaimal, S.E. Larsen; Comment on "A reevaluation of the Kansas mast influence on measurements of stress and cup-anemometer over-speeding". *Boundary Layer Meteorol.* **22** (1977) 245–250.
24. Fairall, C.W., J.A. Edson, S.E. Larsen, P.G. Mestayer; Inertial dissipation air-sea fluxe measurements: A prototype system using realtime spectral coputations. *J. Atmos. Oceanic Technol.* **7** (1990) 425–453.
25. Edson, J.B., C.W. Fairall, P.G. Mestayer, S.E. Larsen; A study of the inertial-dissipation method for computing air-sea fluxes. *J. Geophys. Res.* **96(C6)** (1991) 10689–10711.
26. Webb, E.K., G.J. Pearman, R.Leuning; Correction of flux measurements for density effects due to heat and water vapour transfer. *Quart. J. Roy. Meteorol. Soc.* **106** (1980) 85–100.
27. Larsen, S.E., F.Aa. Hansen, G. de Leeuw, G.J. Kunz; Micrometeorological estimation of fluxes of CO_2, humidity, heat and momentum in the marine atmospheric surface layer during OMEX. Final report, Universite Libre de Bruxelles, 1996.
28. Ohtaki, E., M. Matsui; Infra-red device for simultaneous measurement of atmospheric carbon dioxide and water vapour. *Boundary Layer Meteorol.* **24** (1982) 109–119.
29. Oost,W.A.(ed); *Report of the ASGASEX'94 Workshop.* KNMI Report: TR-174, KNMI, De Bilt 1995, 61pp.
30. Wanninkhof, R. (1992). Relationship between wind speed and gas exchange over the ocean. *J. Geophys. Res.* **97** (1992) 7373–7382.
31. Kunz, G.J., G. de Leeuw, S.E. Larsen, F.Aa. Hansen; Eddy correlation fluxes of momentum, heat, water vapor and CO_2 during ASGASEX. in: W.A.Oost (ed), *Proc. ASGASEX workshop*, Report TR-174, KNMI, deBilt 1995, pp. 12–16.
32. Kunz, G.J., G. de Leeuw, S.E. Larsen, F.Aa. Hansen; Over-water eddy correlation measurements of fluxes of momentum, heat, water vapour and CO_2. in: B.Jähne and E.C.Monahan (eds), *Air-Water Gas Transfer: 3rd Int. Symp. Air-Water Gas Transfer,* Aeon Verlag and Studio, Hanau 1995, pp. 685–701.
33. Keir, R., G. Rehder, H. Erlenkeuser; Inorganic $\delta^{13}C$ and Methane in the European Margin Waters. *Final report for the MAST2 OMEX project,* ULB, Brussels 1996, pp. F83–F98.
34. Larsen, S E, J.B. Edson, P. Hummelshøj, N.O. Jensen, G. de Leeuw, P.G. Mestayer, Dry deposition of particles to ocean surfaces, *OPHELIA* **42** (1995) 193–204.

2.7 Aerosol and Rain Chemistry in the Marine Environment

J.L. Colin[1], R. Losno[1], N. Le Bris[1], M. Tisserant[1], I. Vassali[1], S. Madec[1], S. Cholbi[1], F. Vimeux[1], G. Bergametti[1], H. Cachier[2], C. Liousse[2], J. Ducret[2], B. Lim[2], M.H. Pertuisot[2] and P. Buat Menard[2]

[1]Laboratoire Interuniversitaire des Systèmes Atmosphériques,
Universités Paris 7 et Paris 12, URA CNRS 1404, Faculté des sciences,
61 Av. du général de Gaulle, 94010, Créteil cedex, France
[2]Centre des Faibles Radioactivités, Laboratoire mixte CNRS/CEA,
91198 Gif /Yvette Cedex, France.

In collaboration with :

T. Jickells[3], L. Spokes[3], G. Jennings[4], Michael Schulz[5], A. Rebers[5], W. Maenhaut[6], F. Francois[6], R.V. Grieken[7] and J. Injuk[7]

[3]School of Environmental Sciences, University of East Anglia, Norwich, UK.
[4]University College of Galway (UCG), Galway, Ireland
[5]Institut für Anorganische und Angewandte Chemie (IAAC), Universität Hamburg, Germany
[6]Institut for Nuclear Sciences (INS), University of Gent, Belgium
[7]Department of Chemistry, Universitaire Instelling Antwerpen (UIA), Antwerp, Belgium.

Aims of the research

Within the EUROTRAC programme, the main objectives of our group were devoted to a better understanding of the homogenous and heterogeneous processes which control the atmospheric aqueous phase chemistry and the evolution of particulate aerosol (mineral and carbonaceous) in cloud droplets, before their deposition. These objectives were conducted mainly in marine environment to study atmospheric systems rather homogenous and characterised by long residence times.

The main role of the cloud system is to cover aerosol particles with a layer of water in which aqueous chemical reactions start with acidic species, reducing or oxidising agents, radicals and photons (Fig. 2.7.1).

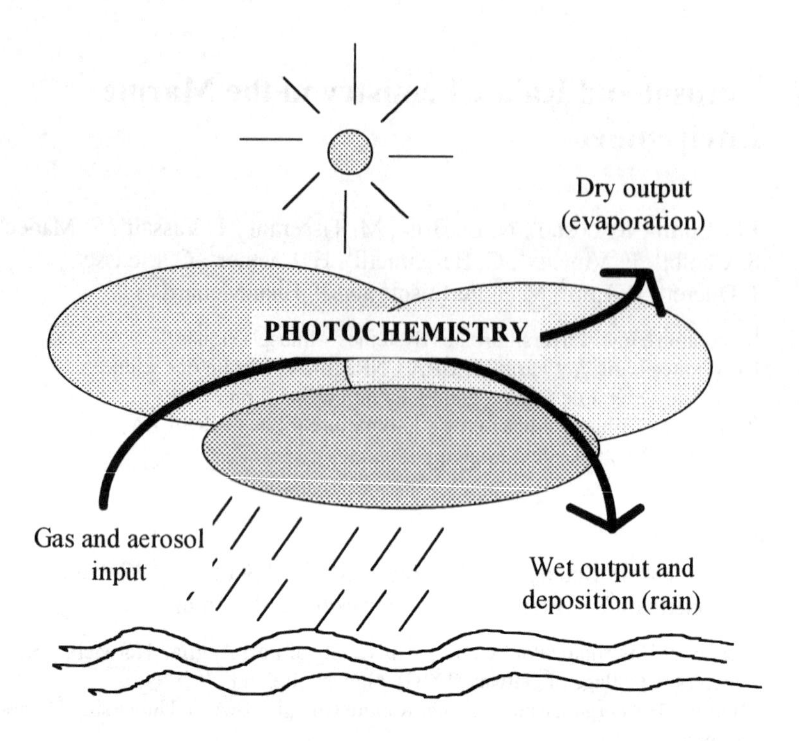

Fig. 2.7.1: Cloud chemistry and its implication for deposition over the oceans.

Dissolution processes release trace metals from the aerosol surface which may be complexed with inorganic ions and organic species thus driving metal speciation in cloud droplets. This speciation is critical to describe more accurately their catalytic effect on aqueous phase chemistry and after several cloud cycles their bioavailability in wet deposit and their potential impact on marine environment.

Research programme

The main goals achieved in this topic were based on field experiments in marine environments, laboratory experiments and chemical modelling.

Field campaigns

A first experiment was conducted in the island of Corsica over the western Mediterranean basin. Daily aerosol samples were collected continuously over one year whilst rain samples were sampled only during short campaigns. The Corsica campaign was not directly developed within EUROTRAC programme, but it is mentioned here because the results obtained in these rainwater samples were of a major interest for further discussion.

A second experiment occurred at Mace Head, on the west coast of Ireland, in November 1990 during a collaborative field experiment with UEA, UCG and the French group (LISA, CFR) with major aims to collect trace metals in rainwater and carbonaceous species in both aerosols and rain.

A third experiment was conducted again at Mace Head in April 1991 with several teams of the ASE group (LISA, UEA, UCG, IAAC, UNS UIA). This campaign was carried out for a general intercalibration of ultra clean procedures for aerosol and rain sampling and analysis. This intercalibration has been discussed at two workshops held in Paris (October 1992) and Arcachon (October 1993).

A wide range of atmospheric conditions are thus covered with these two coastal sites. At Mace Head, Atlantic rain waters are often associated with clean air masses, fairly representative of remote oceans with pH values around 5. By contrast the site in Corsica is representative of contrasting events with strong differences between polluted (pH as low as 3) and non-polluted or acid neutralised air masses with pH values over 7 for events associated with Saharan dust inputs.

Laboratory experiments

We have developed a kinetic reactor to measure dissolution mechanisms in a reaction cell where fine fraction of soils are dispersed in a large amount of aqueous solution. A very small amount of well identified solid powder (clays, soil minerals, loess or Saharan dust) reacts with a controlled aqueous solution, in an open loop where new and clean water solution is continuously injected into the cell. The rate of trace metal dissolution is measured under various pH and oxidant conditions and over short time scales (around the minute). This may give a better view about weathering and transformations of aerosols into a cloud or a rain drop. Models developed can benefit immediately from these laboratory studies.

Modelling

The strongest changes of the aerosol surface occur during cloud processing. We develop modellisation into cloud droplets, involving aerosol. The aerosol contribute to the cloud chemistry by exchanging compounds with the aqueous phase, and especially trace metals. These metals act both as catalyst in the reactivity scheme of oxidising species (OH, HO_2,...), and are modified by these species. For example, Iron(III) is changed into Iron(II) and becomes much more soluble by such processes. Modelling allow us to assess the behaviour of trace metals before their deposition to the oceans.

Rain chemistry

Sampling

Sampling and analysing procedures for pH, major ions and trace metals measurements have been extensively improved during the EUROTRAC programme. Rainwater was manually collected in single-use polyethylene collectors on a wet-only, event-by-event basis. Funnels and bottles were pre-cleaned using ultraclean techniques in a clean room, dried under laminar flow hoods and sealed in polyethylene bags until rain collection. At the end of the rain, the sample is filtered through a 0.4μm pore size Nuclepore™ membrane and two aliquot (60 ml each) are kept. The first for major ions analyses is preserved with chloroform and stored in a freezer. The second for dissolved trace metals is acidified with ultrapure Prolabo™ Normatom® nitric acid. In a final form we have improved this collector with an on line filtration and a direct separation of iron species with a Sep-Pak® cartridge.

The intercalibration of Mace Head in April 1991 was focused on ultraclean procedures for aerosol and rain sampling, analytical procedures for trace metals in aerosol and for major ions and trace metals in rainwater. Our contribution has been focused on major ions and reveals large variations between the results obtained by the different laboratories for a same rain event.

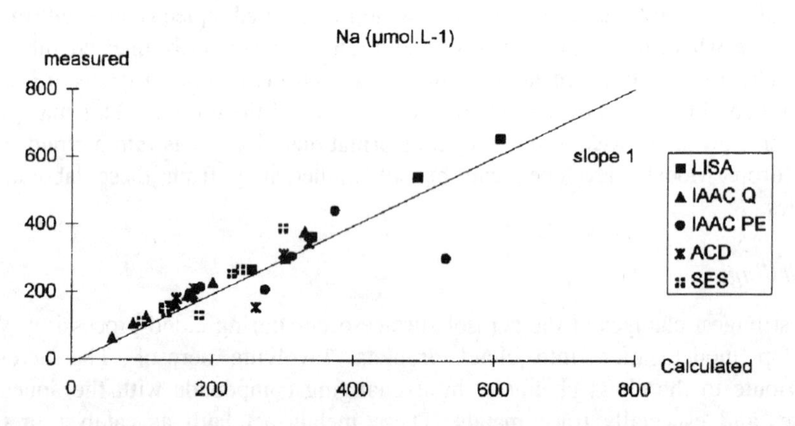

Fig. 2.7.2: Comparison between measured and calculated concentrations of sodium in rainwater collected at Mace Head during an intercalibration campaign "on the field". Samples are labelled with the laboratory name.

The main variation is widely due to sampling methods influenced by non rain deposit (marine brine and sea salt) which is very dependent on the geometry of the sampler and the wind velocity.

Moreover we have shown that at Mace Head and probably in all remote marine areas, the major ion content could be explained by sea salt input and then evaluated by a simple conductance measurement (Fig. 2.7.2) [1].

pH: a major control on metal solubility

The soluble-insoluble partition of trace metals are driven by thermodynamic equilibrium for Al(III), involving mainly hydroxide (OH^-) and sulfate (SO_4^{2-}), and adsorption/desorption mechanisms for Pb, Cu, Zn.

Aluminium

This element shows the same behaviour in rainwater than in river water. At low pH (< 5), Al solubility can be predicted from an equilibrium with an insoluble aluminium hydroxide sulfate salt :

$$Al_4(SO_4)_3(OH)_6\downarrow + 6H^+ = 4\ Al^{3+} + 6H_2O + 3\ SO_4^{2-} \qquad pK= 14.4$$

At pH > 5, Al solubility is controlled by aluminium-hydroxy compounds (Fig. 2.7.3). A special behaviour is observed for very low concentrated sample collected at Mace Head.

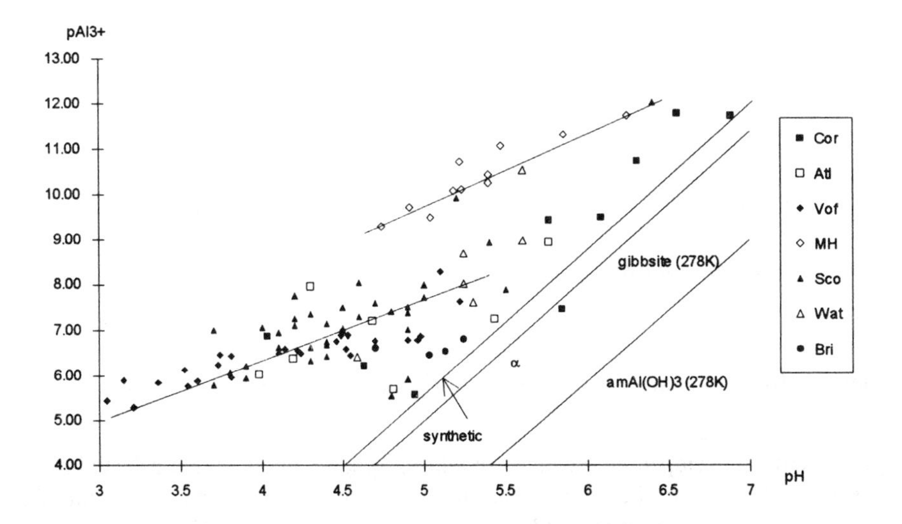

Fig. 2.7.3: Logarithmic concentration of Al^{3+} as function of pH in marine rainwater. (Al^{3+} concentrations for synthetic, α crystalline and amorphous gibbsite equilibria are shown). Cor is Corsica, Atl is middle Atlantic, Vof is Vogues forest, MH is Mace Head, Sco is Scottish snow, Wat is North Atlantic and Bri is Brittany.

Zinc, lead and copper

The solubility of Zn, Pb and Cu are controlled by adsorption/desorption processes in which rainwater particles provide surface-active sorption sites. pH is a critical parameter as illustrated by Zn and Pb behaviour in Fig. 2.7.4. The pH of rainwater ranges between 4 and 7. The soluble zinc ranges between 15 and 99 % of the total zinc, the higher percentage coming for the lower pH. Results suggest that in marine precipitation with pH < 5, more than 80 % of the total Pb, Cu and Zn concentrations are delivered to the surface oceans in the dissolved form [2].

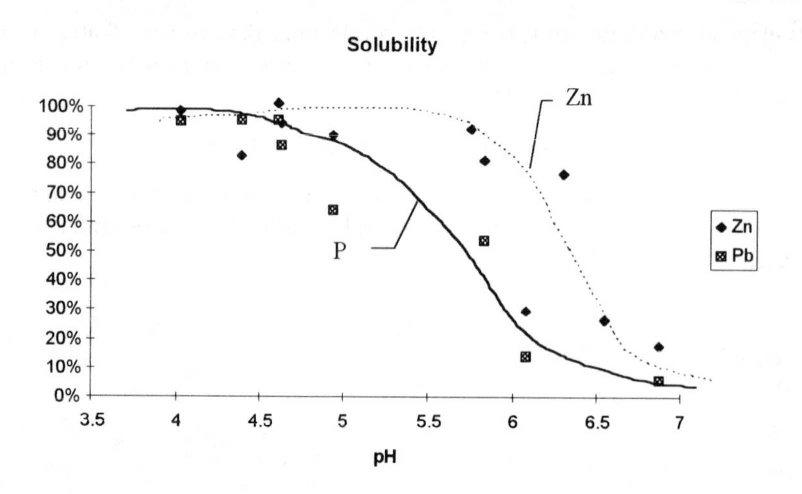

Fig. 2.7.4: Solubility versus pH curves of Zn and Pb in Mediterranean rainwater.

Organic chelating power

Measurements were made at Porspoder, Brittany and Mace Head, Ireland, after a laboratory methodological development. Among various anions, organic may have a strong chelating power, that can change the behaviour of trace metals in the rain water.

The method developed here consist in adding precise amounts of a selected metal (lead was chosen) to the sample, and in measuring the remaining free metal. The difference of slope between added and free metal is considered to be related to available ligands (see legend of Fig. 2.7.5). If ligands were found at Porspoder (Fig. 2.7.5), none were found at Mace Head.

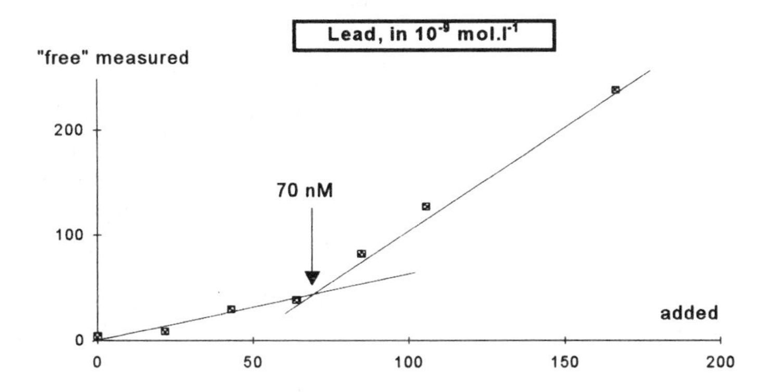

Fig. 2.7.5: Ligand contents measured into a rain water collected at Porspoder (1994). "Free" lead is plotted versus added. The strike of the slope show the ligand concentration and the first slope is related to the complexation constant.

***Red/Ox speciation*: Fe**

To evaluate the potentially biologically labile fraction of iron to marine phytoplankton upon wet deposition, attention has been paid to speciation of Fe(II)/Fe(III) in marine precipitation. Both kinetic and thermodynamic approaches can be used to describe the factors which control the speciation of iron in marine precipitation.

for in cloud processes, we adopted the kinetic approach, since the atmospheric processes that govern the partitioning of Fe(II)/Fe(III) in cloud are rapidly changing. Under light conditions, a steady state involving Fe(II)/Fe(III) redox cycle is reached very quickly (see modelling) with a significant part of iron under Fe(II) state [3].

we hypothesise that the steady state of Fe(II)/Fe(III) in rain is perturbed as rain droplets fall down from cloud level towards the ground and during the rain storage in the rain collector. Using thermodynamic equilibrium we show that Fe(II) is then oxidised in Fe(III) state and freshly precipitated as Fe hydroxide with a stoichiometry of $Fe(OH)_{2.5}$ and a solubility product of 1.8 at 18 °C (Fig. 2.7.6).

To control this iron cycle, we have developed an analytical method to measure directly Fe(II) and Fe(III) species by ionic chromatography coupled with a photometric detection. To be able to detect both species at the nmol L^{-1} level, a cationic resin was used as a precolumn to preconcentrate rain samples so as to get satisfactory detection limits (1.8 nmol/l for iron(II) and 2.1 nmol L^{-1} for iron(III)). But this method tested with freshly collected marine rain samples showed that the concentration levels of iron(II) and iron(III) could not be determined because of quasi-missing recorded signals on chromatograms. Indeed, major cations such as Na^+, K^+, Ca^{2+} or Mg^{2+} were supposed to cause large interference with metallic ions in the mechanism of preconcentration as their concentrations in marine precipitation are much higher than iron concentrations (about ten to hundred µmol L^{-1} versus a few nmol L^{-1}).

Further studies were carried out with synthetic solutions in the laboratory to assess the hypothesis and as a conclusion that the precolumn was either of too low capacity or the competition between major and trace metallic ions for their retention on the sites of the resin induced partial preconcentration. Another point was the delay between the sampling of the rainwater and the analyse latter in the laboratory giving time for the oxidation of Fe(II) to Fe(III) in the sample.

To avoid this artefact we have set up a new analytical protocol allowing a separation of both species on an open column fitted directly to the funnel. This open column (Sep-Pak®, Waters™) is filled with Ferrozine as a stationary phase, which is a specific chelating agent for Fe(II). After the rain collection, Fe(II) is eluted with a water/methanol eluant and then analysed by Graphite Furnace Atomic Absorption Spectroscopy.

These approach will permit to explain the equilibrium reach and observed between soluble iron and its hydroxide in various rainwater (Fig. 2.7.6) which are not collected taking account the reactivity of iron II.

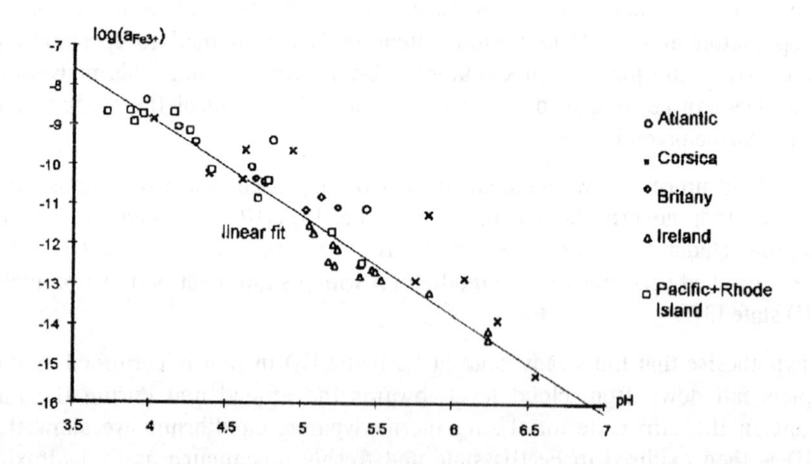

Fig. 2.7.6: Logarithm of Fe^{3+}activity as function of pH. 37 rains sampled over the Atlantic Ocean, Mediterranean Sea and Pacific ocean. The linear fit indicates the existence of an equilibrium between soluble iron and $Fe(OH)_3$.

Dissolution kinetics

Dissolution rates of aerosol particles are measured in an open flow chemical reactor. As the apparatus is simplistic, the main goal of this research is to insure real kinetic conditions far from dissolution equilibrium to explain the dissolution processes. As the solubility of various species are very poor, we must control operating conditions where analytical concentrations are in the range ppt (10^{-12}) to ppb (10^{-9}). These concentrations are consistent with those obtained in marine cloud and rain waters.

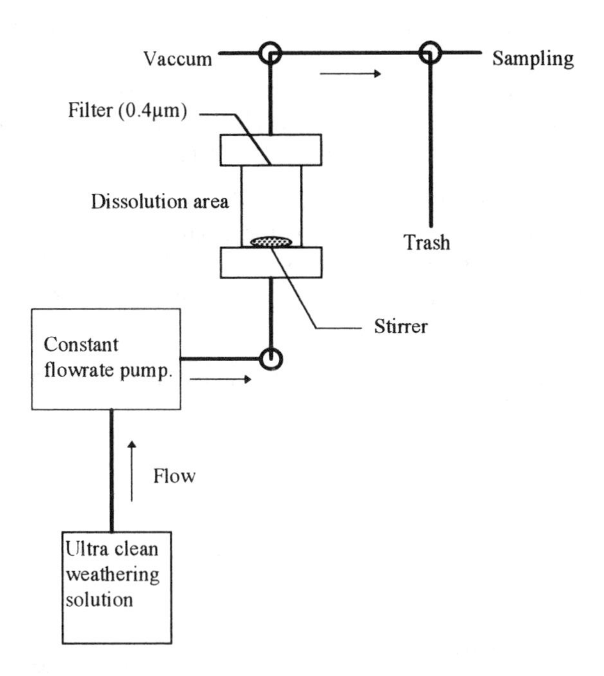

Fig. 2.7.7: The kinetic reactor.

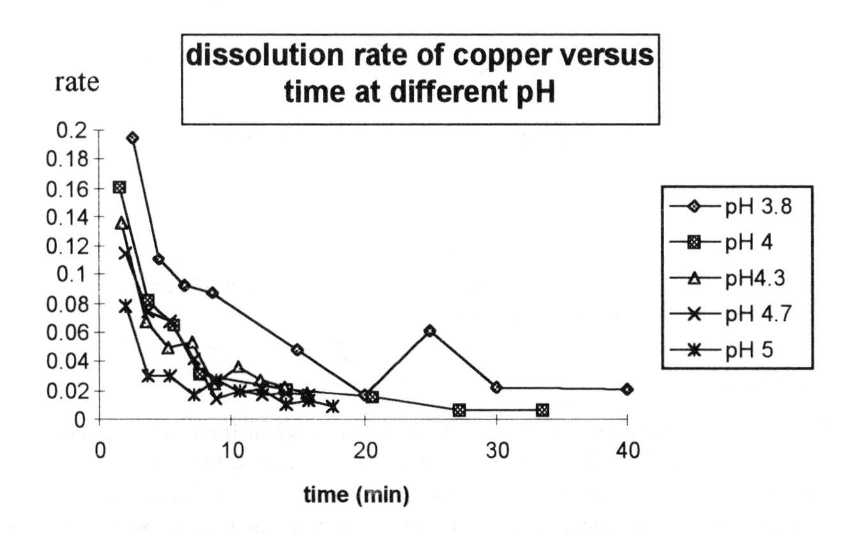

Fig. 2.7.8: Whatever the pH of the weathering solution, the higher dissolution rates are at the beginning of the contact between the water and the solid (less than 10 min). The dissolution process of aerosols must be studied at this scale time.

Soil (20 mg) is introduced into the dissolution area at the beginning of an experiment, then an ultra clean solution at a fixed pH and Redox property is flowed through the cell. The emerging filtered solution is sampled and analysed in trace metals. The volume of the reaction chamber is 34 ml, and the flow rate 18 ml min^{-1}. The average residence time in the stirred cell is about 2 min.

We already measured the dissolution rate of d-block metals as a function of time in a aeolian soil collected on the Cappo Verde Islands. The results show that the dissolution rate is pH dependent, but with different behaviour than equilibrium with hydroxy or oxo-hydroxy salts. As an example, Fig. 2.7.8 shows how the copper dissolves at various pH values.

From those curves, the reaction rate can be plotted at selected positions in the reaction, versus pH (Fig. 2.7.9):

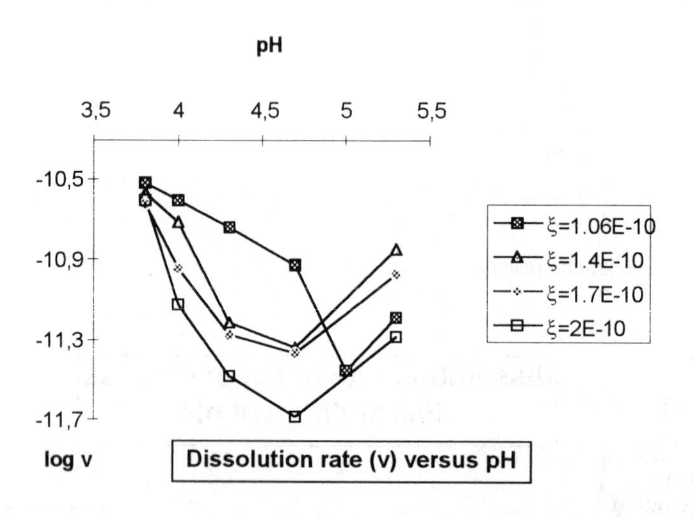

Fig. 2.7.9: Dissolution rate of copper versus pH at selected positions ξ in the dissolution reaction $Cu(II)_{solid} \rightarrow Cu^{2+}_{dissolved}$. The rate reach a minimum at pH around 4.7. At lower pH, H^+ enhance the rate, at higher pH OH^- does it.

Modelling

In marine environment, trace metals are implicated in their biogeochemical fate and also as catalysts in oxidising power of the atmosphere. Both are strongly linked. The following figures (Figs. 2.7.10 and 2.7.11) show that light can produce Fe(II), the most soluble specie of iron, from Fe(III) with a very poor solubility. Alternate step of dark and light are simulated here: 0–5 s light; 5–10 s dark, 10–15 s light, 15–20 s dark. Ordinates are mol L^{-1}, abscises are s.

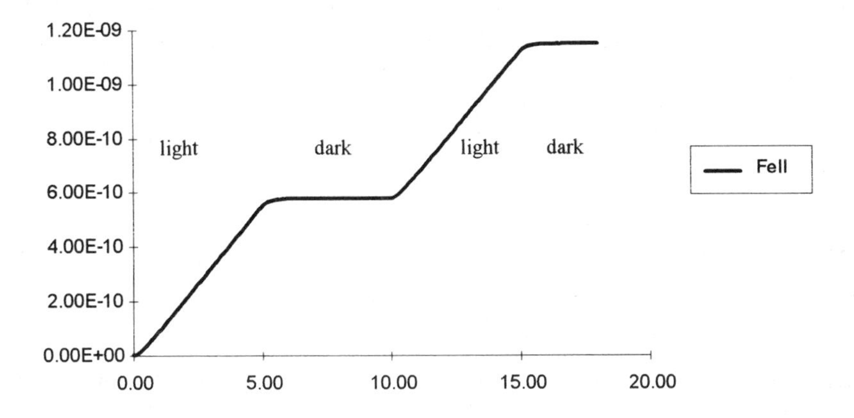

Fig. 2.7.10: Modelled variations of Iron II concentrations in a irradiated cloud drop.

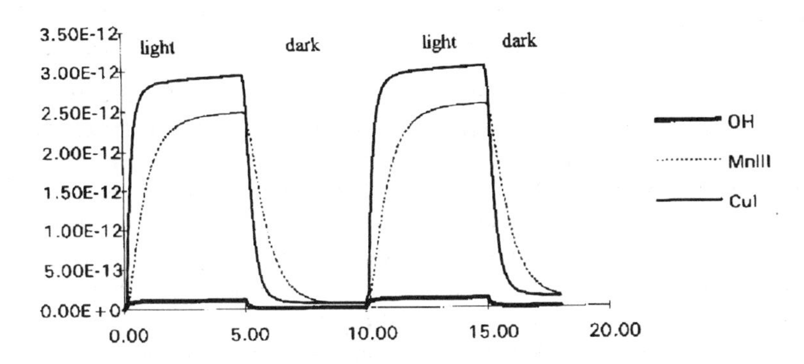

Fig. 2.7.11: Modelled variations of reactive OH, Mn(III) and Cu(I) species under variable irradiation conditions in a marine non polluted cloud. The response time of these species is very fast.

As indicated in Fig. 2.7.11, the response time of the system is very enhanced by the trace metals catalyst, and a steady state is reached after 1 or two seconds. In those conditions, the reactive species concentrations could be well approached by their photostationary values expressed as:

$$[O_2^-] = \frac{2.J_1.[H_2O_2] + J_{19}\,[\alpha FeIII]}{k_{18}.[FeIII] + k_{34}.[CuII] - k_{10}.[O_3] - k_{34}.[CuII]} \qquad [OH] = \frac{2.J_1.[H_2O_2] + J_{19}\dfrac{[\alpha FeIII]}{[H_2O_2]}}{k_4 - \dfrac{k_4.k_{10}.[O_3] - k_4.k_{34}.[CuII]}{k_{18}.[FeIII] + k_{34}.[CuII]}}$$

$$\frac{[FeIII]}{[FeII]} = \frac{k_{18}.[O_2^-]}{k_{25}.[H_2O_2]}\ (\text{slow}) \qquad \frac{[CuI]}{[CuII]} = \frac{k_{34}.[O_2^-]}{k_{31}[H_2O_2]}\ (\text{fast})$$

Reactions and values are given in the Appendix. Reaction (4) is the main HO_2 source, with (16) for several %. Sinks are (10), (13), (18), and (34). Reactions (18) and (34) are the main contribution. For OH, reactions (1), (10), (19), (31) and (4) are the main ones. (31) and (34) build a catalytic cycle for Cu(I)/Cu(II). The rates of (31) and (34) are the same. The response of this system is very fast, around 10^{-10} mol $L^{-1}.s^{-1}$.

Aerosol (carbonaceous particles)

The general aim of this project conducted until 1993 was to study carbonaceous particles behaviour in aerosol and rainwater.

Black carbon (soot) aerosols were measured with optical equipment (aethalometer) during the first Mace Head field campaign in November 1990. At Mace Head, there is a very striking sectorisation of atmospheric concentrations for particulate data. From the clean sector (West) to the dirty sector (East), concentrations vary in a range which can be greater than 2 orders of magnitude. The main value of atmospheric soot carbon in westerly wind conditions (0.015 µg m^{-3}) is very low, only five times the measured value in arctic area.

The two main components of the carbonaceous aerosols (C_t) are the organic fraction (C_o) and the black carbon or soot fraction (C_b). As a tracer of the origin of the carbonaceous aerosols, the C_b/C_t ratio which at Mace Head is on the average of the order of 22 %, attests an industrial origin for the atmospheric particulate material. Accordingly with other results obtained at other sites, a comparison between C_b/C_t ratios in aerosols and in rains points out a systematic relative increase of black carbon in rains which could show a partial disappearance of the organic fraction during its incorporation in hydrometeors (Table 2.7.1). It may be hypothesised that this dissolution of the hydrophilic organic particulate material is accompanied by that of other inorganic species such as sulfates which are likely to be absorbed at the surface of the atmospheric particles. This process could be a key step in the probable CCN role of the carbonaceous aerosols a major part of which is attached to submicron particles. Our future research will focus on that field and will rely primarily on experimental leaching of aerosols.

Scavenging ratio estimates ($S = C_{rain}$ (µgC L^{-1})/ C_{air} (µgC m^{-3})) reinforce the hypothesis that the aerosol acquires a better hydrophilic character during its atmospheric transport (S increases, Table 2.7.1). Also, S values found for black carbon are of the same order as values obtained for soluble species such as sulfur species. Consequently, due to both its chemical inertia and its hydrophilic coating, black carbon could trace the physical processes of the incorporation of atmospheric particulate pollutants.

Table 2.7.1: Incorporation of carbonaceous particles in Mace Head rains : a comparison with results obtained at other representative sites. The increase of the C_b/C_t ratio in rains probably points out a dissolution of the organic fraction of the aerosol.

	C_{black} µgC L^{-1}	C_{total} µgC L^{-1}	C_b/C_t rain (%)	C_b/C_t aerosol (%)	Sca. ratio (S)
Mace Head (n=18)	9–94	47–323 (31)	29 (100)	22	370
Ile de France (n=51)	27–1348 (333)	70–5870 (1190)	30	22	150
Tropical Africa (n=21) (off fires)	75–258	265–745	30	15	100

The idea of an internally mixed aerosol may be corroborated with the surprising correlation often found between excess sulfur and soot-carbon in environments where combustion inputs are dominant. Gases committed during the combustion processes are likely to adsorb onto the carbonaceous nuclei and undergo oxidative conversions leading to a solid coating of the particle. SO_2 is then transformed into SO_4^{2-}. In the particles, the mean C/S ratio value is 1 and may reach 2 in old aerosol phases. This feature could be used in further works to trace with soot-carbon pollution inputs to rain waters. Due to this coating of hydrosoluble substances, the combustion carbonaceous particles have a hydrophilic behaviour, and their capability to be incorporated in hydrometeors is enhanced [4]. This may explain the important sectorisation of the black carbon scavenging ratios observed at the Mace Head site (Table 2.7.1).

The soot-carbon aerosol display optical properties which vary with the size and the coating of the particles [5]. As an example, accordingly with theoretical considerations, we found that a thin coating of sulfates would enhance the light absorption capability of the particles whereas in old particles, a thick surface layer would on the contrary lower the absorption. This had to be taken into account in remote sensing retrieval of aerosol loads or in models of the radiative impact of anthropogenic particulate. This important result was gained primarily with the optical sensor of an aethalometer acquired during the first year of this EUROTRAC contract.

Conclusions

Within this EUROTRAC project, we have documented the impact of mineral and carbonaceous aerosol on marine atmospheric environment. This is of considerable interest since for example some trace metals (*e.g.* Fe, Mn and Zn) can act to enhance marine primary activity while other metals (*e.g.* Cu, Hg) can affect dramatically the initial step of the marine life. Several goals were reached during this programme and are shown in this report. Numerous results are still in the interpretation process and will provide further publication.

Furthermore this EUROTRAC project involving 5 EU research groups experienced on field, laboratory and modelling studies in marine atmospheric environment was of major benefit for international collaboration. Considerable expertise in this area has been thus gained by these groups which have regularly met in annual ASE workshops since 1989.

References

1. Losno R., J.L. Colin, L. Spokes, T. Jickells, M. Schulz, A. Rebers, M. Leermakers, C. Meulemann, W. Bayens; Non-rain deposition significantly modifies rain samples at a coastal site. *Atmos. Environ.* **32** (1998) 3445–3455.
2. Lim B., T.D. Jickells, J.L. Colin, and R. Losno. Solubilities of Al, Pb, Cu and Zn in rain sampled in the marine environment over the North Atlantic Ocean and Mediterranean Sea, *Global Biogeochemical Cycles* **8** (1994) 349–362.
3. Le Bris N., Contribution to the study of trace metals dissolved/particulate partition in wet precipitations (in french) Ph.D. Université Paris 7, 1993.
4. Liousse, C., H. Cachier, S.G. Jennings; Optical and thermal analysis of black carbon aerosol content in different environments: variations of the specific attuenation coefficient, sigma σ; *Atmos. Environ.* **27A** (1993) 1203–1211.
5. Ducret, J., Incorporation of carbonaceous particles into wet precipitations (in french) Ph.D. Université Paris 7, 1993.

Appendix

Reactions used for the modelling, together with the limiting conditions

Compounds and initial concentrations (mol L–1)

O_3:	4.50×10^{-10}	H_2O_2:	5.0×10^{-5}	HO_2:	0.0
O_2^-:	0.0	OH:	1.0×10^{-14}	H^+:	1.0×10^{-5}
hν:	1.0	O_2:	3.0×10^{-4}	Fe(II):	0.0
Fe(III):	5.0×10^{-8}	Mn(II):	3.0×10^{-9}	Mn(III):	0.0
Cu(I):	0.0	Cu(II):	1.0×10^{-9}		

Static

H^+:	1.0×10^{-5}	hν:	1.0	O_3:	4.50×10^{-10}

Acid/Base

HO_2/O_2^- $pK_a = 4.68$

Reactions

(1) $H_2O_2 + h\nu \rightarrow OH + OH$
$k = 5.7 \times 10^{-7}$

(2) $HO_2 + OH \rightarrow O_2$
$k = 7.0 \times 10^{9}$

(3) $O_2^- + OH \rightarrow O_2$
$k = 1.1 \times 10^{10}$

(4) $H_2O_2 + OH \rightarrow HO_2$
$k = 2.7 \times 10^{7}$

(5) $O_3 + OH \rightarrow HO_2 + O_2^-$
$k = 2.0 \times 10^{9}$

(6) $HO_2 + HO_2 \rightarrow H_2O_2 + O_2$
$k = 8.6 \times 10^{5}$

(7) $HO_2 + O_2^- \rightarrow H_2O_2 + O_2$
$k = 1.0 \times 10^{8}$

(8) $H_2O_2 + HO_2 \rightarrow OH + O_2$
$k = 5.0 \times 10^{-1}$

(9) $H_2O_2 + O_2^- \rightarrow OH + O_2$
$k = 1.3 \times 10^{-1}$

(10) $O_3 + O_2^- \rightarrow OH + O_2 + O_2$
$k = 1.5 \times 10^{9}$

(11) $OH + Mn(II) \rightarrow Mn(III)$
$k = 3.4 \times 10^{7}$

(12) $HO_2 + Mn(II) \rightarrow Mn(III) + H_2O_2$
$k = 6.0 \times 10^{6}$

(13) $O_2^- + Mn(II) \rightarrow Mn(III) + H_2O_2$
$k = 1.1 \times 10^{8}$

(14) $HO_2 + Mn(III) \rightarrow Mn(II) + O_2 + H^+$
$k = 2.0 \times 10^{4}$

(15) $O_2^- + Mn(III) \rightarrow Mn(II) + O_2$
$k = 1.5 \times 10^{8}$

(16) $H_2O_2 + Mn(III) \rightarrow Mn(II) + HO_2 + H^+$
$k = 3.2 \times 10^{4}$

(17) $HO_2 + Fe(III) \rightarrow Fe(II) + H^+ + O_2$
$k = 2.0 \times 10^{4}$

(18) $O_2^- + Fe(III) \rightarrow Fe(II) + O_2$
$k = 1.5 \times 10^{8}$

(19) $h\nu + Fe(III) \rightarrow Fe(II) + OH$
$k = 5.9 \times 10^{-4}$

(20) $HO_2 + Fe(II) \rightarrow Fe(III) + H_2O_2$
$k = 1.2 \times 10^{6}$

(21) $O_2^- + Fe(II) \rightarrow Fe(III) + H_2O_2$
$k = 1.0 \times 10^{7}$

(22) $OH + Fe(II) \rightarrow Fe(III)$
$k = 3.0 \times 10^{8}$

(23) $O_3 + Fe(II) \rightarrow Fe(III) + OH + O_2$
$k = 1.7 \times 10^{5}$

(24) $Fe(II) + Mn(III) \rightarrow Fe(III) + Mn(II)$
$k = 2.1 \times 10^{4}$

(25) $H_2O_2 + Fe(II) \rightarrow Fe(III) + OH$
$k = 6.01 \times 10$

(26) $O_2 + Fe(II) \rightarrow Fe(III) + O_2^-$
$k = 7.9 \times 10^{-4}$

(27) $Fe(III) + Cu(I) \rightarrow Fe(II) + Cu(II)$
$k = 1.0 \times 10^{7}$

(28) $OH + Cu(I) \rightarrow Cu(II)$
$k = 3.0 \times 10^{8}$

(29) $HO_2 + Cu(I) \rightarrow Cu(II) + H_2O_2$
$k = 1.5 \times 10^{9}$

(30) $O_2^- + Cu(I) \rightarrow Cu(II) + H_2O_2$
$k = 1.0 \times 10^{10}$

(31) $H_2O_2 + Cu(I) \rightarrow Cu(II) + OH$
$k = 4.0 \times 10^{5}$

(32) $Mn(III) + Cu(I) \rightarrow Cu(II) + Mn(II)$
$k = 2.1 \times 10^{4}$

(33) $HO_2 + Cu(II) \rightarrow Cu(I) + O_2 + H^+$
$k = 1.0 \times 10^{8}$

(34) $O_2^- + Cu(II) \rightarrow Cu(I) + O_2$
$k = 5.0 \times 10^{9}$

2.8 Atmospheric Inputs to Marine Systems

T.D. Jickells, L.J. Spokes, M.M. Kane, A.R. Rendell and R.R. Yaaqub

School of Environmental Sciences, University of East Anglia, Norwich NR4 7TJ, UK

This component of EUROTRAC has concentrated on two themes; the magnitude of atmospheric fluxes to the North Sea and the bioavailablity of such atmospheric inputs of metals as evidenced by their solubility. Good progress has now been made in both areas. The solubility of metals in rain-water is now reasonably well understood, and the atmospheric fluxes of metals and nitrogen to the North Sea have been quantified and the main factors controlling them rationalised. Furthermore much of what has been learnt about inputs to the North Sea is applicable to other coastal areas, providing evidence of several important processes in such environments which need to be included in transport and deposition models.

Aims of the research

Our overall aim within the EUROTRAC project is to better understand the magnitude and effects of atmospheric inputs to marine ecosystems. We have focused particularly on an improved understanding of the factors controlling the solubility of metals in rain-water and studies of the factors controlling deposition processes in the marine atmosphere.

Principal scientific results

Solubility

We have been investigating the controls on trace metal solubility in rain-water for several years. In collaboration with our French colleagues in EUROTRAC we have made considerable progress. Table 2.8.1 illustrates the ranges of solubility found in rain-water from many different locations. For some metals (*e.g.* Zn), solubilities are generally high, while for other metals (*e.g.* Al.) they are much lower. We have now shown that for the metals whose sources are dominated by anthropogenic emissions (*e.g.* Pb, Cu and Zn), solubility can be rationalised in terms of classical adsorption/desorption theories [1] with solubility strongly dependent on rain-water pH. However, this dependence is such that these elements are generally very soluble at rain-water pHs of less than five, as pertain in most ocean areas away from large sources of desert dust. In contrast, elements whose atmospheric source is still predominantly via the resuspension of crustal dust, show low but variable solubility in rain-water which cannot be rationalised fully in terms of simple pH-solubility control. Studies have emphasised the importance of aerosol type in governing dissolution and it now appears that repeated cycling through low pH cloud environments may also, in part, govern the dissolved-particulate distribution of aerosol trace metals in rain-water [2]. Recent work has emphasised the importance of atmospheric inputs of metals (particularly Fe) to the oceans in terms of regulating oceanic productivity [3]. Our results emphasise that it is no simple matter to assume a solubility for these metals in estimates of inputs of bioavailable metals to marine systems. Furthermore, surface water biogeochemistry may further modify solubilities and hence bioavailablity of the metals [4].

Table 2.8.1: Recent estimates of the percentage of some trace metals that are in the dissolved (< 0.4 μm) phase in precipitation collected from rural and remote areas under the influence of maritime air masses. Adapted from Jickells [5].

Element	% Dissolved
Al	5–26
Fe	11–17
Mn	63–64
Cu	27–84
Pb	50–90
Zn	72–96.

Fluxes

Since deposition to marine systems is now recognised as a potentially important control on ocean productivity via the supply of several key nutrients [5], the factors controlling such deposition become critical components of global biogeochemical cycles. Over recent years, atmospheric sampling and subsequent analytical procedures have improved considerably and allow data from many groups to be collated to produce regional maps of aerosol concentrations over the North Sea (*e.g.* Fig. 2.8.1). Such results depend on long sampling campaigns because of large scale daily variability, and the meteorological causes of which are now beginning to be understood [6]. We have been able to study deposition processes of a wide range of species at a coastal site adjacent to the North Sea and on ships over the North Sea itself to try to identify any specific controls on deposition to marine systems which are different to those on land. This data is rationalised in Table 2.8.2 in the form of scavenging ratios which attempt to express the efficiency of aerosol removal by rain-water. Several factors stand out from this table. Firstly the scavenging efficiency differs between elements and this can be rationalised in terms of aerosol particle size. Secondly for many of the components, results from coastal and ship collections are similar. This is not, however, the case for the important nitrogen nutrients nitrate and ammonium. For ammonium this probably reflects local emissions of ammonia from agricultural areas close to the coastal sampling site. In the case of nitrate we believe it arises from the interaction of nitric acid with seasalt which shifts the nitrate aerosol size distribution in marine atmospheres to much larger sizes (8–20 μm) than that seen in terrestrial environments (about 1 μm, [7]). The data in Table 2.8.2 imply that this process increases the scavenging efficiency of rainwater for nitrate by more than three fold over the North Sea compared to the deposition over the neighbouring terrestrial environment. This is clearly a specifically marine process and can occur relatively rapidly since it produces such a large effect over a distance of a few hundred kilometres between the sampling sites. It will thus act to enhance the effect of atmospheric nitrogen deposition on coastal ecosystems.

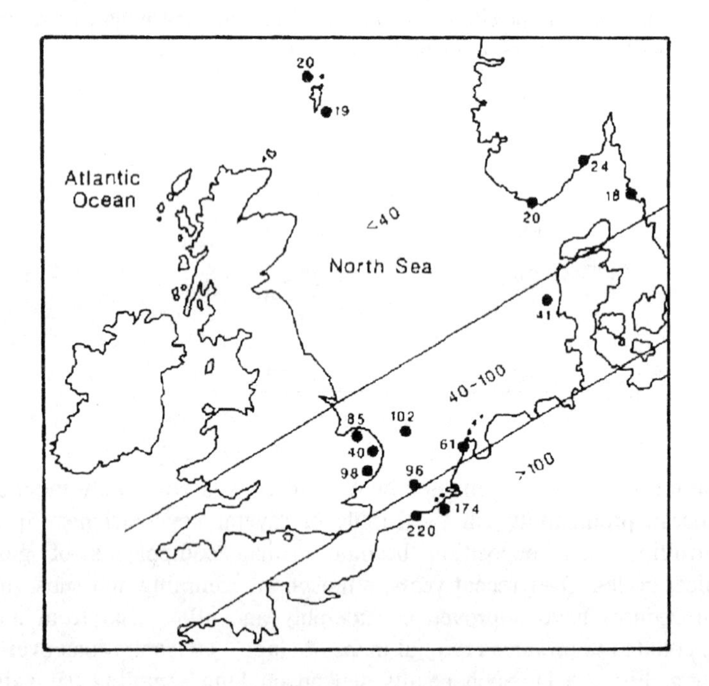

Fig. 2.8.1: Aerosol zinc distribution over the North Sea (ng m^{-3}) based on Yaaqub *et al.* [6].

Overall we estimate that about 30 % of the terrestrial input of nitrogen to the North Sea arises from the atmosphere [8] and thus this process is one which is quantitatively of importance to the marine environment and hence to managers trying to protect coastal ecosystems from eutrophication. Furthermore, we have identified several high deposition episodes in which certain meteorological conditions bring about dramatic local enhancements of nitrogen deposition [9], the effects of which on marine ecosystems are unknown. Atmospheric inputs are similarly important for trace metals with the ratios of the fluxes to the southern North Sea of cadmium, lead, zinc and copper from the atmosphere and from rivers being 1.0, 1.1–2.1, 0.5–0.6 and 0.4–0.8 respectively.

Table 2.8.2: Scavenging ratios for various species at a coastal site on land and over the North Sea. Adapted from Kane *et al.* [10].

Element	Scavenging Ratio	
	Coastal Site	North Sea
Fe	93	105
Mn	249	298
Pb	136	131
Nitrate	204	692
Ammonium	1320	391

$$\text{Scavenging Ratio} = \frac{\text{Concentration in precipitation}}{\text{Concentration in aerosol}} \times \text{Density of air}$$

Concentrations used are volume weighted precipitation concentrations and average aerosol concentrations, both based on samples collected over one year.

Acknowledgements

Various parts of this work have been supported by UK NERC as part of the North Sea Project and via GR3~8404A, UK DoE as part of the EUROTRAC North Sea Experiment and UK MAFF.

References

1. B. Lim, T.D. Jickells, J.L. Colin, R. Losno; Solubilities of Al, Pb, Cu and Zn in rain sampled in the marine environment over the North Atlantic Ocean and Mediterranean Sea. *Global Biogeochem. Cycl.* **8** (1994) 349–362.
2. L.J. Spokes, T.D. Jickells, B. Lim, Solubilisation of aerosol trace metals by cloud processing: A laboratory study. *Geochim. Cosmochim. Acta* **58** (1994) 3281–3287.
3. J.H. Martin *et al.*, Testing the iron hypothesis in ecosystems of the equatorial Pacific. *Nature* **371** (1994) 123–129.
4. R. Chester, G.F. Bradshaw, C.J. Ottley, R.M. Harrison, J.L. Merrett, M.R. Preston, A.R. Rendell, M.M. Kane, T.D. Jickell; The atmospheric distribution of trace metals, trace organics and nitrogen species over the North Sea. *Phil. Trans. Roy. Soc. London A* **343** (1993) 543–556.
5. T.D. Jickells; Atmospheric inputs of metals and nutrients to the oceans: their magnitude and effects. *Mar. Chem.* **48** (1994) 199–214.
6. R.R. Yaaqub, T.D. Davies, T.D. Jickells, J.M. Miller; Trace elements in daily collected aerosols in southeast England. *Atmos. Environ.* **25** (1991) 985–996.
7. C.J. Ottley, R.M. Harrison; The spatial distribution and particle size of some inorganic nitrogen, sulphur and chlorine species over the North Sea. *Atmos. Environ.* **26** (1992) 1689–1699 .
8. A.R. Rendel, C.J. Ottley, T.D. Jickells, R.M. Harrison; The atmospheric input of nitrogen species to the North Sea; *Tellus* **45** (1993) 53–63.

9. L. Spokes, T. Jickells, A. Rendell, M. Schulz, A. Rebers, W. Dannecker, O. Kruger, M. Leermakers, W. Baeyens; High atmospheric deposition events over the North Sea. *Mar. Pollut. Bull.* **26** (1993), 698–703.

10. M.M.Kane, A.R. Rendell, T.D.Jickells; Atmospheric scavenging processes over the North Sea. *Atmos. Environ.* **28** (1994) 2523–2530.

2.9 Transformation of Polluted Air Masses when Transported over Sea with respect to Deposition Processes

M. Schulz, T. Stahlschmidt, A. Rebers and W. Dannecker

Institut für Anorganische und Angewandte Chemie,
Abteilung Angewandte Analytik, Universität Hamburg,
Martin-Luther-King-Platz 6, D-20146 Hamburg, Germany

To improve our understanding of the transformation of air masses during transport over sea we have co-operated in three major field campaigns within ASE:

1) The intercalibration experiment at Mace Head, Ireland, aimed at testing the quality of sampling and analytical procedures of aerosols and rain water in the clean marine atmosphere: Sampling artefacts were found to dominate, though some analytical interference was identified. This emphasises the need of identical, intercalibrated instruments in large scale experiments.

2) The intercalibration thus helped to evaluate the Lagrangian-type North Sea experiment. Changes of atmospheric parameters were measured after transport over sea on two ships being 200 km apart from each other. A significant decrease in concentration was determined for several constituents. The relative composition of these constituents showed however very little variation when comparing upwind and downwind positions. Fine aerosol size distributions of trace elements and major ions showed to be very characteristic in their shape for one 6h measuring interval, but varied considerably in the course of the experiment. This indicates that the aerosol size distribution is emerging rather close to the aerosol sources. For large aerosol particles removal could be explained in part by sedimentation. Substantial entrainment at the top of the marine boundary layer was suggested to be the major reason for the overall concentration decrease. The experiment served thus as a pilot experiment for future Lagrangian experiments in that it emphasises the need for a) well intercalibrated instruments, b) a measurement of inert and slow depositing tracers, and c) a characterisation of the vertical boundary layer structure.

3) Vertical profiles of atmospheric constituents at heights of 2–40 m were measured at the German research platform "Nordsee" in the German Bight. Extent and variability of sea salt vertical distribution was determined. The simultaneous deployment of three impactors revealed, that they do not segregate particles with reproducible precision. Shifts in size distribution of trace elements must be attributed in part to the individual performance of

each impactor. No significant change of size distribution could be detected and thus no size dependent dry deposition velocities could be inferred so far.

Aims of the research

Our work was aimed at the further understanding of deposition processes in the marine atmosphere. In particular we were interested in the deposition of aerosols and associated trace substances. Investigation of these was thought to help interpreting our long-term time series of total deposition and aerosol concentration at marine sites with the ultimate goal of reliable estimates of atmospheric input into European marginal seas. The experiments should allow to:

* Improve and intercalibrate our current techniques to measure aerosols and wet deposition of particulate bound trace metals and nitrogen compounds.
* Identify particle size dependent removal rates and shifts in size distribution due to deposition and transformation in the marine atmosphere.
* Measure the horizontal and vertical distribution of atmospheric pollutants and their deposition features in the marginal seas marine atmosphere.

Principal scientific results and conclusions

Intercalibration experiment Mace Head, Ireland

The experiment was performed at the research station of the University College of Galway at Mace Head, Ireland in April 1991. At this location sampling and trace analysis could be tested most realistically under marine background conditions. We have participated both in the aerosol and rain sampler intercomparison.

With regard to aerosol sampling three devices were deployed: a special PIXE impactor combined to TXRF analysis of the aerosol samples, a SIERRA cascade impactor for high-volume sampling and a bulk aerosol low-volume sampler. Sampling and results of the project partners have been described in detail in Francois *et al.* [1]. Largest differences in bulk aerosol analysis was observed when comparing low and high volume sampling. This was explained by different inlet characteristics. Deviations of less then 25 % were observed among the low volume samplers themselves. Impactor comparison with the PIXE impactors from the University of Gent revealed significant differences for large sized particles for very clean marine air. Although the detailed comparison of size distribution of different trace elements showed the general applicability of both analytical methods, identical impactor performance is still no trivial task. Differences have been explained by a combination of inlet characteristics, precision of impactor nozzle manufacturing and unknown bounce-off on ungreased collection surfaces.

We deployed four rain collectors at the same time, one of them being semi-automatic. While analytical cross-checks with the other project partners were satisfying, it appeared, that also for rain collectors sampling efficiency may be a problem. Rain samples from collectors being inclined into the wind had considerably higher sea salt contents [2]. Dry deposition of sea salt was made responsible for this effect. Trace metal concentrations varied much more, with differences of a factor of 10 being no exception. The very low concentrations were associated with analytical and sampling problems which could not be solved in this project. With regard to the North Sea experiment this experiment clearly pointed to the difficulties of measuring with different devices in different places. This allowed us to evaluate more realistically the results from that study.

North Sea experiment

In the budgeting experiment in the central North Sea two ships have been operated on a circle for two weeks to measure the change in air composition. Our main contribution to this experiment has been the measurement of trace element size distributions [3] and the evaluation of the meteorological measurements during the experiment. The 40–60 % decrease in concentration appeared to be consistent throughout the whole dataset despite uncertainties in sampling errors. The removal velocities, which could be calculated from these measurements, were rather high in the order of 1–2 cm/s even for fine mode trace elements. A discussion of interfering mixing and entrainment processes revealed that only relative deposition velocities can be derived from the current data set (detailed discussion in [4]). Large particles with a diameter of 5μm showed a 5-fold higher removal velocity than those having a diameter of 0.3 μm.

The experiment also allowed us to characterise the spatial extent of nitrogen input events to the North Sea. The sampling network at sea and at the coastal sites showed for one single event a large scale deposition of nitrogen in the eastern part of the North Sea [5, 6]. This event with an extension of a third of the total North sea area was responsible for 1 % of the yearly atmospheric flux to the whole North Sea. Since this deposition event was hit by chance, one can presume that their is a likely chance of more outstanding events. The conclusion from this work was, that especially large wet deposition events should have short term effects on the marine ecosystem and phytoplancton production.

Research platform

Vertical profiles of size-distributions of trace elements in the aerosol were measured at the German research platform in autumn 1992. Three PIXE impactors were deployed at heights between 2 and 40 m. At all levels size distributions of the trace elements had roughly the same shape. By microscopic inspection we found that the nozzles of the impactors are not accurately shaped. This in consequence shifts the cut-off diameter of each individual impactor stage.

Correction procedures and their effectiveness could not be tested so far. With regard to the re-emission of trace elements sea spray constituents of course are most prominent in that they are enhanced near sea level. However two elements, bromine and zinc, seem to have a marine source [7].

Acknowledgements

Thanks are due to the German-Irish co-operation foundation, our encouraging project partners within the EUROTRAC ASE subproject and the University of Hamburg, which enabled the use of other project resources, originally not dedicated to this work.

References

1. Francois F., Maenhaut W., Colin J,-L., Losno R., Schulz M., Stahlschmidt T., Spokes L., Jickells T.; Intercomparison of elemental concentrations in total and size-fractionated aeorosl samples collected during the Mace Head experiment, April 1991. *Atmos. Environ.* **29** (1995) 837–849.
2. Losno R., Colin J,-L., Spokes L., Jickells T., Schulz M., Rebers A., Leermakers M., Meulemann C., Baeyens W.; Non-rain deposits disturb significantly the determination of major ions in marine rainwater. *Atmos. Envion.* **32** (1998) 3445–3455.
3. Haster T., Schulz M., Schneider B., Dannecker W.; The use of short term measurements of trace element size-distributions to investigate aerosol dynamics in a marine Lagrangian-type experiment. *J. Aerosol Sci.* **23** (1992) 703–706.
4. Schulz M., Stahlschmidt T., Francois F., Maenhaut W, Larsen S.E.; The change of aerosol size distributions measured in a Lagrangian-type experiment to study deposition and transport processes in the marine atmosphere. in: P.M. Borrell, P. Borrell, T. Cvitaš, W. Seiler (eds), *Proc. EUROTRAC Symp. '94,* SPB Academic Publishing bv, The Hague 1994, pp. 702–706.
5. Rebers, A., Schulz M., Spokes L., Jickells T., Leermakers M.; A regional case study of wet deposition in the North Sea Area. in: P.M. Borrell, P. Borrell, T. Cvitaš, W. Seiler (eds), *Proc. EUROTRAC Symp. '94,* SPB Academic Publishing bv, The Hague 1994, pp. 621–625.
6. Spokes L., Jickells T., Rendell A., Schulz M., Rebers A., Dannecker W., Krüger O., Leermakers M., Baeyens W.; High atmospheric nitrogen deposition events over the North Sea. *Marine Poll. Bull.* **26** (1993) 698–703.
7. Stahlschmidt T., Schulz M. and Dannecker W.; Vertical profiling of heavy metals in the marine atmosphere to investiagte their air-sea-exchange and dry deposition. in: R.D. Wilken, U. Förstner, A. Knöchel (eds), *Heavy metals in the environment*, 1995.

2.10 Chemical Transformations in the North Sea Atmosphere

Roy M. Harrison, C.J. Ottley, M.I. Msibi and A.-M. Kitto

Institute of Public & Environmental Health, School of Chemistry,
University of Birmingham, Edgbaston, Birmingham B15 2TT, UK

The work described has focused on a number of inter-related areas, mostly concerned with the chemistry of nitrogen compoundsin the marine atmosphere. Initially, concentrations of various individual nitrogen compounds were determined over the North Sea. The averaging of data from many individual cruises permitted the mapping of average concentrations across the southern bight of the North Sea. These measurements were used to infer chemical transformations of nitrogen compounds, particularly the dissociation of ammonium nitrate aerosol and the reaction of nitric acid with sodium chloride particles and to calculate atmospheric nitrogen inputs to the North Sea. The measurements of gaseous ammonia in the North Sea atmosphere were combined with data from ammonium in seawater made during the same series of cruises to estimate surface exchange of ammonia. The work showed that the North Sea system is rather finally balanced with respect to ammonia, with the majority of net fluxes from air-to-sea, but some periods when the sea becomes a net source of ammonia in air. In a Lagrangian experiment, concentrations of a number of gaseous and particulate nitrogen species were measured on two ships located 200 km apart. The data have been used to evaluate the rate of oxidation of nitrogen dioxide in the North Sea atmosphere and to estimate concentrations of the hydroxyl radical in daytime, and to evaluate the mechanism of the night-time oxidation of NO_2. In a variant upon this experiment, neutralisation of aerosol acidity arriving at the English coast in ammonia-depleted North Sea air was followed as the air was advected in ammonia-rich air over eastern England. A pseudo first rate order rate constant for the loss of aerosol H^+ was estimated and found to vary with the degree of neutralisation of the aerosol. In a numerical modelling study collaborative with the Danish National Environmental Research Institute, the North Sea measurement data were compared with the predictions of airborne concentrations from the Danish Eulerian model with non-linear chemistry. After some modifications to the chemical scheme in the model, it was found that model predictions were in extremely good agreement with measured concentrations.

Aims of the research

To improve understanding of the cycling of nitrogen in the polluted marine atmosphere and to evaluate the influence of chemical transformations on the form of the nitrogen and its deposition to the sea.

Principal scientific results

Because of its location between continental Europe and the British Isles the North Sea is subject to substantial pollutant inputs. It has proved peculiarly difficult to provide adequate quantification of the various input pathways, which include riverine discharges, direct discharges to coastal waters and atmospheric deposition. The available estimates suggest that atmospheric deposition is one of the major routes of input for many important pollutant species.

Because of the substantial area covered by the North Sea, it is very difficult to estimate atmospheric pollutant inputs through direct measurements. Most estimates have been based upon measurements of either airborne concentrations or deposition fluxes at coastal, or in more recent studies, island sites with extrapolation subsequently across large areas of the North Sea. Since atmospheric concentrations over the North Sea decline towards the centre and the north where measurement stations are most difficult to place, inputs estimated by such procedures are liable to significant overestimation of the total deposition and are unable to adequately resolve the spatial distribution.

Our initial work on deposition of nitrogen compounds, important as a source of nutrient input, led to a realisation that chemical transformations in the North Sea atmosphere could play a key role in determining deposition fluxes. For example, atmospheric oxidation of nitrogen dioxide to nitric acid could lead to a great enhancement of the deposition velocity to seawater. Additionally, the reaction of nitric acid vapour with the surface of marine sodium chloride could further modify the deposition velocity of the nitrate. In some aspects the marine atmosphere provides an excellent laboratory in which to study the rates of chemical transformations. If measurements are made sufficiently far from the coast, there is relatively high lateral homogeneity in the air mass composition and the trace gases and particles are also well mixed throughout the vertical extent of the boundary layer. It is therefore feasible to carry out Lagrangian-type experiments following the development of air mass composition during advection, without the need to account for additional source inputs.

Estimation of deposition fluxes to the North Sea

As part of the NERC North Sea Programme the opportunity arose to collect samples on the research vessel RRS Challenger which undertook regular cruise tracks of the southern bight of the North Sea, each of two weeks duration, once per month. These were used to define the mean concentration field, and for particulate material the size distribution of relevant trace gases and aerosols. By collecting a sufficiently large number of samples, temporal variations due to different air mass origins were averaged and sensible spatial distributions were generated. An example appears in Fig. 2.10.1.

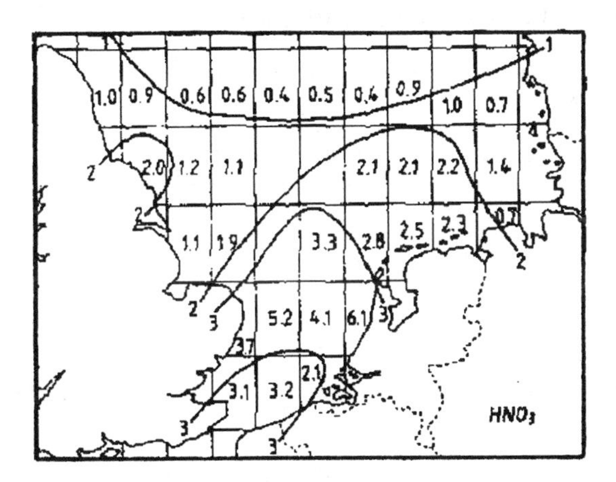

Fig. 2.10.1: Average spatial distribution of HNO_3 over the North Sea

Appropriate values of deposition velocities for the gases were estimated from the model of Hicks and Liss [1] and for aerosols from the model of Slinn and Slinn [2]. Results have been published for trace metals [3] and for nitrogen species [4]. These data have been combined with wet deposition measurements made during the same cruises by Jickells and co-workers in order to calculate total atmospheric inputs from both dry and wet deposition, both for trace metals [5] and for nitrogen compounds [6].

Work in collaboration with Dr Willem Asman of the Danish National Environmental Research Institute sought to use our data for ammonia concentrations in air together with measurements of ammonium in seawater made during the same series of cruises to estimate the surface exchange of ammonia with greater confidence. This work has shown that the North Sea system is rather finely balanced with respect to ammonia, with the majority of net fluxes from air-to-sea, but some periods when the sea becomes a net source of ammonia in air [7]. This is contrary to the situation in remote marine atmospheres where the flux is normally out of the sea, as a result largely of the higher gaseous ammonia concentrations in the North Sea atmosphere. Table 2.10.1 indicates the magnitude of the various atmospheric inputs of oxidised and reduced nitrogen to the southern North Sea; it has been modified from that reported by Rendell *et al.* [6] in the light of the refined estimate of ammonia deposition. One facet of atmospheric chemistry to be revealed by the measurements over the North Sea was that the concentration products $[HNO_3]$ $[NH_3]$ and $[HCl]$ $[NH_3]$ were generally insufficient to permit the formation of NH_4NO_3 and NH_4Cl aerosol. This is contrary to the situation over adjacent land and reflects the lower concentrations of ammonia in the North Sea atmosphere, hence, ammonium nitrate and chloride advected over the North Sea will tend to evaporate rather rapidly. Additionally, the particle size distribution for nitrate aerosol was entirely different from that of non-sea-salt sulfate, being very similar to the sodium surface area distribution. This we interpret as the result of reaction of nitric acid vapour with the surface of marine NaCl.

Table 2.10.1: Atmospheric deposition of nitrogen species to the southern North Sea (south of 56° N) (10^3 t N a^{-1}).

	HNO_3–N	NO_3^-–N	NH_3–N	NH_4^+–N	DON–N	Total–N
dry deposition	20.0	44.7	7.6	22.2		94.5
wet deposition		70.5		47.8	8.1	126.4
total deposition	20.0	115.2	7.6	70.0	8.1	220.9

Estimation of the rates of chemical transformations from Lagrangian experiments

As indicated above, because of the general absence of local sources, the North Sea atmosphere is an excellent laboratory for Lagrangian experiments. One such experiment was designed using two ships, the Alkor and Belgica which were located on opposite sides of the periphery of a circle of diameter 200 km, such that the Belgica was always directly downwind of the Alkor. Measurements were made of a range of chemical constituents including gaseous NO_2, HNO_3 and NH_3 and particulate SO_4^{-2}, NO_3^-, $NH_4^+$and Na^+. Regular radiosondes were used to define the vertical temperature structure and depth of the boundary layer. The results illustrated in Fig. 2.10.2 for nitrogen dioxide showed a considerable loss of some species during the 200 km airmass transport between the two ships. Interpretation of such data proved to be difficult. A trace substance may be lost from the boundary layer by a number of mechanisms including: exchange at the air-sea interface, exchange with the free troposphere at the top of the mixed boundary layer, chemical breakdown and/or scavenging by cloud or rainfall. Rain fell for only a relatively minor proportion of the time during the experiment and is not believed to have had a major influence.

For a species such as nitrogen dioxide, deposition to the sea was believed to be negligible and it was thought initially that the loss observed between the ships was due to oxidation to nitric acid. However, analysis of the data revealed the loss to be unreasonably rapid and it was concluded that exchange with the free troposphere might be accounting for a significant loss. It was therefore necessary to find a marker for this loss rate and we chose to use fine particle sulfate, which would be expected to be affected only slightly by chemical formation from sulfur dioxide and loss by dry deposition.

Hence, losses of fine sulfate were presumed to occur predominantly due to loss from the boundary layer into the free troposphere and hence the rate of loss of this species was taken as a measure of this process, accounting also to some extent for wet deposition losses of pollutants.

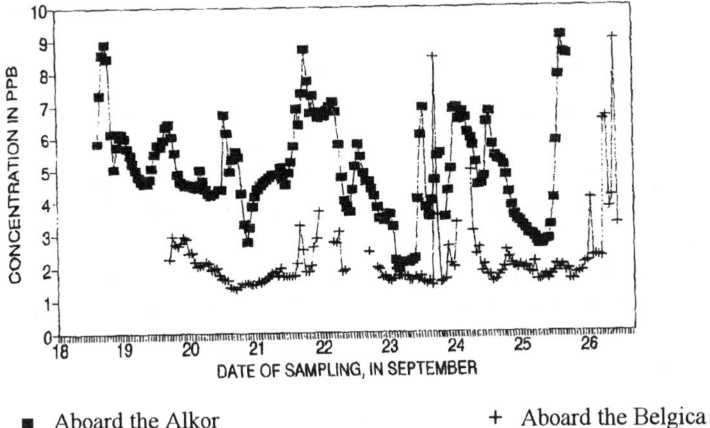

Fig. 2.10.2: Concentrations of NO_2 on the upwind and downwind ships in the Lagrangian Experiment

Daytime chemical loss of nitrogen dioxide is expected to be by reaction with the hydroxyl radical:

$$NO_2 + OH \rightarrow HNO_3.$$

Using the rate constant for this process and assuming first order loss of nitrogen dioxide from the boundary layer at the same rate as fine particle sulfate allows estimate of the concentration of hydroxyl radial. The calculated concentration for hydroxyl of $1\text{–}2 \times 10^6$ cm^{-3} is highly consistent with that expected for September at the latitude of the North Sea [8]. At night-time the main loss rate for NO_2 is expected to be via reaction with ozone:

$$NO_2 + O_3 \rightarrow NO_3 + O_2$$

$$NO_3 + NO_2 \leftrightarrow N_2O_5$$

$$N_2O_5 + H_2O \rightarrow 2HNO_3$$

The NO_2 loss rate at night-time was consistent with the expected rate of the first reaction of NO_2 with ozone without subsequent amplification by further reaction of NO_3 with NO_2. Since there are other pathways open for reaction of NO_3 this is not a surprising finding.

Loss rates for HNO_3 and aerosol NO_3^- and NH_3 and aerosol NH_4^+ are more difficult to interpret as there is the possibility of exchange between the gaseous and particulate species as well as loss processes. The measurements were therefore interpreted in terms of the loss of (HNO_3 + aerosol NO_3^-) and of (NH_3 + aerosol NH_4^+).

In the case of nitrate species there appeared to be no net loss and hence formation from NO_2 is approximately balanced by deposition processes. In the case of ammonia and ammonium there was a substantial net loss [9].

In a variant upon this experiment the neutralisation of aerosol H^+ arriving at the English coast in ammonia-depleted North Sea air was followed as the air was advected for 1–2 hours in ammonia rich air over eastern England [10]. The rate of decrease of aerosol H^+ concentration was measured and a pseudo-first-order rate constant for the loss reaction was calculated in the range 4×10^{-6} s^{-1} to 4.1×10^{-4} s^{-1} consistent with model results and experimental data from the literature derived by other methods. As expected, the rate constant decreased as the aerosol became less acidic. These pseudo-first-order rate constant varied with the H^+/NH_4^+ mole ratio according to the expression

$$k = 23 \times 10^{-5}\ (H^+/NH_4^+) + 4 \times 10^{-5}\ (r = 0.65;\ n = 24), \text{ for } k \text{ in units of } s^{-1}$$

These experiments are again subject to numerous inaccuracies, but given the relatively short travel times, neither dry deposition nor exchange with the free troposphere is expected to be a significant loss mechanism compared to chemical neutralisation by ammonia.

Modelling study

The spatially resolved measurements of airborne concentrations over the North Sea provided an ideal basis for a numerical modelling study which was carried out in collaboration with Dr Zahari Zlatev using the Danish Eulerian model with non-linear chemistry [11]. The model, which uses data collected by the Norwegian Meteorological Institute in connection with EMEP is able to calculate airborne concentrations corresponding to specific individual sampling intervals within North Sea cruises. The estimates are therefore both highly spatially and temporally resolved. The two-dimensional space domain considered in the model contains Europe together with parts of the Atlantic Ocean, Asia and Africa discretised into a 32 by 32 grid with space increments of 150 km by 150 km at 60° N latitude. The model correctly predicted many features of the data and gave a generally good performance [12]. It was subsequently modified in two ways designed to better describe heterogeneous chemical processes. These involved the inclusion of kinetic constraints in the $NH_3/HNO_3/NH_4NO_3$ reaction system and use of the rate data from Harrison and Kitto [10] reported above for the neutralisation of acid sulfate by ammonia, previously assumed to be instantaneous. These modifications lead to a better model performance as estimated by the parameter DEV listed in Table 2.10.2 in which DEV = 100 × (calculated-measured)/measured. After modifications, the model predictions were extremely good as evaluated by the procedure suggested by Hanna and Heinold [13].

Table 2.10.2: Comparison of measured and model-predicted mean concentrations ($\mu g\ m^{-3}$) before (*a*) and after (*b*) modifications to model

Species	Measured	Calc–*a*	Dev–*a*	Calc–*b*	Dev–*b*
Particulate ammonium	2.0	1.9	5%	2.2	10%
Particulate sulfate	4.5	3.7	16%	4.1	9%
Particulate nitrate	4.0	3.3	17%	3.2	20%
Nss sulfate	4.2	3.7	12%	4.1	2%
Gaseous ammonia	0.4	0.8	200%	0.5	25%
Gaseous nitric acid	2.3	3.1	35%	2.5	9%

Conclusions

The work has succeeded in better quantifying the input of nitrogen species to the North Sea. It has demonstrated the great importance of dry and wet deposition of ammonium and nitrate ions and the significance of atmospheric chemistry in influencing concentrations of nitric acid and ammonia which contribute significantly to nitrogen inputs. The system is rather finally balanced with respect to ammonia, but the predominant flux is from the atmosphere to the sea. Chemical reactions involving dissociation of ammonium nitrate and ammonium chloride and reaction of nitric acid with sea-salt aerosol are important in influencing the deposition characteristics of the pollutant nitrogen. The model study demonstrates that these and other chemical processes can successfully be accommodated in a numerical model which well describes the chemistry of nitrogen in the North Sea atmosphere. Lagrangian experiments can provide useful information upon transformation rates if exchange with the free troposphere at the top of the mixed boundary layer can be quantified. This can provide useful information on the concentrations of reactive species such as the hydroxyl radical responsible for promoting chemical change.

Overall, the project has been successful in addressing its objectives and has benefited immensely from the collaborations developed through EUROTRAC.

Acknowledgements

Thanks are due to the U.K. Department of Environment and Natural Environment Research Council for funding of the research. The collaborative work by C.J. Ottley (University of Lancaster), T. Jickells (University of East Anglia), Z. Zlatev and W. Asman (Danish National Environmental Research Institute) has been essential to the success of this work and is gratefully acknowledged.

References

1. Hicks, B.B., Liss, P.S., *Tellus*, **28** (1976) 348–354.
2. Slinn, S.A., Slinn, W.G.N., *Atmos. Environ.* **14** (1980), 1013–1016.
3. Ottley, C.J., Harrison, R.M., *Atmos. Environ.* **27A** (1993) 685–695.
4. Ottley, C.J., Harrison, R.M., *Atmos. Environ.* **26A** (1992) 1689–1699.
5. Chester, R., Bradshaw, G.F., Ottley, C.J., Harrison, R.M., Merrett, J.L., Preston, M.R., Rendell, A.R., Kane, M.M.,Jickells, T.D., *Phil Trans R. Soc. Lond. A* **343** (1993) 543–556.
6. Rendell, A.R., Ottley, C.J., Jickells, T.D.,Harrison, R.M., *Tellus* **45B** (1993) 53–63.
7. Asman, W.A.H., Harrison, R.M.,Ottley, C.J., *Atmos. Environ.* **28** (1994) 3647–3654.
8. Ehhalt, D.H., Dorn, H.P., Poppe, D. *Proc. R. Soc. Edinburgh*, **97B** (1991) 17–34.
9. Harrison, R.M., Msibi, M.I., Kitto, A.-M.N., *Atmos. Environ.* **28**, (1994) 1593–1599.
10. Harrison, R.M., Kitto, A.-M.N., *J. Atmos. Chem.* **15** (1992) 133–143.
11. Zlatev, Z., Christensen, J., Hov, O., *J. Atmos. Chem.* **15** (1992) 1–37.
12. Harrison, R.M., Zlatev, Z., Ottley, C.J. *Atmos. Environ.* **28** (1994) 497–516.
13. Hanna, S.R., Heinold, D.W. *Air Pollution Modelling and its Application*, **V** (1985) 441–452.

2.11 Elemental Composition and Sources of Atmospheric Aerosols above and around the North Sea

Willy Maenhaut and Filip François

Institute for Nuclear Sciences, Proeftuinstraat 86, B-9000 Gent, Belgium

Our research within the EUROTRAC subproject ASE focuses on the study of size-fractionated atmospheric aerosols at different sites above and around the North Sea and in the Arctic. The work involves continuous samplings at (near) coastal stations and participation in intensive sampling campaigns (usually on sea and together with other ASE groups). In the Mace Head 1991 campaign, we were involved in an intercomparison of aerosol sampling and analytical techniques. In the North Sea experiment 1991 we contributed to the study of the aerosol transport and deposition processes by collecting samples at different locations during passage of an air mass over the North Sea and by comparing the atmospheric elemental concentrations and size distributions. In the 1992 joint experiment on the German research platform Nordsee, we examined the alterations in aerosol compositions with the origin of the air mass. During the International Arctic Ocean Expedition 1991, we studied the atmospheric aerosol in the high Arctic during summer. In our long-term continuous samplings at two locations in southern Norway, we investigate the long-range transport of continental aerosols over the North Sea and make inferences about the deposition of these species into the sea. Continuous samplings are also performed at the Zeppelin background station in Spitsbergen, with the aim to assess the sources of the non-sea-salt sulfate and the particulate trace elements at various times of the year, and to examine changes in elemental mass size distributions.

Aims of the research

Our contribution to the EUROTRAC subproject ASE aims at studying the chemical (multi-elemental) composition and the sources (source regions) and transformation and deposition processes of atmospheric aerosols above and around the North Sea and in marine Arctic regions. Aerosol samples are collected in two size fractions (coarse and fine) with PM10 stacked filter units (SFUs) [1] and with a modified Sierra-Andersen Hi-Vol cascade impactor [2]. In order to obtain more detailed size fractionation, we use Battelle-type PIXE cascade impactors (PCIs) [3] and occasionally also other cascade impactors [4]. The aerosol collections are performed in long-term continuous samplings and as part of intensive field campaigns (usually joint EUROTRAC ASE experiments). Continuous samplings are done at Birkenes and Skreådalen in southern Norway and at the Zeppelin mountain station in Spitsbergen. The southern Norwegian locations are excellent sites to study the long-range transport of continental and anthropogenic species over the North Sea, whereas the Zeppelin station allows us to examine this transport further to the Arctic basin.

The samples are analysed by gravimetry (for measuring the particulate mass (PM)), by a light reflectance technique (for black carbon), by particle-induced X-ray emission (PIXE), instrumental neutron activation analysis (INAA), and ion chromatography (IC) [4–6]. This results in data sets with concentrations and size distributions for up to 40 components, which can be divided into several categories, depending on their main sources. Sea-salt derived elements are Na, Mg, S, Cl, K, Ca, Br and Sr, mineral dust elements are Al, Si, Ti, Sc, Fe, the rare earth elements and Th, and the anthropogenic elements (species) comprise black carbon (BC), S, V, Mn, Ni, Cu, Zn, As, Se, Mo, Cd, In, Sb, W and Pb. The main ionic components that are measured by IC are SO_4^{2-}, methanesulfonate (MSA), NO_3^-, and NH_4^+.

Interpretation of the multi-sample, multi-element data sets yields information about the sources (source types, source regions), transport, transformation and deposition processes of the atmospheric aerosol constituents. Several techniques are used in interpreting the data. The detailed elemental size distributions from the PCI samples (or even the fine/coarse concentration ratios from the SFUs) and the enrichment factors relative to average crustal rock or sea water composition already give a very good idea about the possible source processes for the various elements. For a more thorough interpretation and for assessing the contribution from the source types or source regions, we make use of various receptor models [2, 7–9] (occasionally also transport models [10]) and/or we relate the concentration data to air mass trajectories, wind sectors or meteorological data [11-13]. To assess the extent of physical transformation processes during the atmospheric transport, the detailed elemental size distributions (or fine/coarse ratios) at the various sites along the transport routes are compared [14]. The detailed elemental mass size distributions are also used to calculate dry deposition velocities [15], and by multiplying airborne concentrations by dry deposition velocities or scavenging ratios, fluxes to sea and oceans are obtained [16].

Principal scientific results

a. Mace Head Experiment, Ireland, April 1991

As part of this joint ASE experiment we were involved in an intercomparison exercise for aerosol collection and analysis. Besides ourselves, three other ASE groups participated. A variety of samplers was used, including both high- and low-volume devices, with different types of collection substrates: Hi-Vol Whatman 41 filter holders, single Nuclepore filters and stacked filter units, as well as PIXE cascade impactors. The samples were analysed by each participating group, using in-house analytical techniques and procedures. For the majority of the elements, a good agreement was observed among the low-volume samplers. With the Hi-Vol Whatman 41 filter sampler, on the other hand, much higher results were obtained, in particular for the sea-salt and crustal elements. The discrepancy was dependent upon the wind speed and was attributed to a higher collection efficiency of the Hi-Vol sampler for the very coarse particles, as compared to the low-volume devices, under high wind speed conditions. The results of the intercomparison are discussed in detail elsewhere [17]. A brief account on our other work in the Mace Head experiment is given by François *et al.* [14].

b. North Sea Experiment (NOSE), 16th-26th September, 1991

During this Lagrangian transport experiment aerosol samples were collected by (or for) us at 6 different locations, *i.e.* aboard 2 research vessels (FS Alkor and RV Belgica) and at 4 (near) coastal land-based sites (the Dutch isle of Texel, Rörvik in Sweden, and the two southern Norwegian stations Birkenes and Skreådalen). The two research vessels were aligned in wind direction and at about 200 km from each other, with the Alkor always upwind of the Belgica. Furthermore, the start and end of each sampling period aboard the two ships were determined on the basis of the ship positions and the wind speed, so that the "parallel" collections at the two ships sampled the same air mass. Aboard the two ships we collected both SFU and PCI samples, but the samplings at the (near) coastal sites were only done with SFUs or a total filter sampler. The prevailing wind direction during NOSE was south to west for the entire duration of the experiment, so that the two southern Norwegian sites were essentially downwind of the two research vessels. The average SFU fine/coarse ratios for the two ships and the various land-based sites were in general rather similar. Furthermore, the SFU fine/coarse ratios observed at the Belgica agreed very closely with those at the Alkor. This indicates that the size distribution of the mineral dust and anthropogenic elements (in the particle size range < 10 μm equivalent aerodynamic diameter (EAD)) remained virtually unchanged during the 200 km transport over the North Sea. This was confirmed by the detailed size distribution data derived from the parallel PCI samples at the two ships. From a detailed comparison of the data from the parallel SFU samples at the two ships (see Fig. 2.11.1), it appeared that the concentration ratio Belgica/Alkor was very close to one for the aerosol mass and the sea-salt elements (both for the coarse and the fine size fraction). For the mineral dust and anthropogenic elements, the average Belgica/Alkor concentration ratios were typically 0.6–0.7, and this for both size fractions. Adopting an Alkor-Belgica transport time of 5 hours and a mixed layer height of 1000 m, this corresponds to an average total deposition velocity (sum of wet plus dry) of 2 cm/s. The ratio Belgica/Alkor varied from sample to sample, though, and was clearly lower for samples collected during rainy periods than for dry weather conditions. However, it is quite likely that not only deposition, but also dilution and mixing processes are at the origin of the observed large concentration differences between the Belgica and Alkor. The fact that the detailed size distribution is virtually the same on the two ships provides support for the importance of these other processes. More discussions on this work and on the changes in the size distribution between the two research vessels are given by François *et al.* [14] and Schulz *et al.* [18].

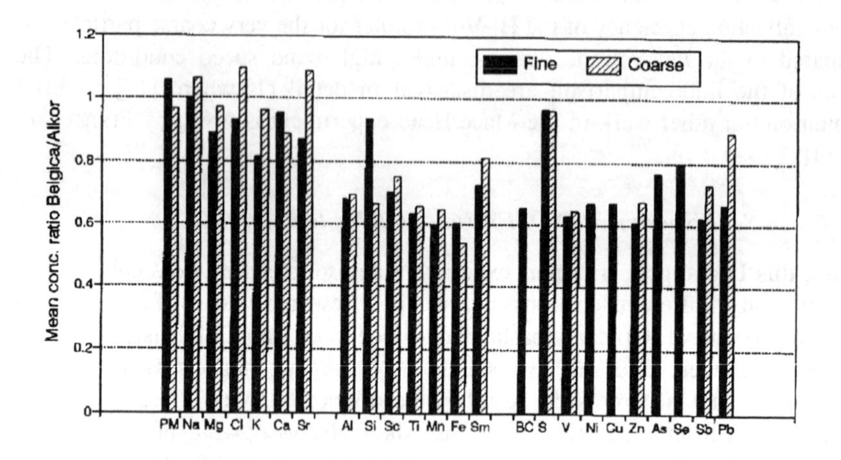

Fig. 2.11.1: Average concentration ratios Belgica/Alkor in the fine and coarse size fractions for PM, the sea-salt, mineral dust and anthropogenic elements (based on 13 SFU sample pairs). Although K is grouped with the sea-salt elements, its fine fraction is to a large extent anthropogenic. Similarly, S is grouped with the anthropogenic elements, but this is only correct for its fine fraction; the coarse S is mainly from sea-salt origin.

c. *Experiments aboard the research platform Nordsee, September 1992*

During this three-week campaign-type experiment, we collected aerosol samples by means of SFUs and PCIs at two different heights (27 and 45 m) above sea level. The grouping of the data into wind sectors showed that there was a negative concentration gradient, spanning more than one decade, with decreasing continental influence (SE > SW > W > NW) for the fine PM, the fine anthropogenic elements and the coarse mineral dust elements. This suggests that, for that area of the North Sea, the transport from eastern Europe and Germany is much more important for the input of particulate pollutants into the sea than the transport from other source regions. Also the mass size distributions (or fine/coarse ratios) of various particulate species seemed to depend upon the wind sector. Furthermore, it was found that fogs cause drastic changes in the mass size distributions. More details on this work are given by François and Maenhaut [13].

d. *International Arctic Ocean Expedition, August-October 1991 (IAOE-91)*

It was found that the concentrations of the PM, non-sea-salt (nss) sulfate and various natural and anthropogenic aerosol constituents were quite low in the high Arctic (north of 80°). For many species, the levels were 5-10 times lower than summer concentrations observed at Ny Ålesund in Spitsbergen. With the exception of elemental carbon, S and I, all species and elements measured, including the typical anthropogenic metals (*e.g.* Zn, As, and Sb), were predominantly associated with the coarse size fraction. The highest levels of the anthropogenic metals and of the mineral dust elements were found in air, which in general had not been in contact with large continental or anthropogenic source regions during the last 5 days prior to its arrival, and in an area of the pack ice, which was influenced by continental river run-off from the Siberian coast. It was therefore suggested that the elevated concentrations of both those metals and the crustal elements were the result of local mechanical windblown generation of coarse aerosols from the river effluent materials which were present on or at the surface of the ice. As to the sources of the nss-sulfate in the high Arctic, it was concluded that this was mainly from anthropogenic origin near the end of the IAOE-91, but that the marine biogenic sulfur source must have dominated during the first month and a half of the expedition. The results from this research are described in more detail elsewhere [12].

e. Long-term aerosol study in Birkenes and Skreådalen, southern Norway

Since February 1991, SFU samples are continuously being taken according to a 2-2-3 day sampling scheme at these two Norwegian Institute for Air Research (NILU) sampling stations. The results obtained for more than two years of samplings indicate that the fine fraction concentration data from the two sites are very well correlated with each other. This is clearly so for the particulate mass (PM) and for most anthropogenic species (*e.g.* BC, S, Zn). Furthermore, these latter anthropogenic species, and in particular S, were very well correlated with the fine PM at each of the two sites. This suggests that the fine PM is to a large extent attributable to anthropogenic emissions. The good correlation between the two sites, on the other hand, suggests that this anthropogenic aerosol is advected to southern Norway by long-range transport of polluted air masses from central, eastern or western Europe. The coarse fraction crustal elements at the two sites also exhibited a fairly good correlation, so that a major fraction of the crustal aerosol is probably also advected by long-range transport. However, for a few elements, especially fine Mn, the correlation between the two sites broke down, which indicates that local (and regional) southern Norwegian contributions are also important for certain species.

f. Long-term aerosol study at Ny Ålesund, Spitsbergen

In a co-operation with NILU, aerosol samples are continuously collected for us since early 1991 at the Zeppelin background station in Ny Ålesund, Spitsbergen. The aerosol samplers consist of a modified Sierra-Andersen Hi-Vol cascade impactor and of a PCI. The time series of the fine (< 2.5 μm EAD) nss sulfate and MSA atmospheric concentrations, as derived from the back-up filters of the 1992 Hi-Vol samples, is presented in Fig. 2.11.2. During the summer months (June through August), the two sulfur species were for more than 90 % associated with the fine size fraction. In order to resolve the relative contributions from marine biogenic dimethylsulfide (DMS) emissions and from anthropogenic sources to the fine nss sulfate during summer, various multiple linear regression (MLR) calculations were applied with nss sulfate as dependent variable and with MSA (as indicator for the biogenic contribution) and an anthropogenic tracer (As or Sb) as an independent variable [19]. These MLR analyses for the 1991, 1992 and 1993 summer data sets indicated that, on the average, about one third of the fine sulfate was from marine biogenic origin and about two thirds from anthropogenic origin. However, there was considerable variability in the relative contribution from both source types from sample to sample and even from season to season. During the summer 1993 the average contribution from biogenic sources was only 20 % versus about 40–50 % in 1991 and 1992. On the other hand, the ratio of (fine MSA/fine biogenic sulfate) showed no variation from summer to summer and was about 0.4–0.5.

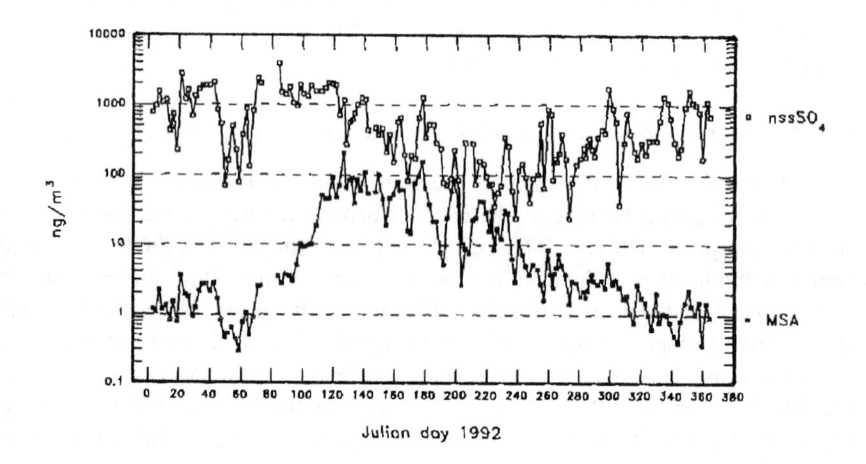

Fig. 2.11.2: Time trend of the nss sulfate and MSA concentrations at the Zeppelin mountain station during 1992. The data apply to the size fraction < 2.5 μm EAD.

Conclusions

Our ASE work provided extensive data sets of the size-fractionated aerosol composition above and around the North Sea and in the Arctic. Several aspects of the sources (source types), transport, transformation and deposition processes were examined. A surprising finding was the rather large decrease in boundary layer concentrations of the mineral dust and anthropogenic elements, and at the same time the virtually unchanged size distributions, during the Lagrangian transport experiment over the North Sea. Of important significance is our result that during summer at Spitsbergen one third of the fine nss sulfate is on the average attributable to marine biogenic sources. More work needs to be done with regard to the source regions of the various aerosol constituents for Birkenes/Skreådalen and Ny Ålesund, and with regard to the calculation of the fluxes of the constituents to the North Sea. This work is currently in progress.

Acknowledgements

We are very grateful to the Norwegian Institute for Air Research, and in particular to Dr. H. Dovland and J.E. Hanssen, for their co-operation in the Birkenes/Skreådalen and Ny Ålesund samplings. G. Ducastel and K. Beyaert participated in the research on the Arctic aerosol. J. Cafmeyer helped in the collection of the Mace Head and NOSE samples and in the sample analysis. Further assistance in the analysis was given by C. Gilot. Financial support is provided by the Belgian "Nationaal Fonds voor Wetenschappelijk Onderzoek", the "Interuniversitair Instituut voor Kernwetenschappen", and an EUROTRAC ASE grant from the Belgian State-Prime Minister's Service, Federal Office for Scientific, Technical and Cultural Affairs.

References

1. W. Maenhaut, F. François, J. Cafmeyer, Applied Research on Air Pollution using Nuclear-Related Analytical Techniques, *Report 1st Research Co-ordination Meeting, Vienna, April 1993*, NAHRES-19, IAEA, Vienna 1994, p. 249.
2. W. Maenhaut, P. Cornille, J.M. Pacyna, V. Vitols, *Atmos. Environ.* **23** (1989) 2551.
3. S. Bauman, P.D. Houmere, J.W. Nelson, *Nucl. Instr. Meth.* **181** (1981) 499.
4. W. Maenhaut, *Belgian Impulse Programme "Global Change", Symposium 17 & 18 May 1993, Proceedings*, Vol. 1, Belgian Science Policy Office 1993, p. 5.
5. W. Maenhaut, A. Selen, P. Van Espen, R. Van Grieken, J.W. Winchester, *Nucl. Instr. Meth.* **181** (1981) 399.
6. P. Schutyser, W. Maenhaut, R. Dams, *Anal. Chim. Acta* **100** (1978) 75.
7. G.E. Gordon, *Environ. Sci. Technol.* **22** (1988) 1132.
8. R.C. Henry, C.W. Lewis, P.K. Hopke, H.J. Williamson, *Atmos. Environ.* **18** (1984) 1507.
9. W. Maenhaut, J. Cafmeyer, *J. Trace Microprobe Techn.* **5** (1987) 135.
10. J.M. Pacyna, A. Bartonova, P. Cornille, W. Maenhaut, *Atmos. Environ.* **23** (1989) 107.
11. W. Maenhaut, I. Salma, J. Cafmeyer, H.J. Annegarn and M.O. Andreae, *J. Geophys. Res.* 101 (1996) 23631.
12. W. Maenhaut, G. Ducastel, C. Leck, E.D. Nilsson, J. Heintzenberg, *Tellus* **48B** (1996) 300.
13. F. François, W. Maenhaut, in: P.M. Borrell, P. Borrell, T. Cvitaš, W. Seiler (eds), *Proc. EUROTRAC Symp. '94,* SPB Academic Publishing bv, The Hague 1994, p. 472.
14. F. François, J. Cafmeyer, C. Gilot and W. Maenhaut, in: P.M. Borrell, P. Borrell, T. Cvitaš, W. Seiler (eds), *Proc. EUROTRAC Symp. '92,* SPB Academic Publishing bv, The Hague 1993, p. 788.
15. M.H. Bergin, J.-L. Jaffrezo, C.I. Davidson, J.E. Dibb, S.N. Pandis, R. Hillamo, W. Maenhaut, H.D. Kuhns, T. Mäkelä, *J. Geophys. Res.* **100** (1995) 16275.
16. R.A. Duce and 21 co-authors, *Global Biogeochem. Cycles* **5** (1991) 193.
17. F. François, W. Maenhaut, J.-L. Colin, R. Losno, M. Schulz, T. Stahlschmidt, L. Spokes, T. Jickells, *Atmos. Environ.* **29** (1995) 837.

18. M. Schulz, T. Stahlschmidt, F. François, W. Maenhaut, S.E. Larsen, in: P.M. Borrell, P. Borrell, T. Cvitaš, W. Seiler (eds), *Proc. EUROTRAC Symp. '94,* SPB Academic Publishing bv, The Hague 1994, p. 702.
19. W. Maenhaut, G. Ducastel, K. Beyaert, J.E. Hanssen; in: P.M. Borrell, P. Borrell, T. Cvitaš, W. Seiler (eds), *Proc. EUROTRAC Symp. '94,* SPB Academic Publishing bv, The Hague 1994, p. 467.

2.12 Study of Individual Particle Types and Heavy Metal Deposition for North Sea Aerosols using Micro and Trace Analysis Techniques

René Van Grieken, J. Injuk, L. De Bock and H. Van Malderen

Department of Chemistry,University of Antwerp (U.I.A.),
B-2610 Antwerp-Wilrijk, Belgium

In the period under review (1991–1994), our group has participated in three sampling programs that have been organised in the framework of the EUROTRAC ASE project: the "Mace Head Intercomparison Experiment" (8th–20th April 1991), the "North Sea Experiment" (15th–27th September 1991) and the ASE experiment on the research platform "Nordsee" (1st–22nd September 1992). Also, an extensive aerosol sampling program was accomplished during several research cruises on the North Sea with the R.V. Belgica (1991, 1992, 1993, 1994) and at a fixed station at the Belgian coast (1992–1994). Our research was intended to estimate the amount of atmospheric pollutant fluxes to the North Sea and to identify the nature of the particles responsible for this flux. In such a complex study, the combined use of different analytical techniques namely: energy dispersive X-ray fluorescence (EDXRF), total reflection X-ray fluorescence (TXRF), micro-proton induced X-ray emission (micro-PIXE) and electron probe X-ray micro-analysis (EPXMA) was effectively implemented.

The atmospheric supply of material determined indicates the extreme importance of the atmosphere as a pathway by which the pollutants are reaching the North Sea environment. When compared with other input sources such as rivers, direct discharges and incineration, the atmospheric input equals approximately:

34 % (Ni), 17 % (Zn), 18 % (Cu), 32 % (Pb).

The single particle analysis was mainly focused on the determination of the inorganic composition of giant particles with diameters above 1 μm. Several thousands of individual particles were identified to determine their nature, composition, origin, physical size, shape and surface characteristics. Multivariate techniques were used for the reduction of the data set and for source apportionment. The compositional heterogeneity for major and trace elements within the single particles were determined and based on the relative abundances, found by hierarchical cluster analysis according to the Ward error sum method, three to eight different aerosol types were distinguished. The composition changes in air masses crossing the North Sea appeared as a decrease in the abundance for the aluminosilicate particles and a relative increase for NaCl and sea water crystallisation products. Principal factor analysis revealed four different aerosol types: aluminosilicates, NaCl and sea water crystallisation products, and two aerosols types from industrial sources.

Aims of the research

A first aim of our work was in an improved determination of elemental aerosol size distributions above the North Sea, essential for a more reliable atmospheric estimation of the heavy metal inputs into the marine environment. Since parameterisation of both wet and dry deposition by direct flux measurements has been generally inadequate, our semi-empirical deposition calculations were based on the experimental values of aerosol concentrations as a function of aerosol size and meteorological conditions. A novel approach for the isokinetic sampling conditions was applied by means of sampling in a wind tunnel, which avoids the loss of very large particles (> 1 μm) at the sampler inlet. Sampling was done on Nuclepore filters and quartz discs by cascade impactors (May and Battelle) and Millipore filter holders.

The major purpose of this work was, however, to identify quantitatively the different individual particle types in North Sea aerosols and rain water suspensions, using two complementary micro-analysis techniques, EPXMA and micro-PIXE.

Principal scientific results

The average airborne heavy metal concentrations for different sampling campaigns on the North Sea (from research vessels, a platform, and a fixed station at the Belgian coast) were all in the same ranges. A clear decreasing trend was observed from the southern to the northern parts of the North Sea. There is a strong dependence on the wind direction; the heavy metal concentrations originating from the south-southeast sector are about 20 times higher than for the north-northwest sector, when the air masses are from purely marine origin. Based upon the size-segregated airborne concentrations, the dry and wet deposition fluxes to the North Sea area were calculated by the model of Slinn and Slinn [1]. We have used the assumption that over the North Sea, a relatively humidity of 98.3 % exists at the air-sea interface in which hygroscopic particles (all except Fe-rich particles) instantly reach their equilibrium hydrated size. For Fe-containing particles, the assumption was made that these particles are predominantly hydrophobic. Since dry depositions velocities are mass dependent, normalising aerosol mass and particle size is necessary. In this study, we used the mass-weighted dry deposition velocities given by the Battelle cascade impactor. The mass-weighted dry deposition velocities were calculated from each cascade impactor data set for each element. In the calculation of dry fluxes to the sea, the total airborne concentration of an element within each cascade impactor was multiplied by the corresponding mass-weighted deposition velocity. Here we assume that the measured concentrations are representative for the concentration over the whole North Sea. Wet deposition calculations were carried out using the scavenging factors as reported by GESAMP [2], an annual rainfall above the North Sea of 677 mm y^{-1} and the average elemental atmospheric concentrations of the whole data set. The average heavy metal airborne concentrations, wet and dry fluxes together with corresponding standard deviations, are given in

Table 2.12.1. All reported values are weighted for different wind-frequency directions, observed above the North Sea. Inferring the removal of aerosol from the differences in airborne concentration above two research vessels positioned 200 km downwind from each other (Langrange type experiment) yielded similar values.

Table 2.12.1: Average airborne concentrations, dry, wet and total atmospheric fluxes, with associated standard deviations determined from various sampling campaigns above the North Sea in a period from 1991–1994. All values are weighted for the wind-frequency directions. The North Sea area is assumed to be 5.3×10^6 km^2.

Element	Concentration (ng m^{-3})	Dry flux (kg km^{-2} y^{-1})	Wet flux (kg km^{-2} y^{-1})	Total (ton y^{-1})
Cr	4.4 ± 0.5	1.10 ± 0.52	1.33 ± 0.44	1,270
Mn	7.8 ± 0.5	0.89 ± 0.53	1.73 ± 1.21	1,390
Fe	152 ± 2	22 ± 8	89 ± 52	59,300
Ni	3.7 ± 0.3	0.33 ± 0.17	0.88 ± 0.33	650
Cu	3.1 ± 0.2	0.42 ± 0.23	0.89 ± 0.28	690
Zn	17.8 ± 15.4	1.47 ± 0.43	5.03 ± 1.93	3,480
Pb	11.2 ± 1.1	0.49 ± 0.20	3.20 ± 1.47	1,970

It was observed that the deposition velocities varied much from day to day and that they were particularly high under high wind speeds. The atmospheric fluxes were evaluated relative to the other input sources: those borne by rivers, dumping and incineration. It is clearly recognised that substantial inputs of metals are introduced into the North Sea through rivers and the atmosphere. The atmospheric deposition balanced 34 % (Ni), 18 % (Cu), 17 % (Zn), 32 % (Pb) among all input sources. The determined total deposition fluxes point out the importance of the atmosphere as a passage by which the pollutants are reaching the North Sea environment. It was found that the "giant" particles are by far the most important in this flux.

Detailed elemental compositions of the individual aerosol particles responsible for determined fluxes, down to absolute masses of 50 fg, were determined by micro-PIXE [3]. Such single particle analysis combined with multivariate numerical analysis reveals three major particle types:

(1) particles dominated by (Na)Cl, S, K, and Ca, apparently sea salt, aged sea salt and gypsum,

(2) particles with relatively high contents of Cr, Fe, Ni, Zn, V and Ti, probably fly ash and

(3) mixed or aggregated marine/continental particles, which are internally heterogeneous and contain *e.g.* Cr and sea salt elements in spatially separated parts of the same giant particle (Table 2.12.2).

Collision of fly ash particles with abundant sea spray might partially explain the increased deposition velocities during stormy conditions. However, more detailed classification was obtained by results based on EPXMA [4].

Table 2.12.2: The range of elemental concentrations (pg) for the three representative particle groups found in the North Sea troposphere - based on micro-PIXE analysis.

Particle group	Elemental amounts (pg)	
Sea salt	(Na)Cl 20–400	
Sea salt +gypsum	S 10–33	K 5–17
	Cl 20–390	Ca 1–4
Agglomerated aerosols	S 2–20	K 1–16
	Cl 8–100	Ca 1–4
	Ti 0.1–3.8	Fe 0.2–1.1
	V 0.1–0.3	Ni 0.01–0.08
	Cr 0.2–2.6	Zn 0.01–0.52

The single particle aerosol data set from EPXMA measurements contained the results of 22,500 analyses of individual particles. In a such huge data set, the interrelation of 13 variables (Na, Mg, Al, Si, P, S, Cl, K, Ca, Ti, Fe, Ni and Zn) was investigated by principal factor analyses (PFA) with orthogonal varimax rotation. This multivariate technique revealed four major different sources of giant aerosols: aluminosilicate dispersion, including both soil dust and fly ash particles, and a marine source producing fresh sea spray particles and sea water crystallisation products as a marine source and two industrial sources, characterised by particles rich in Ni and Zn and in Ni and P. Elements like Cu, Pb, Mn, Cr, Ba, Pt and V were occasionally detected in some particles. which contributed less than 0.5 % of the total aerosol abundance and therefore they were further excluded from the data matrix. Particle classification into different groups based on their chemical composition was achieved by hierarchical cluster analysis. Correct interpretation of the cluster analysis results was possible by taking into account the exact ship positions and meteorological data like wind speed, wind direction and relative humidity during sampling periods. Depending on the aerosol sample, three to eight different aerosol types were distinguished. The abundance of sea salt particles was very high in all the impactor samples, which were generally collected under stormy conditions, ranging from 10–90 % of the total aerosol abundance. High abundances of aged sea salt where found in some of the samples but their average abundance was low. All samples were also characterised by significant amounts (3–33 %) of Ca and S-rich particles; several hypotheses have been put forward to explain their origin. Sea water crystallisation products such as KCl, $MgCl_2$, $MgSO_4$, K_2SO_4 are found in 43 % of the samples and their abundances are below 19 %. Aluminosilicates as fly ash and soil dust account sometimes for up to 30 % of the total aerosol abundance. A few times Si-rich and Fe-rich particles were observed and organic particles were present in often significant concentrations. Especially for the largest particle fraction, the particle diameters were observed to be below the cut-off aerodynamic diameters of the May impactor. In a marine environment, this can be attributed to four different possible effects: a particle density above unity; splintering of large particles into pieces upon impaction; collection of wet particles which are reduced in volume after drying; or fractional crystallisation occurring after evaporation of the water present, leading to many small particles. The appearance of NaCl in low concentrations in all the hierarchical clustering tables can be explained as a NaCl coating.

By investigating possible changes between the samples collected during six corresponding sampling periods on board the F.S. Alkor and the R.V. Belgica (positioned at 200 km downwind from each other), the following conclusions were drawn:

(1) during the air mass trajectory, no significant decrease in aerosol particle diameters is observed; the average diameter of particle "types" for each stage remains more or less constant;

(2) the abundance of sea water crystallisation products increases during travelling across the sea for the first two stages (particles with diameters above 8 mm), with a factor of *ca.* 2.3. The discrimination between different element combinations seems to be more complicated for the downwind R.V. Belgica samples;

(3) in stages with particle diameters above 4 μm, a decrease in aluminosilicate abundance is noticed;

(4) 48 % of the analysed samples does not contain organic particles. No significant changes in their abundance can be discovered between samples from the two research vessels;

(5) during the air trajectory, an increase in freshly produced sea salt particles is observed at the first three stages; aged sea salt almost disappears at large particle stages and it is replaced by pure NaCl.

The results of hierarchical cluster analysis of the rain water suspension samples from both vessels also showed some differences. Firstly a significant difference in composition was seen. Samples collected on board the upwind vessel F.S. Alkor contained, besides aluminosilicates and Si-rich particles, also high abundances of particles rich in S, P, Fe and Cr, which were not seen on the R.V. Belgica. In addition, higher abundances of organic material were detected in the Alkor samples. Secondly, some differences in particle diameters could be noticed.

More than 200 particles were subjected to manual EPXMA followed by ZAF corrections. This type of analysis disclosed seven major particle groups: NaCl, aluminosilicates, Si-rich, S- and K-rich, Fe-rich, Ca- and S-rich particles and particles with a high contribution of heavy metals. The shapes were compared with those described in literature. Irregular shapes were found for Si-rich, S- and K-rich, Ca- and S-rich, Fe-rich particles and also for particles with increased heavy metal concentration. Aluminosilicates could be found both in irregularly and spherical shapes, respectively soil dust and fly ash. Besides the cubic form described in the literature for halite, also elongated NaCl particles were found. Spherically shaped Fe particles were seldom found. $CaSO_4$ particles were detected in an irregular shape but also in a rhombic and hexagonal shape (gypsum).

Conclusions

The biogeochemical cycle of particulate matter in the North Sea environment was studied by different trace and micro-analysis techniques in a context that have not been systematically applied before. The characterisation of the composition, origin and size of the particulate matter in the North Sea environment will certainly improve our understanding of the chemical processes that are occurring due to the air/sea interface.

The participation in such a multi-purpose experiments was a valuable experience for developing new techniques for aerosol sampling, for the optimisation of sample loading times and treatment, and for aerosol analysis methodology.

Acknowledgements

This work was partially supported by the Belgian Prime Minister's Service - Science Policy Office, under contract EU7/08, until August 31st, 1993.

References

1. S.A. Slinn, W.G.N. Slinn, Predictions for particle deposition on natural waters. *Atmos. Environ.* **14** (1980) 1013.
2. GESAMP-IMPO/FAO/UNESCO/WHO/IAEA/UN/UNEP/ Joint Group of Experts on the Scientific aspects of Marine Pollution. *The Atmospheric Input of Trace Species to the World Ocean*, Rep. Stud. GESAMP (38) 1989.
3. J. Injuk, L. Breitenbach, R. Van Grieken, U. Watjen, Performance of a nuclear microprobe to study giant marine aerosol particles, *Mikrochim. Acta.* **114/115** (1994) 345.
4. L. A. De Bock, H. Van Malderen and R. Van Grieken, Individual Aerosol Particle Composition Variations in Air Masses Crossing the North Sea, *Environ. Sci. Technol.* **28** (1994) 1513.

2.13 Air-Sea Exchange Fluxes of Mercury and Atmospheric Deposition to the North Sea

W. Baeyens, M. Leermakers, C. Meuleman and Q. Xianren

Department of Analytical Chemistry, Free University of Brussels (VUB), Pleinllan 2, 1050 Brussels, Belgium

Our research in the EUROTRAC ASE framework was focused on the distribution and behaviour of atmospheric mercury above and around the North Sea as well as studying parameters influencing mercury speciation in water and air-sea exchange of mercury.

Atmospheric mercury was analysed during the North Sea Experiment (September 1991), the North Sea Platform Experiment (September 1992) and on the North Sea and English Channel (November 1993). A preliminary experiment at Mace Head, Ireland, was carried out in April 1991 in order to evaluate sampling and analytical equipment. Estimations of deposition fluxes were made.

From 1992 till 1994 surface waters of the Scheldt estuary were sampled the year round, in order to assess the behaviour of the various Hg species. On the basis of these data annual and seasonal fluxes (evasion fluxes, formation rates of Hg^0 and MeHg, *etc.*) were calculated.

Aims of the research

The general aim of the research is to understand the processes involved in the transformation of mercury species in atmosphere and surface waters and to model evasion processes as well as deposition processes. Questions which remain to be answered are:

* What is the deposition flux of mercury species to the North Sea?
* Is the atmosphere an important source of MeHg to the North Sea compared to riverine fluxes?
* What is the seasonal variation in the concentration of mercury species (dissolved inorganic mercury, dissolved methylmercury, dissolved gaseous mercury, particulate inorganic and methylmercury) in surface waters?
* Which abiotic and biotic factors influence the production of methylmercury and dissolved gaseous mercury in surface waters? What are the evasion fluxes of dissolved gaseous mercury in supersaturated waters and how does this flux vary seasonally? How important is this evasion flux in relation to the deposition flux?

Principal scientific results

The North Sea Experiment (NOSE 1991)

The North Sea Experiment (NOSE 1991) was carried out from September 15th to 27th, 1991. In order to measure changes in concentrations in air and precipitation in a moving air mass (Lagrangian experiment), samples were collected on board of two research vessels arranged to lie 200 km apart, aligned in wind and along air-parcel forward trajectories. In a Lagrangian experiment variations in concentrations of a given component during transport are either due to internal processes (such as oxidation or adsorption on particles) or external processes such as sources and sinks. Whereas over land anthropogenic activities represent a variety of atmospheric emission sources, over sea only natural emission sources, which are much more homogeneous than those over land, have to be considered. Fluxes can be derived using a mass balance box-model approach. Recent atmospheric mercury measurements above the North Sea are scarce and limited to coastal stations and near shore measurements [1, 2].The aim of this experiment was to obtain more information about: 1) the atmospheric mercury concentrations above the North Sea, and 2) the processes that may affect these concentrations. The results of this experiment have been published in *Water, Air and Soil Pollution* [3].

TGM varied from 0.7 to 1.9 ng m^{-3} on the R.V. Belgica with an average of 1.2 ng m^{-3} while on the F.S. Alkor TGM varied from 0.7 to 2.6 ng m^{-3} with an average concentration of 1.5 ng m^{-3}. These concentrations are lower than previously reported data for the northern hemisphere [4–6] and correspond to background concentration over sea for winds deriving from relatively unpolluted wind sectors (westerly winds). An average 20 % decrease in TGM from the up-wind to the down-wind ship was found. A substantial decrease in almost all trace metal concentrations on size segregated aerosols from the luff to lee ship was found [7–9]. An average decrease of 30–40 % was found for anthropogenic elements, which was much larger than could be expected for dry deposition. Important factors which may effect removal are entrainment with the free troposphere by vertical dilution, wet deposition and air-sea exchange. An overall removal velocity can be calculated,

$$v_r = -\ln (C_{lee}/C_{luff})\, z_i/D_t$$

where C_{lee} and C_{luff} are the Hg concentrations on down-wind and upwind ships, z_i is the mixing layer height and D_t is the travel time of the air mass. Calculated removal velocities varied between 0 and 5.5 cm s^{-1} with an average of 0.95 cm s^{-1} for the whole experiment. The average removal velocity for dry periods is 0.5 cm s^{-1}, while during rain events removal velocities ranged from 1 to 5.5 cm s^{-1}. Entrainment with the free troposphere is probably the dominating reason for the depletion in the boundary layer. For nitrogen species entrainment velocities were estimated to be typically 1 to 2 cm s^{-1} and up to 5 cm s^{-1} or more during frontal passages [10].

Mercury concentrations in rain water varied from 5 to 25 ng L^{-1} and are comparable on both ships. They are in good agreement with data found in coastal and marine environments [11–13]. The calculated wash-out ratios, defined as the ratio of Hg concentrations in rain water to those in air, vary from 1×10^4 to 5×10^4.

The North Sea Platform Experiment (NSP 1992)

The aim of this experiment was to obtain information of mercury speciation (besides gaseous mercury particulate mercury and methylmercury were sampled) in the atmosphere as well as in rain and to investigate the cycling of dissolved gaseous mercury (Hg^0 and DMeHg) at the air-sea interface.

Total gaseous mercury concentrations ranged from 0.7 to 3.25 ng m^{-3}. No day to night variations were observed. The highest concentrations were found with easterly and south-easterly winds. Particulate concentrations ranged from 9 to 310 pg m^{-3}. A good correlation was found between gaseous mercury and particulate mercury concentrations. Organomercury species were all below the detection limit of 5 pg m^{-3}. Rain water concentrations ranged from 19.7 to 170 ng L^{-1}. These concentrations are relatively high for marine rain. Methylmercury concentrations ranged from 0.1 to 0.2 ng L^{-1}. These values are comparable to continental unpolluted regions but significantly higher than found above the Atlantic ocean revealing the importance of wet deposition of methylmercury to coastal regions.

Speciation of mercury in the Scheldt estuary

Surface waters of the Scheldt Estuary were sampled on various occasions between 1991 and 1994. Total Hg (TM) concentrations ranged from 20 to 200 ng L^{-1}, of which over 90 % is bound to suspended matter. Total dissolved Hg (TDM) concentrations ranged from 0.5 to 5.2 ng L^{-1} and are strongly influenced by removal, mobilisation and mixing processes in the estuary. Total dissolved mercury is predominantly bound to strong complexing ligands (organic substances) in the upper estuary, but this fraction generally decreases with increasing salinity. Significant seasonal variations were observed in dissolved MeHg (DMeHg) concentrations with concentrations ranging from 11 to 120 pg L^{-1} in the winter and 80 to 600 pg L^{-1} in summer and autumn. 20 to 40 % of the total MeHg (TMeHg) was found in the dissolved phase during a winter cruise. Supersaturation of Hg^0 is observed throughout the whole estuary with concentrations ranging from 10 to 130 pg L^{-1}. An overview of the concentrations of the mercury species is shown in Table 2.13.1. A detailed description of the speciation results can be found in Leermakers *et al.* [3] and the budget and flux calculations are described in Baeyens and Leermakers [14].

Table 2.13.1: Mercury speciation in the Scheldt estuary

species	zone 1 0–2 ‰	zone 2 2–10 ‰	zone 3 10–20 ‰	zone 4 20–27.5 ‰
Total dissolved Hg (ng L^{-1})				
summer	1.05	1.89	1.16	1.16
winter	2.65	1.72	1.57	1.45
Reactive Hg (ng L^{-1})				
summer	0.45	0.78	0.55	0.55
winter	0.55	0.49	1.06	1.33
Dissolved gaseous Hg (pg L^{-1})				
summer	60	60	60	80
winter	50	60	50	45
Dissolved MeHg (pg L^{-1})				
summer	183	335	181	382
winter	105	83	23	39

The behaviour of TDM in the Scheldt estuary is very sensitive to salinity and redox conditions, In summer, a large part of the upstream area becomes anoxic resulting in the precipitation of mercury sulfides, while in winter the undersaturation of oxygen does not lead to the presence of sulfides in the water column. In the downstream area, reoxydation of these sulfides results in a temporarily increase of dissolved mercury.

In winter DMeHg concentrations were much lower than in summer and autumn (0.011–0.127 ng L^{-1}). Maximum concentrations are found in the upper estuary (salinities 1–5 ‰) and remain relatively constant in the lower estuary. The percentage of DMeHg ranged from 0.7 to 8 % of the TDM concentrations. In summer DMeHg concentrations ranged from 0.1 to 0.6 ng/L accounting for 7 to 60 % of the TDM. Comparable to the variation in TDM, DMeHg concentrations are low at the lowest salinities and increase at salinities ranging from 8 to 12 ‰, followed by a decrease in the mid estuary and a second peak in the lower estuary at 25 to 30 ‰ . At the same time high plankton population densities prevailed in the lower estuary. PMeHg concentrations range from 1 to 8 ng g^{-1} in the winter cruise. These concentrations are comparable to the concentrations found in the surface sediments of the intertidal flats of the Scheldt estuary [15]. In contrast to TDM which only accounts for a few % of TM, DMeHg accounts for up to 20 to 40 % of the TMeHg.

Besides the formation of MeHg, the physicochemical conditions prevailing in the estuary do not only promote methylation, but also the competitive demethylation and volatilisation reactions. The formation of Hg^0 can be the result of either a) bacteriological reduction of inorganic Hg and MeHg [16], b) reduction of dissolved Hg by phytoplankton [17] or c) an electron transfer reaction in Hg humic complexes; a mechanism which is also enhanced by solar radiation [18]. Higher concentrations are found in summer than in winter. Hg^0 concentrations range from 20 to 130 pg L^{-1}. The latter high value was observed at the mouth of the estuary, in the summer, when high plankton concentrations were found at the same place. Hg^0 is supersaturated throughout the whole estuary, in summer as well as in winter. The behaviour and magnitude of the dissolved gaseous fraction is of a major importance, since it is directly related to the loss of Hg in the estuary.

Modelling the evasion flux of mercury, which takes into account wind speed and wave spectra was proposed by Baeyens *et al.* [12]. A model appropriate to gas transfer at a sharp interface contains viscous and diffusive sublayers on both sides of the interface as well as outer boundary layers in which turbulence reduces the resistance dramatically This causes the diffusive layers to generally form the largest resistance for gas exchange. Because the molecular diffusion of weakly soluble gases such as Hg^0 is much lower in liquids than in the gas phase, the gas fluxes are normally under liquid phase control. Gas phase control only occurs when: 1) Hg^0 is rapidly scavenged by particles in highly turbid environments [19], and 2) Hg^0 is produced in the surface microlayer [20]. The transfer of Hg^0 across the air-sea interface can in a general way be described by Fick's first law: $F = -r\ V\ (C_z - C_s)$ where F is the Hg flux; C_z the Hg concentration in the fluid with the largest resistance to the flux (dissolved gaseous mercury), and C_s the Hg^0 concentration at the interface between both fluids (surface microlayer); r is the density, and V is the transfer velocity. The transfer velocity under liquid phase control can be described by a shear turbulence model [21]. Breaking waves can, however, dramatically change the character of the turbulence just below the surface or in the viscous sub-layer. Calculations of the transfer velocity under wave breaking conditions corresponds to the shear induced turbulence under extreme storm conditions [12].

The evasion fluxes were calculated for four different salinity zones, in summer and in winter. An average wind speed of 29 km h^{-1} was assumed resulting in an evasion velocity of 1.3 m day^{-1}. An average gaseous Hg^0 concentration of 3 ng m^{-3} was used; this corresponds to an equilibrium concentration of 10 pg L^{-1} at the water surface.

We also estimated the Hg^0 formation rates. Based on the dissolved Hg burden contained in the depth integrated water column (between 15 and 20 µg m^{-3}), conversion rates of 0.25 to 0.42 % day^{-1} in winter (Table 2.13.2) are required to balance the estimated evasion fluxes of 17 to 24 µg m^{-2} yr^{-1}. In summer the integrated Hg burden in the water column varies from 8 to 17 µg m^{-3}, hence conversion rates of 0.40 to 0.83 % day^{-1} are required to balance the evasion fluxes of 24 to 33 µg m^{-2} yr^{-1}. These Hg^0 formation rates are of the same order as those estimated by Mason *et al.* [22, 23] for open ocean waters (0.2 to 1 % day^{-1})

but much lower than the values they found (2 to 4 % day^{-1}) in the Upper Mystic Lake [24].

Table 2.13.2: Hg exchange fluxes at the atmosphere-water interface and Hg^0 formation rates.

box	supersat- uration	transfer velocity	flux	surface	total flux	formation rate
‰	ng m^{-3}	m day^{-1}	μgHg $m^{-2}yr^{-1}$	km^2	g Hg yr^{-1}	% yr^{-1}
			Fluxes in summer			
0-2	50	1.3	24	4.8	110	305
2-10	50	1.3	24	15	360	141
10-20	50	1.3	24	66	1500	188
20-27.5	70	1.3	33	88	2900	220
			Fluxes in winter			
0-2	40	1.3	19	4.8	90	95
2-10	50	1.3	24	15	360	155
10-20	40	1.3	19	66	1200	11
20-27.5	35	1.3	17	88	1500	90

Estimations of deposition fluxes to the North Sea

Based on compiled data of the mentioned experiments wet and dry deposition fluxes to the North Sea were calculated. Dry deposition of particulate mercury was estimated to be 1.11 ng Hg cm^{-2} yr^{-1}. Wet deposition accounts for a flux of 0.46 ng Hg cm^{-2} yr^{-1} of total mercury and a flux of MeHg of 0.0057 ng cm^{-2} yr^{-1}.

NATO ARW

From June 10th–14th, 1995, A NATO advanced research workshop on "Global and regional mercury cycles; sources, fluxes and mass balances" was held in Novosibirsk, Russia. The main aim of the workshop was to evaluate available data and identify gaps and uncertainties in our knowledge on of the sources, fluxes, and mass balances of Hg on regional and global scales by bringing together specialists on this matter from NATO and the Cooperation Partner Countries. The themes that were focused on were analytical methods, atmospheric mercury, oceanic mercury and Hg in terestrial systems and inland waters.

Conclusions

Wet and dry deposition fluxes of mercury above the North Sea were calculated. The fluxes for total Hg are comparable to previous estimations [25], whereas for MeHg no previous data are available. The relative importance of the atmospheric deposition flux of MeHg compared to riverine inputs is difficult to evaluate given the limited amount of atmospheric data and the uncertainties on the behaviour of MeHg during estuarine mixing. Biological activity (plankton activity in the lower estuary and bacterial activity in the upper estuary) plays a dominant in the formation of MeHg as well as Hg^0 in estuarine waters. Supersaturation of Hg^0 is found throughout the whole estuary, resulting in an important evasion flux, which is dependant on the wind speed. Speciation data in North Sea waters are currently under investigation.

Acknowledgements

This work was sponsored by the Ministry of Scientific Policy (FDWTC). C. Meuleman obtained a grant from the Institute fro Research and Industry and Agriculture (IWONL) and Q. Xianren from the Brussels Agglomeration.

References

1. Ebinghaus R., Hintelmann H., Wilken R.D.: *Frez. J. of Anal. Chem.* **350** (1994) 21–29
2. Lindqvist O.: *Water, Air and Soil Poll.* **55** (1991).
3. Leermakers M., Meuleman C., Baeyens W.: *Water, Air and Soil Poll.* **80** (1995) 641–652.
4. Slemr F., Langer E.: *Nature* **355** (1992) 434–437.
5. Fitzgerald W.F., Gill G.A., Hewitt A.D.: in: Wong (ed), *Trace Metals in Seawater*, Plenum Press, New York 1983, pp. 297–315.
6. Schroeder W. H.: in: C. Watras, J. Huckabee (eds), *Mercury Pollution, Integration and Synthesis*, Lewis Publishers, Boca Raton 1994, Ch. II. 10, pp. 281–290.
7. Schulz M., Stahlschmidt T., Francois F., Maenhaut W., Larsen S.: in: P.M. Borrell, P. Borrell, T. Cvitaš, W. Seiler (eds), *Proc. EUROTRAC Symp. '94*, SPB Academic Publishing bv, The Hague 1994, pp. 702–706.
8. Francois F., Cafmeyer J., Gilot C., Maenhaut W.: in: P.M. Borrell, P. Borrell, T. Cvitaš, W. Seiler (eds), *Proc. EUROTRAC Symp. '92*, SPB Academic Publishing bv, The Hague 1993, pp. 788–791.
9. Injuk J., Van Malderen H., Van Grieken R.: *X-Ray Spectrometry* **22** (1993) 220–228.
10. Larsen S. E.: *EUROTRAC Annual Report 1993, part 3, ASE*, EUROTRAC ISS, Garmisch-Partenkirchen 1994, pp. 86–93.
11. Gill G.A., Fitzgerald, W.F.: *Global Biogeochmical Cycles*, **1(3)** (1987) 199–212.
12. Baeyens W., Leermakers M., Dedeurwaerder H., Lansens P.: *Water, Air and Soil Poll.* **56** (1991) 731–744.
13. Mason, R.P., Fitzgerald W. F., Vandal G: *J. Atmos. Chem.* **14** (1992) 489–500.

14. Baeyens W.. M. Leermakers, Particulate, dissolved and methylmercury budgets for the Scheldt estuary (Belgium and the Netherlands), in: W. Baeyens, R. Ebinghaus, O. Vasiliev (eds), *Global and regional mercury cycles: sources, fluxes and mass balances*, NATO ARW Series, Kluwer Publishers 1996, pp. 285–381.
15. Muhaya B, M. Leermakers, W. Baeyens; *Water, Air and Soil Pollution* **94** (1996) 109–123.
16. Nakamura K., M. Sakamoto, H. Uchiyama,O. Yagi; *Applied Environmental Microbiology* **1** (1990) 304–305.
17. Mason R. P., W. F. Fitzgerald, J. Hurley, A. K. Hanson Jr., P. L. Donaghay, J. M. Sieburth; *Limnolgy and Oceanography* **38** (1993) 1227–1241.
18. Iverfeldt A.; PhD thesis. University of Gothenburg, Sweden 1984.
19. Baeyens, W., G. Decadt, F. Dehairs, L. Goeyens; *Oceanologica Acta* **5** (1982) 261–266.
20. Iverfeldt, A., O. Lindqvist; in: W. Brusaert, G. Jirka (eds), *Gas transfer at water surfaces*, D. Reidel Publishing Company 1984, pp. 533–538.
21. Kitagorodskii S.A., Donelan M.A., *John Hopkins University Int. Report* 1984, p. 27.
22. Mason R.P., O'Donnell J., Fitzgerald W.F.: in: (1994), In: C. Watras, J. Huckabee (eds.) *Mercury as a global pollutant: Towards Integration and synthesis* , Lewis Publishers, 83-98.
23. Mason R.P., Rolfhus K.R., Fitzgerald W.F.: *Water, Air and Soil Poll.* **80** (1995) 665–677.
24. Mason R.P., Morel F.M.M., Egmond H.F.: *Water, Air and Soil Poll.* **80** (1995) 775–787.
25. Decadt G., *PhD thesis*, Free University of Brussels, 1986.

2.14 Deposition of Nitrogen Compounds to the Danish Coastal Waters

Willem A.H. Asman[1], Ruwim Berkowicz[1], Ole Hertel[1], Erik H. Runge[1], Lise-Lotte Sørensen[1], Kit Granby[1], Peter Klint Jensen[1], Jesper Christensen[1], Hans Nielsen[1], Bjarne Jensen[1], Sven-Erik Gryning[2], Anna Maria Sempreviva[2], Niels A. Kilde[2], Henning Madsen[3], Peter Allerup[3], Søren Overgaard[3], Jess Jørgensen[3], Flemming Vejen[3] and Klaus Hedegaard[3]

[1]National Environmental Research Institute (NERI), Frederiksborgvej 399, 4000 Roskilde, Denmark
[2]Risø National Laboratory (RISØ), Risø, 4000 Roskilde, Denmark
[3]Danish Meteorological Institute (DMI), Lyngbyvej 100, 2100 Copenhagen, Denmark

Methods, parameterisations and models have been developed, which give insight in processes that are important for deposition of nitrogen compounds to the sea. This knowledge has been applied to the Kattegat Strait for which a nitrogen deposition of 960 kg N km^{-2} yr^{-1} was calculated with an atmospheric transport model.

Aims of the research

The aim of this project is to determine the deposition of nitrogen compounds to the Kattegat Strait between Denmark and Sweden. The reason for this is that nitrogen is the limiting factor for algae growth in this area, where regularly very low oxygen concentrations are measured, caused by decomposition of algae.

Principal scientific results

It is difficult to measure dry deposition continuously. Moreover, there are only a few islands in this area, which makes it not so easy to get representative measurements of the wet deposition. It was therefore decided to develop an atmospheric transport model to calculate the deposition and to investigate some of the processes that were not very well known experimentally. A large part of this research is reported in Danish in 6 reports of the Danish Environmental Protection Agency. In this contribution reference is only made to reports and papers in English.

Emission inventories and preliminary calculations

As no emissions on a fine scale were available for Denmark, a land use map was made [1], that was used as a basis for emission inventories on 1×1 km^2 for SO_2, NO_x, NH_3 and VOCs [2–4]. Moreover, an emission inventory for NH_3 was made for Europe, as such an inventory was not available [5]. Preliminary calculations of the nitrogen deposition to the Kattegat Strait were made with the TREND model, developed by the National Institute of Public Health and Environmental Protection (RIVM), the Netherlands [6, 7]

Development of a NH_3 *monitor*

One of the subprojects within the project was the determination of the dry deposition velocities of nitrogen compounds by the gradient method. To be able to measure low concentrations of NH_3, a diffusion scrubber method was developed [8, 9]. The diffusion scrubber consists of an outer tube with a water saturated microporous membrane tube centred along the axis of the tube. The outer tube is made of PFA (perfluoralkoxy). The air is sucked through this tube with a laminar flow. During the passage through the tube part of the NH_3 will diffuse through the membrane and will be collected in the water in the membrane tube. The water is pumped to a fluorescence system where NH_3 is determined by reaction with OPA (*o*-phthaldehyde) and sulfite. As only part of the NH_3 diffuses through the membrane it is necessary to calibrate the system with NH_3 in known concentrations. The sampling resolution time of this method is 10 min. and the detection limit is estimated to be about 0.01 ppb, depending on the background present in the system. The results of the method compared reasonably well with the results of the AMANDA wet annular denuder system and a filterpack [8]. The method has the advantage that it can be used on board ships.

Dry deposition velocities at sea

It was tried to measure the dry deposition velocity of NH_3, HNO_3, SO_2 and NH_4^+, NO_3^- in aerosol at the beach on the island of Anholt in the Kattegat by the gradient method. Large gradients were found for HNO_3 and SO_2, higher than can be accounted for by turbulent transport alone. It is supposed that this could have something to do with an uptake by sea spray. Moreover, the NH_4^+ concentration in sea water was measured. It appeared that the NH_4^+ concentration in sea water near the beach was a factor 15 higher near the beach than at 100 m from the beach. The results of these experiments could not be used to evaluate model parameterisations of the dry deposition velocity at sea. It was therefore decided to use a theoretical approach that takes into account reactions in sea water [10, 11].

Some related work based upon UK measurements in the North Sea lead to the conclusion that the dry deposition of NH_3 to the southern bight of the North Sea was only half the expected value, because NH_3 was present in the sea water. During some periods even emission from the sea was observed [12]. A method was developed to calculate the NH_3 flux at sea from measurements [12].

Precipitation at sea

The amount of precipitation to the Kattegat Strait was measured by a network of automatic rainfall recorders and precipitation gauges at coastal stations and on islands. The amount of precipitation to this area appeared to be 84% of the amount found for the Jutland peninsula. During modelling it became evident that the amount of precipitation at the Kattegat Strait and its seasonal variation could be very different from the amount observed on land. For this reason it is absolutely necessary to know the amount of precipitation at sea. The precipitation fields used in transport models, *e.g.* the EMEP model, are often based on land-based precipitation stations and may not be valid for sea areas.

Mixing height

Transport calculations are sensitive to changes in the mixing height. Variations in the mixing height over the Kattegat are in general not routinely known because most radio soundings are made from land-based stations. It was therefore decided to measure the height of the lowest mixed layer occurring in the atmosphere over the Kattegat within the framework of this project. This was done by launching 400 radio soundings from the island of Anholt (in the middle of the Kattegat). The mixing height at Anholt did not show the diurnal variation, which is characteristic of mixed layers over land. It appeared to be climatologically constant over a 24-hour period with an average value of 800 m. This constancy is caused by dependence of the variation in mixing height on the temperature difference between the air and the surface. The sea water temperature is constant and does not show a diurnal variation as does the soil temperature on land; this leads to a nearly constant mixing height over sea. Additional information on this climatology and modelling of the mixing height can be found in [13, 14].

Below-cloud scavenging

During below-cloud scavenging many reactions and transfer processes play a role. Every time a reaction is added, many of the equations in a below-cloud scavenging model have to be revised. This is rather time-consuming. For that reason a below-cloud scavenging model using general reaction simulation was developed, that only needs small changes to include more reactions. This is, however, on the expense of CPU time. It was shown that the NH_4^+ concentration in precipitation is only to a minor degree influenced by local NH_3 sources [15].

In a related project a parameterisation of the below-cloud scavenging coefficient for highly soluble gases was developed, taking into account evaporation of the drops. This parameterisation needs only relative humidity and temperature at ground level as an input, as well as the diffusivity of the gas at 25 °C [16]. This information is normally available.

Numerical method for modelling chemistry

A numerical method was developed to solve the stiff systems of ordinary differential equations (ODEs) that are found in atmospheric transport models. The method, Euler backward iterative method was in general more efficient with a constant number of iterations than the QSSA method that is often used [17]. This method was used in the ACDEP model developed within this project.

Transport model

An Langrangian photochemical atmospheric transport model was developed (ACDEP) that consists of 10 vertical layers and that uses trajectories which go 4 days back in time. CBM-IV chemistry was used and in- and below-cloud scavenging were parameterised with scavenging coefficients. The model includes a special parameterisation of the dry deposition to the sea. The model results compare mostly within 30 % with measurements [18].

Conclusions

The model results show that the total nitrogen deposition by NH_x and NO_y is 960 kg N km^{-2} yr^{-1}, of which wet deposition of NH_x contributes 42 %, wet deposition of NO_y 30 %, dry deposition of NH_x 17% and dry deposition of NO_y 11 % [19]. Model results show that Denmark contributes 35 % of the total NH_x deposition to the Kattegat Strait, but only 7 % of the NO_y deposition. In fact, Germany is the largest contributor to the NO_y deposition with about 35 %.

The project has not only led to an estimate of the nitrogen deposition to the Kattegat Strait, but also to methods, parameterisations and models that can be used for other areas or for other types of atmospheric chemistry problems.

Acknowledgements

The Danish Environmental Protection Agency and the Netherlands Foundation for Atmospheric Chemistry supported this research partially. This research was part of the Danish Marine Research Programme 90. One of the authors (A.M.S.) acknowledges funding from the EEC project Human Capital and Mobility for her stay in Denmark. Another author (O.H.) acknowledges the Danish Research Academy for their support of his Ph. D. study. The trajectories and precipitation data for Europe were made available to this project by The Meteorological Synthesising Centre - West, European Monitoring and Evaluation Programme (EMEP-MSC/W) at the Norwegian Meteorological Institute, Oslo, Norway.

References

1. E.H. Runge, W.A.H. Asman, Land use in Denmark. *Report DMU LUFT*-B135, National Environmental Research Institute, Roskilde, Denmark 1989, 23 pp.
2. E.H. Runge, W.A.H. Asman, N.A. Kilde, A detailed emission inventory of sulfur dioxide for Denmark, *Report Risø* M-2937, Risø National Laboratory, Roskilde, Denmark 1992, 33 pp.
3. E.H. Runge, W.A.H. Asman, N.A. Kilde, A detailed emission inventory of nitrogen oxides for Denmark. *Report Risø* M-2929, Risø National Laboratory, Roskilde, Denmark 1991 68 pp.
4. W.A.H. Asman, A detailed ammonia emission inventory for Denmark. *Report DMU LUFT A132*, National Environmental Research Institute (NERI), Roskilde, Denmark 1990.

5. W.A.H. Asman, Ammonia emission in Europe: updated emission and emission variations. *Report* 228471008, National Institute of Public Health and Environmental Protection, Bilthoven, The Netherlands 1992, 88 pp.

6. W.A.H. Asman, E.H. Runge, Atmospheric deposition of nitrogen components to Denmark and surrounding sea areas, a preliminary estimate. *Report* R-91-1, Netherlands Foundation for Atmospheric Chemistry, Utrecht, The Netherlands 1991, 46 pp.

7. W.A.H. Asman, J.A. van Jaarsveld, A variable transport model applied for NH_x in Europe, *Atmos. Environ.* **26A** (1992) 445-464.

8. L.L. Sørensen, H. Nielsen, K. Granby and W.A.H. Asman, Diffusion scrubber technique for measuring ammonia, *Technical Report* 99, National Environmental Research Institute, Roskilde, Denmark, 54 pp.

9. L.L. Sørensen, K. Granby, H. Nielsen, W.A.H. Asman, Diffusion scrubber technique used for measurements of atmospheric ammonia, *Atmos. Environ.* **28** (1994) 3619–3628.

10. W.A.H. Asman, R. Berkowicz, Atmospheric nitrogen deposition to the North Sea. *Marine Poll. Bull.* **29** (1994) 426–434.

11. M.A. Sutton, W.A.H. Asman, J.K. Schjørring, Dry deposition of reduced nitrogen, *Tellus* **46B** (1994) 255–273.

12. W.A.H. Asman, R.M. Harrison, C.J. Ottley, Estimation of the net air-sea flux of ammonia over the southern bight of the North Sea. *Atmos. Environ.* **28** (1994) 3647–3654.

13. E. Batchvarova, S.E. Gryning, An applied model for the growth of the daytime mixed layer, *Boundary-Layer Meteorol.* **56** (1991) 261–274.

14. E. Batchvarova, S.E. Gryning, Applied model of the height of the daytime mixed layer including the capping entrainment zone. In: S.E. Gryning and M.M. Millán: Air pollution modeling and its application X, Plenum Press, New York, USA 1994, pp. 253–260.

15. P.K. Jensen, W.A.H. Asman, General chemical reaction simulation applied to below-cloud scavenging. *Atmos. Environ.* **29** (1995) 1619–1625.

16. W.A.H. Asman, Parameterization of below-cloud scavenging of highly soluble gases under convective conditions. *Atmos. Environ.* **29** (1995) 1359–1368.

17. O. Hertel, R. Berkowicz, J. Christensen, Ø. Hov, Test of two numerical schemes for use in atmospheric transport-chemistry models, *Atmos. Environ.* **27A** (1993) 2591–2611.

18. O. Hertel, J. Christensen, E.H. Runge, W.A.H. Asman, R. Berkowicz, M.F. Hovmand, Ø. Hov, Development and testing of a new variable scale air pollution model ACDEP, *Atmos. Environ.* **29** (1995) 1267–1290.

19. W.A.H. Asman, O. Hertel, R. Berkowicz, J. Christensen, E.H. Runge, L.L. Sørensen, K. Granby, H. Nielsen, B. Jensen, S.E. Gryning, A.M. Sempreviva, S. Larsen, P. Hummelshøj, N.O. Jensen, P. Allerup, J. Jørgensen, H. Madsen, S. Overgaard, F. Vejen; Atmospheric nitrogen input to the Kattegat Strait. *Ophelia* **42** (1995) 5–28.

Chapter 3

Biological Processes and Air–Sea Exchange of Biogases

3.1 Oceans as Sources or Sinks of Biologically Produced Gases relevant to Climatic Change

S. Rapsomanikis

Max Planck Institute for Chemistry, Biogeochemistry Dept., P.O.Box 3060, 55020 Mainz, Germany

Gases may be produced in the world's oceans largely by biological and chemical processes in sediments at the sediment water interface, water column, photic zone, surface waters and at the water atmosphere interface. Some of these gases play an important role in the forcing of global climate (*e.g.* CO_2, CH_4, $(CH_3)_2S$, COS, and N_2O) [1]. Their eventual transfer to the atmosphere is driven by supersaturation of ocean waters in comparison with the water-air equilibrium concentration for each individual gas. Conversely, the oceans may act as sinks for some of these gases, under conditions of undersaturation. Tropospheric and stratospheric chemistry are influenced by these gases and the net result is either direct positive radiative forcing (*e.g.* CH_4, CO_2, N_2O; "greenhouse gases"), or indirect negative radiative forcing (*e.g.* $(CH_3)_2S$) [2]. Increases in the concentrations of the former gases result in global warming. The oxidative products of the latter two gases form condensation nuclei which in turn form marine clouds affecting the earth's albedo and hence resulting in a negative radiative forcing (global cooling) [3]. In relation to the ASE EUROTRAC subproject, the fluxes of all the above gases have been studied by different groups and in various contexts. Below, a short state of the art is elaborated, for the role of the oceans as sources or sinks for the above named gases.

Carbon dioxide, CO_2

CO_2 is considered as the main greenhouse gas because of its high atmospheric concentration in comparison with the other greenhouse gases. Its atmospheric concentration has been increasing steadily since the industrial revolution, from values of approximately 280 ppm to present day 353 ppm. The current rate of annual atmospheric concentration increase is 1.8 ppm.

The oceans represent the largest carbon reservoir of our planet and one can conclude that they regulate the CO_2 concentrations in the remote (nonperturbed by human activities) atmosphere. The concentration of CO_2 in sea surface waters, $p(CO_2)$, is strongly influenced by biological activities. Marine biota may act as a biological pump. They transport organic carbon to deeper waters in the form of particles and CO_2 is upwards transported from deeper CO_2 richer waters. It should be noted, however, that this "biological pump" does not help in the cycling of anthropogenically produced CO_2 since marine biota do not respond directly to increased CO_2 concentrations in the top 200–300 meter ocean waters. Climatic changes may affect the utilisation or removal of CO_2 in the oceans because they may alter the nature and population concentrations of marine biota. Estimates of fluxes of CO_2 from and to the ocean waters are based on scarce measurements of $\Delta p(CO_2)$ *i.e.* the difference between atmospheric and ocean surface concentrations. The mass transfer coefficient of CO_2 across the ocean atmosphere interface is also required, to balance the flux equation:

$$F(CO_2) = K_w \cdot \Delta p(CO_2)$$

Alternatively short term variations in the flux of CO_2 may be directly measured by using the eddy correlation micrometeorological technique. In this case the other two terms of the above equation are not useful, or one may also measure $\Delta p(CO_2)$ and evaluate K_w. Then this value of K_w can be related to other prevailing parameters and conditions at the time of the measurements.

Principally, the same equation is used for the calculation of the fluxes of the other gases considered below [6]. These estimates of each variable of the above simplified equation may help modelling the fluxes of CO_2 to and from the oceans by relating other easily and commonly measured parameters (*e.g.* wind speed). Currently using global circulation models, the oceanic uptake of CO_2 is estimated to be 2.0 ± 0.8 Gigatons per year.

Methane, CH_4

Methane is a greenhouse gas with radiative forcing 23 times that of CO_2. It appears that although global land based budgets of methane may be balanced, oceanic sources and sinks are considered of minor importance. Continental shelf waters and upwelling areas may be supersaturated with methane which is mainly produced in anoxic environments as a final step in the anaerobic degradation of organic matter. It may also be produced in the water column by yet unknown biological or chemical processes [4]. A very small number of world-wide measurements (*ca.* 100) exist for the flux of methane across the ocean atmosphere interface and for the upwelling areas of the world's oceans, measurements are indeed scarce [5]. It appears that oceanic emissions may represent only 2 % of the global CH_4 budget but an annual atmospheric increase of ~40 Teragrams of CH_4 needs to be accounted for (Table 3.1.1). More data from the world's ocean are needed in order to account for spatial and temporal variations of CH_4 emissions. Despite the fact that shelf areas and estuaries represent a small part of the world's ocean, they contribute about 75 % to the global oceanic CH_4 emissions. We reassessed these emissions on the basis of our data and a literature compilation, using the concept of regional gas transfer coefficients. Our estimates computed with two different air-sea exchange models lie in the range of 11–18 Tg CH_4/year [7]. These values are not different from the ones in Table 3.1.1 [1] but they narrow down the uncertainty range.

Table 3.1.1: Estimated sources and sinks of CH_4 [1].

	Range of annual change (Teragram CH_4)		
Sources			
Natural wetlands	100	-	200
Rice paddies	25	-	70
Enteric fermentation	65	-	100
Gas drilling, venting, transport	25	-	50
Biomass burning	20	-	80
Termites	10	-	100
Landfills	20	-	70
Coal mining	19	-	50
Oceans	5	-	20
Fresh waters	1	-	25
CH_4 hydrate destabilisation	0	-	100
Sinks			
Removal by soils	15	-	45
Reaction with atmospheric OH	400	-	600
Atmospheric increase	40	-	48

Nitrous oxide, N_2O

Nitrous oxide is also a greenhouse gas with an important contribution from the oceans to its estimated global budget. Its annual atmospheric concentration is currently increasing by 0.2–0.3 %, indicating an imbalance of 30 % in its global budget (Table 3.1.2). If this is due to underestimating emission sources and not due to overestimating the stratospheric sinks of N_2O (reaction with singlet oxygen), then shelf and upwelling areas may be the missing source. Nitrous oxide production in oceanic waters has been associated with oxygen minima and nitrification processes but also with denitrification processes. Our own measurements of N_2O oceanic fluxes indicate that there exist a necessity to improve on the spatial and temporal data pool. The imbalance in the global N_2O

budget remains an unanswered question. However after a compilation of literature data including our own, we conclude that the mean saturations in coastal regions (with the exception of estuaries and upwelling regions) are only slightly higher than in the open ocean. When estuarine and coastal upwelling regions are included, a computation of the global oceanic N_2O flux indicates that a considerable portion (approximately 60 %) of this flux is from coastal regions, mainly due to high emissions from estuaries. We estimate, using two different parametrisations of the air-sea process, an annual global sea-to-air flux of 11–17 Tg N_2O [8]. Our results suggest a serious underestimation of the flux from coastal regions in previous estimates as in Table 3.1.2 [1].

Table 3.1.2: Estimated sources and sink of N_2O [1].

	Range of annual change (Teragram N as N_2O)
Sources	
Oceans	1.4 – 2.6
Soils (tropical forests)	2.2 – 3.7
Soils (temperate forests)	0.7 – 1.5
Combustion	0.1 – 0.3
Biomass burning	0.02 – 0.2
Fertiliser	0.01 – 2.2
Sinks	
Removal by soils	?
Photolysis in the stratosphere	7 – 13
Atmospheric increase	4 – 4.5

Volatile sulfur compounds, VSC

The direct emission of $(CH_3)_2S$ (dimethyl sulfide; DMS) by certain phytoplankton species and its emission via the enzymatic cleavage of its precursor compound dimethyl sulfonium propionate (DMSP), result in dissolved DMS within the photic zone. DMS is ubiquitous and the predominant volatile sulfur compound in marine surface waters. The hypothesis that climate and phytoplankton activity are connected, rests on the fact that cloud condensation nuclei (CCN) in marine remote areas are formed by the atmospheric oxidation products of DMS. The cloud droplet size and hence the cloud albedo depends on the number concentration of CCN. One may argue that DMS has, indirectly, an opposite effect on the climate change than the other greenhouse gases [3].

The DMS fluxes from different biogeographical regions of the oceans to the atmosphere are summarised in Table 3.1.3 [2].

Table 3.1.3: Global oceanic DMS concentrations and annual fluxes [2].

Biogeographic region	Mean concentration (nM)	Total flux (Teramole S)
Oligotrophic	2.4	0.2–0.6
Temperate	2.1	0.1–0.3
Upwelling	4.9	0.2–0.7
Coastal/shelf	2.8	0.1–0.2
	Mean: 3.0	Total: 0.6–1.7

The production or consumption of DMS and its emission to the atmosphere in upwelling areas is not well documented. It appears that climatic General Circulation Models (GCMs) show an atmospheric cooling above upwelling areas. However, a direct function relation between DMS ocean water concentrations and atmospheric temperatures or cloud coverage, remains to be established. Carbonyl sulfide (COS) is another significant volatile sulfur compound of high atmospheric concentration (~500 ppt). Its oxidation in the stratosphere may be responsible for sustaining the stratospheric sulfate layer [2]. COS is present in the surface sea waters and it has been shown that there exists a relationship between photochemical activity and COS diel production cycle. Estimates of the global gaseous sulfur compounds emissions are listed in Table 3.1.4[1].

Table 3.1.4: Estimates of global, annual, gaseous sulfur compounds emissions [1].

	Flux (Teragram S)
Sources	
Antropogenic (mainly SO_2) from fossil fuel combustion	80
Biomass burning (SO_2)	7
Oceans (DMS)	40
Soils and plants (H_2S, DMS)	10
Volcanoes (H_2S, SO_2)	10
Total:	147

Conclusions

The uncertainties about the fluxes of greenhouse gases from oceanic regions can be summarised as follows.

1. The mass transfer coefficients across the sea-air interface require more accurate estimates for each of the gases considered above. It is also required that these mass transfer coefficients are related to other easily measured parameters like wind speed [6] and atmospheric and sea water temperature.
2. A large number of measuring campaigns would be necessary to establish the contribution of the different biogeographical regions of the world's oceans to the global fluxes of the gases considered. Seasonal variations of these fluxes should also be taken into account.
3. Models directly relating concentrations and fluxes of these gases with observed temperature, cloud coverage or other remotely sensed parameters may help in estimating the impact of oceanic fluxes of greenhouse gases to the global climatic change.

References

1. *Climatic Change: The IPCC Scientific Assessment*, Cambridge University Press 1990.
2. Moore B., D. Shimel (eds), *Trace Gases and the Biosphere*, UCAR, Office for Interdisciplinary Earth Studies, Boulder, Colorado 1992.
3. Charlson, R.J., J.E. Lovelock, M.O. Andreae, S.G. Warren; Oceanic phyto-plankton, atmospheric sulfur, cloud albedo and climate, *Nature* **326** (1987) 6651-6661.
4. Conrad, R., W. Seiler; Methane and hydrogen in sea water (Atlantic Ocean), *Deep Sea Res.* **35A** (1988) 1903-1917.
5. Owens, N.J.P., C.S. Law, R.F.C. Mantoura, P.H. Burkill, C.A. Llewellyn; Methane flux to the atmosphere from the Arabian Sea, *Nature* **354** (1991) 293–296.
6. Liss, P., L. Merlivat; Air-sea exchange rates, in: P. Buat-Menard (ed), *The Role of Air-Sea Exchange in Geochemical Cycling*. Reidel, Dordrecht 1986, pp. 113–117.
7. Bange, H.W., S. Rapsomanikis, M.O. Andreae; Methane in the Baltic and North Seas and a reassessment of the marine emissions of methane. *Global Biogeochem Cycles* **8** (1994) 465–480.
8. Bange, H.W., S. Rapsomanikis, M.O. Andreae; Nitrous oxide in coastal waters. *Global Biogeochem. Cycles* **10** (1996) 197–207.

3.2 Influence of Photochemical and Biological Processes on the Selenium Cycle at the Ocean-Atmosphere Interface

D. Amouroux, C. Pecheyran, M.M. De Souza Sierra and O.F.X. Donard

Laboratoire de Photophysique et Photochimie Moléculaire, URA CNRS 348, Groupe de Chimie Bio-Inorganique et Environnement, EP CNRS 132, Université de Bordeaux I, 351 Cours de la Libération, 33405 Talence Cedex, France

A study of the selenium biogeochemical cycle at the ocean-atmosphere interface has been performed. The speciation of volatile selenium forms in sea water has been established by the development and the utilisation of an analytical method involving a thermal molecular trapping and a separation coupled with a specific atomic detector (absorption, fluorescence). The role of the biological productivity and the photochemical reactivity on the distribution of the volatile selenium compounds in the photic zone has been investigated by field measurements and laboratory experiments. Finally, the fluxes of selenium from the ocean to the atmosphere have been determined for two marine ecosystems (eastern Mediterranean, Gironde estuary). A generalisation of the results obtained demonstrates that gaseous phase marine emissions represent certainly the major source for atmospheric selenium. These volatilisation processes contribute then significantly to remove the selenium from the aquatic environments.

Aims of the work

The study of the selenium biogeochemical cycle at the ocean-atmosphere interface involves many different fundamental and applied scientific concerns [1]. The identification and the characterisation of the sources and fluxes of the exchanges between the ocean and the atmosphere may allow us to estimate the global equilibrium of selenium distribution in the marine environment. These results may also allow a prediction of the influence of natural or anthropogenic disturbance when selenium is deficient or in excess.

The motivations which contributed to study the formation and the emission of volatile selenium species from marine waters to the atmosphere are described in the following points.

* The high selenium enrichment in marine aerosols, without continental or sea salts inputs suggested that the selenium incorporates tropospheric aerosols from the gaseous phase by photochemical processes [2]. Kinetics of dimethyl selenide (DMSe) photo-oxidation by atmospheric oxidants were found much higher than those for dimethyl sulfide (DMS) [3]. This allows us to postulate a rather fast transfer of selenium between the gaseous phase and the aerosols.

* The production of organic selenium compounds at the sea surface and in the estuarine and coastal environment has been observed [4]. The formation of these organic species (*i.e.* amino and organic acids) seems to be related to the biological productivity and sometimes presents the major fraction of total dissolved selenium. Similarly to the sulfur cycle, inorganic selenium (*i.e.* selenate and selenite) is taken up by some living species, reduced and methylated into the intracellular medium. Some volatile methylated selenium species were then determined from continental aquatic ecosystems [5].

Marine environments are then potential source of selenium to the atmosphere by a gaseous phase transfer between the two compartments. The volatile selenium species are coming from the biological activity after release of the organic methylated compounds synthesised by biochemical pathways.

From these above postulate, we then fixed some objectives in order to determine qualitatively and quantitatively the different processes involved:

* to study the factors which are controlling the distribution of the volatile selenium species in sea water in order to confirm the role of the biological productivity and examine the impact of the photochemical reactivity;
* to perform the speciation of the volatile selenium species in different marine environments; and
* to quantify the fluxes of volatile selenium from marine waters to the atmosphere and estimate the global consequences on the selenium biogeochemical cycle.

Major results and discussion

The overall results obtained during this work improved the comprehension of the selenium cycle at the ocean-atmosphere interface, as shown in Fig. 3.2.1.

This work also affirmed the perception of the physico-chemical processes involved in the transformations and the fates of selenium at the sea surface. The principal results rising from this study are described in the following section.

Analytical developments

The development of an analytical technique for the determination of hydrogen peroxide (H_2O_2) in natural waters permits to study the impact of marine photochemical processes in the field and during experimentation in the laboratory [6].

The development of a sensitive and selective analytical technique for the speciation of volatile selenium compounds allowed a study of the distribution of the different species in marine surface waters and troposphere, despite the low concentrations encountered (0.1–10 pmol/l and 0–2 pmol/m^3, respectively) [7].

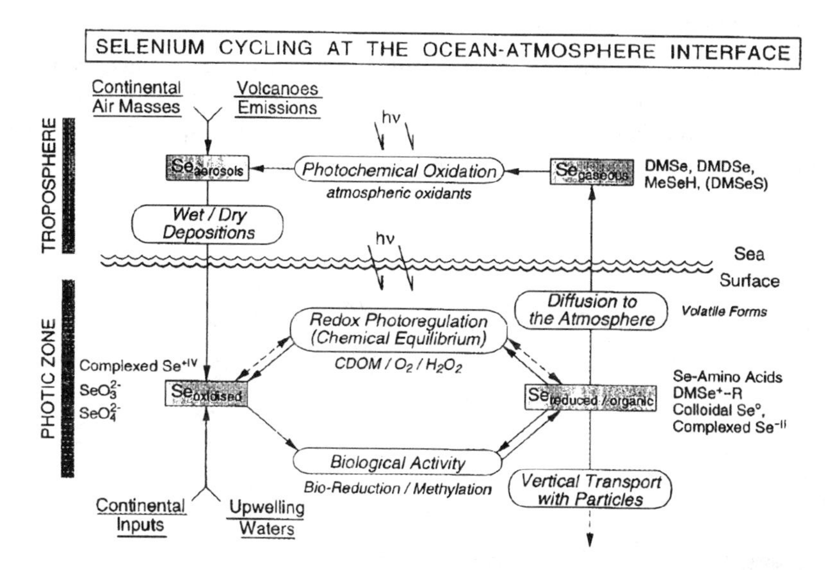

Fig. 3.2.1: Comprehensive selenium biogeochemical cycle at the ocean-atmosphere interface.

Atomic fluorescence spectrometry, preceded by a cryo-condensation step of the volatile species, was found to be a useful detection technique for trace selenium measurements in the environment (Fig. 3.2.2). The different methods developed were adapted and proved their utilisation in field conditions.

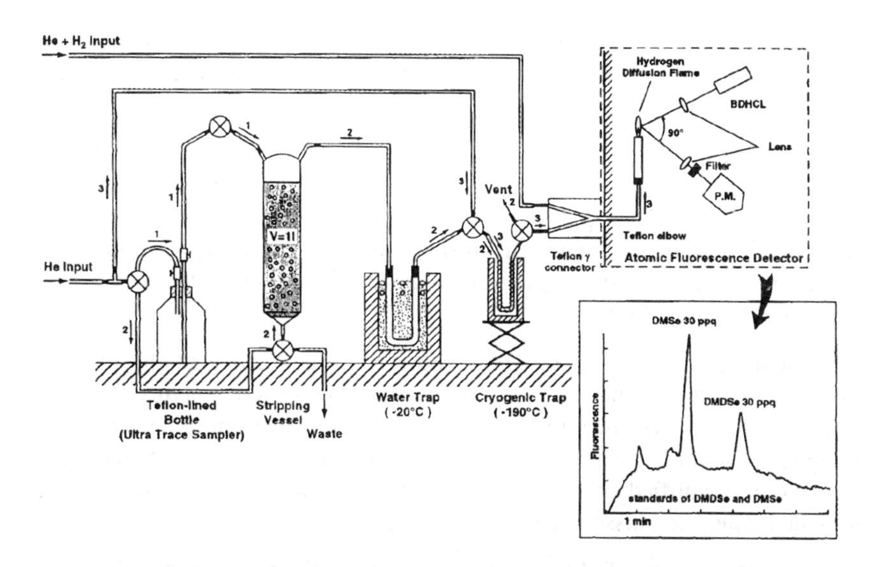

Fig. 3.2.2: On-line purge and cryogenic trapping, gas chromatography and atomic fluorescence spectrometry (PCT-GC-AFS) for the determination of volatile selenium species in the environment.

Study of the photochemical reactivity

Different photochemical parameters of marine waters have been studied in order to determine the role of the photochemical activity in marine environments. The impact of the photochemical reactivity was followed during experiments in the laboratory under simulated sunlight conditions (Fig. 3.2.3) [8, 9].

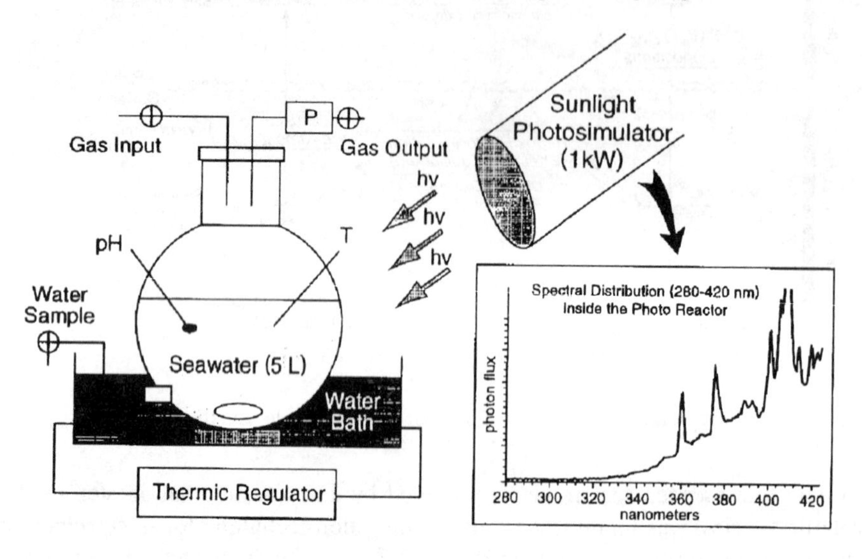

Fig. 3.2.3: Experimental device for sunlight simulated photochemical experiments in marine waters.

The fluorescence spectroscopy of dissolved organic matter (DOM) was used to characterised different type of chromophores constituent of the DOM by specific spectral signatures [10, 11].

The measurements of H_2O_2 in marine waters allow to determine the recent photochemical activity in the photic zone [12]. In the natural medium, the physical and biological phenomena interacting within the water column may disturb this type of interpretation. In the laboratory, the H_2O_2 accumulation rate obtained during experiments permit a direct evaluation of the photochemical reactivity in the medium [8, 9].

A complete processing of the H_2O_2 photochemical kinetics in the presence of dissolved selenium species, lead us to propose different type of chemical mechanism. No evidence for either reduction or methylation mechanisms of inorganic selenium species in marine waters was observed. The photoproduction of H_2O_2 is amplified, probably by a catalytic cycle in the presence selenite, organic ligands and dissolved oxygen.

The photoproduction of H_2O_2 is also amplified in the presence of seleno-amino-acids (*e.g.* Se-methionine) and results in the production of volatile selenium species assisted by the light. The volatile forms observed were mainly dimethyl diselenide (DMSe) and the heterogeneous dimethyl selenyl sulfide (DMSeS).

The presence of selenium compounds and light in marine waters also increases the kinetics of H_2O_2 disappearance by assisted oxidation of selenite and some organic selenides.

Photochemistry and H_2O_2 production may then be directly or indirectly involved in the selenium recycling (oxidation) and in the formation of volatile selenium species in the marine environment. These results also demonstrate that the main compound encountered in natural waters, dimethyl selenide (DMSe), is produced by a specific precursor such as an organic selenonium ion ($DMSe^+$–R).

Study of volatile selenium sea-to-air exchanges

The speciation of volatile selenium species was performed during oceanographic cruises over different marine systems such as the Gironde estuary and the eastern Mediterranean [13, 14]. DMSe was the major fraction of volatile selenium, but significant quantities DMDSe and methane selenol (MeSeH) were also observed in these environments.

Table 3.2.1: Comparison of the type and intensity of volatile selenium emission to the atmosphere between the different marine environments studied in this work and different previous studies; (a) coastal salt marsh (USA; Velinsky and Cutter, [15]), (b) anthropogenic activity (27 European countries; Pacyna, [16]), (c) contaminated ponds (USA, Cooke and Bruland, [5]).

Sources	Volatile selenium compounds[a]	Av. concentrations (pmol/l)	Selenium Fluxes ($nmol/m^2.yr$)
Eastern Mediterranean (été 1993)	**DMSe** MeSeH,DMDSe	0.353	10–1052
Gironde Estuary (1992 –1993)	**DMSe** (MeSeH,DMDSe)	2.480	13–1405
(a) Great Marsh Delaware (1985–86)	**DMSe**	(0.9–7.5)[b]	400–3800
(b) European anthropogenic emissions (1979)	**SeO_2** (Se^0, Me^{II}–Se)	(0.2–3.0)[c]	950[d]
(c) Keterson Reservoir California (1985–86)	**DMSe** DMDSe	$3–30 \times 10^4$	$2.4–2 \times 10^6$

[a] major compound in bold;
[b] concentration range in nmol/g of total organic selenium in the sediment;
[c] concentration range in ppm in oil and coal (their combustion provides the major source);
[d] fluxes normalised to the total area of the 27 countries.

The seasonal study performed in the Gironde estuary showed that the hydrodynamic conditions and the biological production are the main factors controlling the formation and release of volatile selenium species [13]. DMSe concentrations present a direct correlation with the temperature of the aquatic medium. This suggests the contribution of active planktonic and microbial biomethylation processes.

In the eastern Mediterranean, the concentrations of volatile selenium species measured showed a strong relation with the planktonic biomass [14]. The concentrations of DMSe seem also related with the type of water mass (oligotrophic, eutrophic) and the associated plankton species, as demonstrated for DMS.

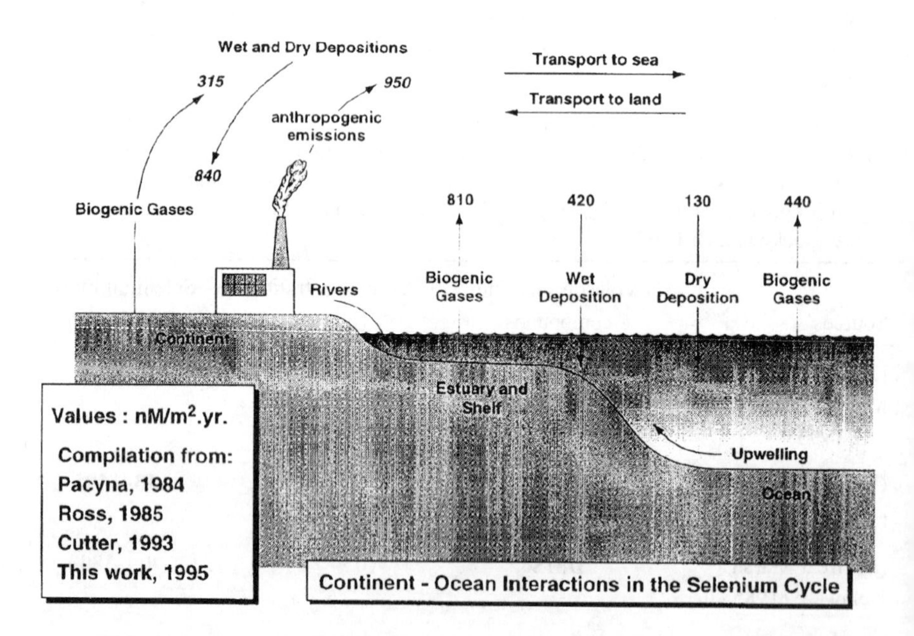

Fig. 3.2.4: Selenium biogeochemical cycle in coastal environments and compilation of the different flux intensities with the atmosphere.

In both marine systems studied, we were able to determine the gaseous selenium fluxes from the ocean to the atmosphere. The principal results are synthesised in Table 3.2.1 and displayed in Fig. 3.2.4. The flux calculations were validated after the determination of the DMSe Henry's law constant by comparing its molecular structure and physico-chemical properties with DMS.

The intensity of the selenium emissions obtained in the marine environment during this work are comparable to the emission factors of some wetlands previously studied (a). These emission factors appear to be in the same range than those reported for the anthropogenic emissions of industrialised European countries (b). However, the impact of contamination seems to stimulate the production of volatile selenium species in natural ecosystems and increase significantly the selenium emission to the atmosphere (c).

Conclusions and perspectives

Emissions of gaseous selenium from marine ecosystems to the atmosphere is a major part of the exchanges in the biogeochemical cycle of selenium, as was postulated previously. An accurate evaluation of the global selenium emission to the atmosphere can only be performed by additional field measurements in marine and continental environments. The direct determination of gaseous forms of a metalloid in marine water and in the atmosphere suggests that this kind of concept must be considered and extended to other metalloids (As, Sb, Te) and metals (Hg, Pb, Sn). For most of them, the volatilisation has been examined or considered after field or laboratory measurements. The contribution of natural emissions of some trace elements to the atmosphere compared to the anthropogenic emissions would then need to be re-estimated in the global budget of these elements in the environment.

Acknowledgements

We wish to thank Prof. M.O. Andreae and Dr. S. Rapsomanikis for inviting us on two cruises. We acknowledge T. Besson, F. Martin-Lecuyer, R. Munoz and B. Thomas for their help in this work. We gratefully thank the crews and the captains of the *R/V Meteor*, *Aegaio* and *Côte d'Aquitaine*. PSA Analytical and Perkin Elmer are also acknowledged for their instrumental assistance.

References

1. Amouroux, D.; Study of the selenium biogeochemical cycle at the ocean-atmosphere interface. Ph. D. Thesis, University of Bordeaux I, 1995.
2. Ross, H. B., 1990. Biogeochemical cycling of atmospheric selenium. *NATO ASI Ser.* **G23** (1990) 523–543.
3. Atkinson, R., S. M. Aschmann, D. Hasegawa, E. T. Thompson-Eagle, W. T. Frankenberger Jr.; Kinetics of atmospherically important reactions of dimethyl selenide, *Environ. Sci. Technol.* **24** (1990) 1326–1332.
4. Cutter, G. A., K.W. Bruland; Marine biogeochemistry of selenium : a re-evaluation. *Limnol. Oceanogr.* **29** (1984) 1179–1192.
5. Cooke, T. D., K.W. Bruland; Aquatic chemistry of selenium : evidence of biomethylation. *Environ. Sci. Technol.* **21** (1987) 1214–1219.
6. Amouroux, D., O.F.X.Donard; Determination of hydrogen peroxide in estuarine and marine waters by flow injection with fluorometric detection. *Oceanol. Acta* **18** (1995) 353–361.
7. Pecheyran, C., Amouroux, D., O.F.X. Donard,; Determination of volatile selenium species in natural waters by cryofocusing and detection by non-dispersive atomic fluorescence spectrometry. *J. Anal. Atmos. Spectr.*

8. Amouroux, D., Pecheyran, C., Uher, G., Ulshofer, V., Donard, O.F.X., M.O. Andreae; Experimental study of the photochemical reactivity in seawater; in: P.M. Borrell, P. Borrell, T. Cvitaš, W. Seiler (eds), *Proc. EUROTRAC Symp. '94,* SPB Academic Publishing bv, The Hague 1994, pp. 1060–1063.

9. Amouroux, D., Pecheyran, C., O.F.X. Donard. Photochemical reactivity of hydrogen peroxide in marine waters with humic substances and algae's exudates under simulated sunlight conditions. *Water Reserch*.

10. De Souza Sierra, M.M., Donard, O.F.X., Lamotte, M., Belin, C., M. Ewald; Fluorescence spectroscopy of coastal and marine waters. *Mar. Chem.* **47** (1994) 127–144.

11. Donard, O.F.X., De Souza Sierra, M.M., Ewald, M., Belin, C., Amouroux, D., H. Etcheber; Fluorescence of continental and marine waters. *Annales Inst. Océanogr.* 69 (1993) 189–191.

12. Andreae, M.O., Amouroux, D., Andreae, T., Donard, O.F.X., Meyerdierks, D., Rapsomanikis, S., Thiel, C., G. Uher; Biogeochemical and atmospheric investigation by the international air-sea exchange group during the Meteor North Atlantic Cruise 1992. *EUROTRAC annual report, part 3 ASE,* EUROTRAC ISS, Garmisch-Partenkirchen 1993, pp. 39–44.

13. Amouroux, D., O.F.X. Donard; Evasion of selenium to the atmosphere via biomethylation processes in the Gironde estuary, France. *Mar. Chem.*

14. Amouroux, D., O.F.X. Donard; Maritime emission of selenium to the atmosphere in eastern Mediterranean seas. *Geophys. Res. Lett.*, **23** (1996) 1777–1780.

15. Velinsky, D.J., G.A. Cutter; Geochemistry of selenium in a coastal salt marsh. *Geochim. Cosmochim. Acta* **55** (1991) 179–191.

16. Pacyna, J.M.; Estimation of the atmospheric emissions of trace elements from anthropogenic sources in Europe. *Atmos. Environ.* **18** (1984) 41–50.

3.3 Biological Production of Trace Gases in Surface Sea Water and their Emission to the Atmosphere

P.S. Liss, W.J. Broadgate, A.D. Hatton, R.H. Little, G. Malin, N.C. McArdle, P.D. Nightingale and S.M. Turner

School of Environmental Sciences, University of East Anglia, Norwich, NR4 7TJ, UK

Measurements of a variety of biogenic trace gases have been made in several European coastal, shelf and deep waters. Gases examined include dimethyl sulfide and its biochemical precursor dimethyl sulfoniopropionate, organo-halogens, and non-methane hydrocarbons. These measurements have been used to estimate the fluxes of these gases from the sea to the atmosphere. Complementary measurements of sulfur isotopic ratios on atmospheric aerosols have been used to assess the relative importance of man-made versus marine biogenic sources of non-sea-salt sulfur to the aerosols.

Aim of the research

A major aim of this contribution is to understand the role of marine organisms (and in some cases sea surface photochemistry) in the production of volatile forms of sulfur (mainly dimethyl sulfide - DMS), organo-halogens (particularly CH_3X and CHX_3, where X = Cl, Br or I) and non-methane hydrocarbons (including isoprene) in European shelf and coastal waters, as well as in some selected open ocean areas. A second aim is to estimate the magnitude of the fluxes of these compounds from sea water to the atmosphere, and to assess the role these fluxes have for the chemistry of the overlying atmosphere and adjacent land masses. These aims have been pursued via field and laboratory studies. Measurements of stable sulfur isotopic ratios have been used to quantitatively differentiate between man-made (fossil fuel) and natural marine (DMS) sources of non-sea-salt sulfur in atmospheric aerosols and rain.

Principal scientific results

Sulfur, DMS and related compounds

There is much evidence [1, 2] to suggest that DMS emitted by phytoplankton plays significant roles in a number of environmental concerns, *e.g.* pH control of rain and climate control via the formation of cloud condensation nuclei (CCN). It is necessary to understand the production and loss mechanisms of DMS and its precursor, dimethylsulfoniopropionate, to be able to predict how man's perturbation of the environment may ultimately affect albedo and acidity of rain.

In order to investigate some of these important areas, a large number of field observations, in coastal and shelf seas, have been made and hypotheses tested using simplified laboratory systems as part of ASE.

From field measurements in the North Sea and N.E. Atlantic, it has been found that the ability of phytoplankton to produce DMSP varies considerably, depending on species, phase of the life cycle and possibly nutrient availability [3–6]. The species which produce most DMSP are coincidentally those which in some parts of the North Sea have flourished at the expense of the low DMSP producers. Initial field and laboratory studies suggest that more DMSP is produced when nitrate is limited, supporting the hypothesis that DMSP becomes the preferred osmoregulant rather than its nitrogen analogue, glycine betaine. These observations have implications for the effects of increasing eutrophication. Preliminary experiments with a large number of North Sea bacterial isolates suggest that the ability to cleave DMSP and produce DMS is common, possibly suggesting that bacteria play a significant role in the production of DMS. First estimates of the rates of DMSP and DMS production have been made using data from a series of cruises, in which we participated, in the UK NERC North Sea Project [3, 7].

As previously found, the variation in DMS and DMSP concentrations followed the development of the phytoplankton. During the winter, the mean DMS level was very low, at about 4 ng S l^{-1}. As spring progressed, DMS levels increased to a maximum in May, when *Phaeocystis pouchetii* was predominant off the coasts of Holland and Belgium. Concentrations then decreased, with a slight elevation in August. During the nine months studied, a number of 'hotspots' of DMS were identified, which were associated with blooms of particular species of phytoplankton. The ratio of DMSP to DMS varied spatially and seasonally; during the summer, the mean ratio was 14, a value found in previous years, and in winter the ratio was much higher, at about 60.

The mean (spatially averaged) concentration of DMS from April to September inclusive was close to 300 ng S l^{-1} which represents a flux of sulfur to the atmosphere equivalent (on a unit area/time basis) to about 20 % of the emission rate of anthropogenic, European sulfur (as SO_2) over the spring and summer months. In winter the emission from the North Sea is insignificant compared with man-made inputs to the atmosphere in Europe.

A study conducted in June 1991 in an intense and extensive *coccolithophore* bloom in the N.E. Atlantic showed that the flux of DMS to the atmosphere from this region was, per unit area, at least as great as the equivalent flux from the southern North Sea in summer [8].

An analytical chemistry advance made as part of the work has led to the development of a reliable, straightforward and, most importantly, specific enzymatic method for measuring dimenthylsulfoxide (DMSO) in sea water [9]. This oxidation product of DMS was difficult to determine accurately by previous techniques since some fraction of the biochemical precursor of DMS, dimenthylsulfoniopropionate (DMSP), is generally returned as DMSO. Results obtained using the new method indicate that concentrations of DMSO are intermediate between those of DMS and DMSP. All three compounds exhibit rather similar vertical profiles, except that DMSO concentrations tend to remain high to greater depths in the water column than for the other compounds.

Sulfur isotopic ratios

The relative importance of anthropogenic versus natural sulfur emissions to the remote marine atmosphere is uncertain but of great interest. The aim of this part of the research was to examine the potential of stable sulfur isotope composition to act as a tracer for marine biogenic sulfur, thus enabling it to be distinguished from anthropogenic sulfur. Fractionation of the four stable isotopes of sulfur takes place during various naturally occurring transformations, with the result that some sources of sulfur have characteristic isotopic signatures. For example, sea-water sulfate has a $\delta^{34}S$ value of +20 ‰, which is consistent throughout the world. If important sources of sulfur have characteristic signatures which are sufficiently different from each other then a simple mass balance calculation can be used to find the relative strength of each source.

Atmospheric sulfate derived from the oxidation of dimethylsulfide produced by marine phytoplankton, is though to have a $\delta^{34}S$ value close to that of sea water, whereas sulfur derived from fossil fuel burning has been found to have $\delta^{34}S$ values in the range 0 to 5‰.

Aerosol samples were collected over a two-year period (1990–1992) at Mace Head, a remote site on the west coast of Eire. To complement the Mace Head data a suite of samples were also obtained for Ny Ålesund, Spitsbergen. The samples from both sites have been analysed for a selection of anions, cations and trace metals, as well as for stable sulfur isotopes [10, 11]. Very similar results were obtained for both sites, with a range in $\delta^{34}S$ values of +4 to +23 ‰. A strong seasonality was observed in the isotope results, with the highest values occurring in spring and summer. The concentrations of methanesulfonate (MSA, another oxidation product of DMS) displayed a very similar seasonal cycle. The signature for marine biogenic sulfate was estimated to be +23 ‰ at both sites. The typical $\delta^{34}S$ value for anthropogenic sulfate is considered to be +5 ‰ at Mace Head and +6 ‰ at Ny Ålesund. The estimated isotopic signatures were used to calculate the relative contributions of these two sources to non-sea-salt (NSS) sulfate. It was found that ~ 30 % NSS sulfate sampled at the two sites during summer was derived from the oxidation of DMS. The maximum concentration of biogenic sulfate of 26 nmol/m^3 and summer mean value of ~5 nmol/m^3 found for both locations, indicate that both the Arctic and Atlantic oceans produce large amounts of DMS during the spring and summer.

Statistical analysis of the chemical data supported the findings from sulfur isotopes. Air mass back trajectory data were obtained for many of the samples and confirmed that low isotopic signatures are associated with continentally influenced air masses, whilst high $\delta^{34}S$ occur when essentially marine air is sampled.

In addition to these two series of samples, a number of isolated samples were collected at a selection of sites around Britain. These samples, along with three samples of flue gases from power stations, support the value of $\delta^{34}S$ estimated for the anthropogenic end member at Mace Head.

Organo-halogens

As with DMS, low molecular weight organo-halogens produced by marine algae (and possibly photochemically) are proposed to play significant roles in atmospheric chemistry on emission from surface sea waters. Examples of such reactions are the oxidation of DMS to DMSO in the presence of methyl iodide and ozone, and the possible effect of organo-halogens on the redox chemistry of the atmosphere.

Our investigation in the N.W. Mediterranean [12] has shown that in contrast to the North Sea, there is little anthropogenic influence on levels of halocarbons in the Ligurian Sea from either the French or Italian coasts. Levels of halocarbons along transects from Nice towards the centre of the sea showed only a small coastal gradient. Depth profiles have indicated that methyl iodide is produced, most probably indirectly, by biological activity. A distinct maximum in concentrations was observed above the chlorophyll maximum and close to the thermocline. Profiles of dibromomethane and the trihalomethanes possess sub-surface maxima. Levels appear to be more directly associated with biological activity than does methyl iodide. Indicators of biological activity were also bimodal, for example bacterial activity, heterotrophic flagellates and cyanobacteria. It was not possible to ascribe production to a particular group of these organisms. An unknown compound appears to be produced biogenically and indeed possessed the largest sub-surface maximum. Both this compound and methyl iodide would also appear to be removed in the water column.

Carbon tetrachloride concentrations declined rapidly with depth. A removal mechanism for this compound may also be present. Composite plots of methyl chloroform and tetrachloroethene suggest that levels of these compounds might be influenced by biological activity. Finally, there would appear to be little diurnal variation in halocarbon concentrations. An investigation to observe a 24 hour cycle was influenced only by changing water masses, although this did indicate that elevated halocarbon concentrations were associated with increased biological activity.

Non-methane hydrocarbons

Non-methane hydrocarbons (NMHCs) have an influence on the oxidising capacity of the atmosphere via the reactions with OH, ozone and NO_3. The importance of anthropogenic emissions of NMHCs and biogenic emissions of monoterpenes and isoprene from the terrestrial biosphere has long been recognised and extensively studied. However, the ocean as a source of reactive species such as isoprene has only recently been identified. The emission of isoprene and other NMHCs from the ocean may provide a significant source of reactive organics to the atmosphere over the remote oceans.

Monthly measurements of isoprene over a full seasonal cycle in the North Sea indicate that the concentration range is over two orders of magnitude with a summer maximum and winter minimum [13]. The concentration of isoprene is correlated with chlorophyll levels indicating a production mechanism which is linked to phytoplankton activity [14]. High concentrations of isoprene were observed in a bloom of *Phaeocystis pouchetii* in the North Sea and a spring bloom dominated by diatoms in the Southern Ocean.

Conclusions

The main conclusions of this work are the following.

i) There are considerable interspecies differences in ability of phytoplankton to form dimethlysulfoniopropionate (DMSP) and that nitrate limitation may favour the production of DMSP over its nitrogen analogue, glycine betaine. Bacteria appear to play a major role in the cleavage of DMSP to produce dimethyl sulfide (DMS).

ii) In the southern North Sea the seasonal cycle of DMS and DMSP concentrations closely follows development of the phytoplankton. In winter, DMS and DMSP concentrations are very low, rising to levels in spring and summer which predict a sea-to-air flux (per unit area per unit time) of DMS which is up to 20 % of the equivalent emission from land areas in Europe. Extensive *coccolithophore* blooms in the north-east Atlantic produce fluxes at least as great as those from the southern North Sea.

iii) A new enzymatic method for measuring dimethyl sulfoxide (DMSO) in sea water has shown that concentrations of DMSO tend to be between those of DMS and DMSP, and that elevated levels of DMSO persist to greater depths in the water column than DMS or DMSP.

iv) Sulfur isotope ratio measurements at Mace Head (Eire) and Ny Ålesund, Spitsbergen have been used to differentiate between man-made (fossil fuel) and DMS-derived sulfur in atmospheric aerosol samples. At both sites approximately 30% of the non-sea-salt sulfate in summer is derived from atmospheric oxidation of marine biogenic DMS.

v) Organo-halogen, such as methyl iodide, dibromomethane and trihalomethanes, measured in the north-west Mediterranean, show depth profiles which strongly suggest that production of these compounds is via biological processes.

vi) Monthly measurements of a variety of non-methane hydrocarbons in the North Sea indicate a concentration range over two orders of magnitude between the summer maximum and the winter minimum. Concentrations of isoprene in the water show a good correlation with chlorophyll levels in both the North Sea and the Southern Ocean, indicating a biological production mechanism.

Acknowledgements

Support for this work has come from the UK's Natural Environment Research Council and Department of the Environment, and the Commission of the European Communities.

References

1. Liss, P.S. The Sulfur Cycle. In: J.C. Duplessy, A. Pons, R. Fantechi (eds), *Climate and Global Change*, CEC, Bruxelles 1991, p. 75.
2. Fell, N., P.S. Liss. Can algae cool the planet? *New Scientist* **1887** (1993) 34.
3. Liss, P.S., G. Malin & S.M. Turner. Production of DMS by marine phytoplankton. In: G. Restelli, G. Angeletti (eds), *Dimethylsulfide: Oceans, Atmosphere, and Climate*, Kluwer, Dordrecht 1993, p. 1.
4. Liss, P.S., G. Malin, S.M. Turner, P.M. Holligan. Dimethyl sulfide and *Phaeocystis*: a review. *J. Marine Systems* **5** (1994) 41.
5. Malin, G., S.M. Turner, P.S. Liss. Sulfur: the plankton/climate connection. *J. Phycol.* **28** (1992) 590.
6. Malin, G., S.M. Turner, P.S. Liss. Dimethyl sulfide: production and atmospheric consequences; in: J.C. Green, B.S.C. Leadbeater (eds), *The Haptophyte Algae,* Clarendon Press, Oxford 1994, p. 303.
7. Liss, P.S., A.J. Watson, M.I. Liddicoat, G. Malin, P.D. Nightingale, S.M. Turner, R. Upstill-Goddard. Trace gases and air-sea exchanges; *Phil. Trans. Roy. Soc. Lond.* **A343** (1993) 531.
8. Malin, G., S.M. Turner, P.S. Liss, P. Holligan, D. Harbour. Dimethylsuphide and dimethylsulfoniopropionate in the Northeast Atlantic during the summer coccolithophore bloom. *Deep-Sea Research* **40** (1993) 1487.
9. Hatton, A.D., G. Malin, P.S. Liss, A.C. McEwan. Determination of dimethyl sulfoxide in aqueous solution by an enzyme-linked method. *Anal. Chem.* **66** (1994) 4093.
10. McArdle, N.C.; The use of stable isotopes to distinguish between nautral and anthropogenic sulphur in the atmosphere. Ph. D. thesis, University of East Anglia, 1993.
11. McArdle, N.C., P.S. Liss; Isotopes and atmospheric sulfur. *Atmos. Environ.* **2** (1995) 2553.
12. Nightingale, P.D.; Low molecular weight halocarbons in seawater. Ph. D. thesis, University of East Anglia, 1991.
13. Broadgate, W.J.; Non-methane hydrocarbons in the marine environment, Ph. D. thesis, University of East Anglia, 1995.
14. Broadgate, W.J., P.S. Liss, S.A. Penkett; Ocenanic emissions of non-methane hydrocarbons, *CACGP/IGAC Conf. on Global Atmospheric Chemistry*, Fuji-Yoshida, Japan 1994.

3.4 Origin and Importance of Particulate Dimethylsulphoniopropionate (DMSPp) in the Marine Cycle of Dimethylsulphide (DMS)

S. Belviso[1], M. Corn[1], U. Christaki[1], J.C. Marty[2], C. Cailliau[2] and P. Buat-Ménard[3]

[1]Centre des Faibles Radioactivités, CNRS-CEA, Av. de la Terrasse, 91198 Gif-sur-Yvette, France
[2]Laboratoire de Physique et Chimie Marines, Observatoire Océanologique de Villefranche, UPMC-CNRS-INSU, BP28, 06230 Villefranche-sur-Mer, France
[3]Département de Géologie et Océanographie, URA CNRS 197, Université Bordeaux 1, av. des Facultés, 33 405 Talence Cedex, France

The large scale variability of dissolved DMSP-DMS and chlorophyll *a* is explored to determine the feasibility of a predictive relationship. Changes in dissolved DMSP-DMS and particulate DMSP levels were examined during the 3rd and 4th Eumeli cruises in the subtropical northeastern Atlantic Ocean, the 2nd Antares cruise in the Austral Ocean and during the Dyfamed time-series programme in the Mediterranean Sea, areas where the intensity of the phytoplankton biomass varied within a 1 to 100 ratio. The free DMSP and DMS pool of sea water appears to be predictable from two parameters: the phytoplankton abundance (using chlorophyll *a* as a proxy) and the microplanktonic DMSP standing stock (using peridinin as a proxy and microzooplankton abundance for which no proxy is presently available).

Aims of the research

The aim of this research is to determine the controlling factors of the spatio-temporal variability of dissolved DMSP and DMS concentration in contrasted marine ecosystems.

Principal scientific results

This study has investigated the occurrence and abundance of the dissolved compounds DMSP and DMS in 3 contrasting trophic regimes of the subtropical northeastern Atlantic Ocean (Eumeli programme), in the Mediterranean during the Dyfamed time-series programme, and in the southern Indian Ocean along a transect at 62 °E (Antares programme).

Water samples were collected for particulate DMSP (DMSPp), dissolved DMSP and DMS, chlorophyll *a* and carotenoid analysis using bottle samplers fixed to a CTD system. Integrated quantities of DMSPp, dissolved DMSP and DMS, chlorophyll *a* and carotenoids in the top 50 m water column were calculated from vertical profiles.

No significant relationship was found between dissolved DMSP + DMS and chlorophyll *a* . The sulfur compounds were covarying with the total particulate-DMSP standing stock of sea water (Fig. 3.4.1). It is made from two distinct pools of DMSP-containing particles. DMSPp in the size class > 10 µm always accounted for an important fraction (> 20 %) of the total DMSPp pool and was predominant in up-welled waters where it covaried with peridinin, a biomarker of dinoflagellates. A weaker peridinin to DMSPp relationship was observed in the two other areas where DMSP could originate from microzooplankton (ciliates, copepod larvae). DMSPp in the size class < 10 µm covaried with diadinoxanthin, a non photosynthetic carotenoid with photoprotective properties, except in the central up-welling area where diatoms low in DMSP but high in diadinoxanthin are predominating, and during spring time in the Mediterranean Sea for a reason presently unknown. This relationship provides *in-situ* indication of light-dependent DMSP accumulation in prymnesiophytes. So, the large scale variability of DMSPp appears to be driven from the standing crop of prymnesiophytes and dinoflagellates and from heterotrophs (microzooplankton and flagellates possibly).

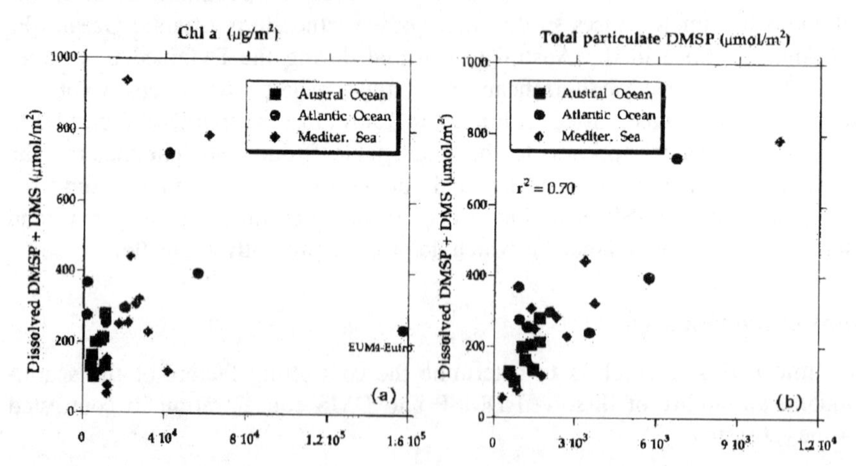

Fig. 3.4.1: Dissolved DMSP + DMS versus (a) chlorophyll *a*, and (b) total particulate DMSP concentrations from the three oceanic areas investigated.

Although DMSPp in the Austral Ocean was carried mainly by particles in the size class < 10 µm, no significant relationship was found between concentration of dissolved DMSP + DMS and this DMSP fraction. Dissolved levels were significantly correlated with levels of DMSPp in the size class > 10 µm. It is suggested that dinoflagellates, microzooplankton and/or microscopic material (aggregates, fecal pellets) could play a key role in controlling the concentrations of dissolved DMSP and DMS in these waters. For the 3 oceanic areas investigated, the (DMSPp 10–200 µm)-to-(dissolved DMSP + DMS) ratio is shown to be closely positively related to phytoplankton biomass (Fig. 3.4.2).

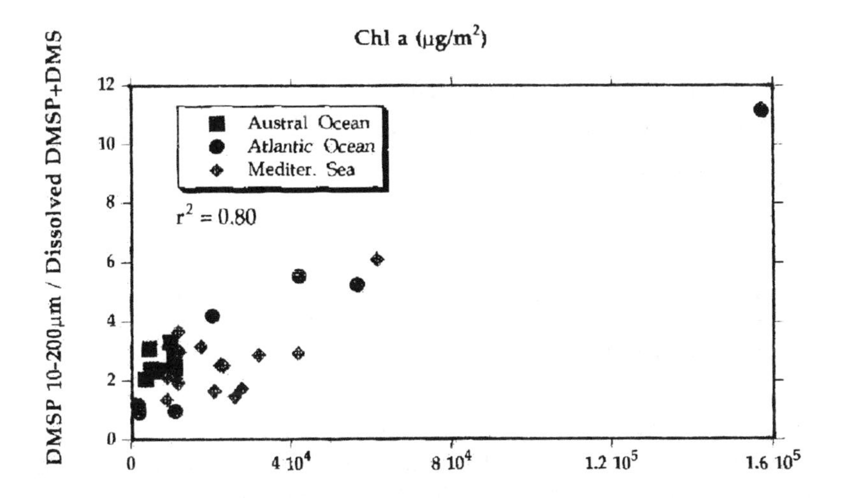

Fig. 3.4.2: The relationship of chlorophyll *a* with the (DMSPp 10–200 µm)-to-(dissolved DMSP + DMS) ratio for the three oceanic areas investigated.

Conclusions

In the framework of the EUROTRAC subproject ASE, we have evaluated the contribution of biological processes to the atmospheric fluxes of DMS from surface sea water 1-4. We conclude that the free DMSP and DMS pool of sea water appears to be predictable from two parameters: the phytoplankton abundance (using chlorophyll *a* as a proxy) and the microplanktonic DMSP standing stock (using peridinin as a proxy and microzooplankton abundance for which no proxy is presently available).

Acknowledgements

This work was supported by the French Centre National de la Recherche Scientifique (CNRS), Institut National des Sciences de l'Univers (INSU), Institut Français de Recherche pour l'Exploitation de la Mer (IFREMER), Commissariat à l'Energie Atomique (CEA) and Ministère de L'Environnement.

References

1. S. Belviso, S.-K. Kim, F. Rassoulzadegan, B. Krajka, N. Mihalopoulos, P. Buat-Ménard, *Limnol. Oceanogr.* **35** (1990) 1810.
2. S. Belviso, P. Buat-Ménard, J.-P. Putaud, B.C. Nguyen, H. Claustre, J. Neveux, *Marine Chem.* **44** (1993) 55.
3. J.-P. Putaud, S. Belviso, B.C. Nguyen, N. Mihalopoulos, *J. Geophys. Res.* **98** (1993) 14863.
4. M. Corn, S. Belviso, P. Nival, A. Vigot, P. Buat-Ménard, *Océanol. Acta.* **17** (1994) 233.

3.5 Seasonal and Latitudinal Variations of Dimethylsulfide Emissions from the North-East Atlantic Ocean

B.C. Nguyen[1], J.-P. Putaud[1] and N. Mihalopoulos[2]

[1]Centre des Faibles Radioactivitis, Laboratoire Mixte CNRS-CEA, LMCE, CEN Saclay Bat. 709, Orme des Merisiers, 91191 Gif-sur-Yvette Cedex, France
[2]Crete University P. 0. Box 1470 Iraklion Crete

Dimethylsulfide (DMS) and dimethylsulfoniopropionate (DMSP) concentrations in sea water have been measured during the Darwin cruise between Las Palmas, Canaries, and Barry, Wales, in December 1993–January 1994, the Belgica cruise between Bordeaux, France, Cork, Ireland, and Zeebrugge, Belgium, in April–May 1994 and the Poseidon cruise between Reykjavik, Iceland, and Lisbon, Portugal, in August and September 1995.

Significant correlations were observed between DMS and DMSP in these campaigns. Clear latitudinal gradients were observed with highest DMS and DMSP concentrations in the high latitude between 55 °N and 65 °N. In addition a clear seasonal signal was observed with higher DMS and DMSP concentrations during spring and summer than in winter time. These results can be to some extent explained by the variations in the biological parameters (chlorophyll *a*, phaeopigments, nutrients ...) measured simultaneously.

From a sea-air exchange model and the meteorological data recorded during these campaigns, seasonal and latitudinal variations of DMS fluxes varying from 4 to 18 μmol m^{-2} d^{-1} with the highest flux were observed at the high latitude between 55 °N and 65 °N and in the vicinity of continental margins.

Introduction

Dimethylsulfide (DMS), the most abundant volatile sulfur compound in surface sea water, is a by-product of plankton activity, yielded by cleavage of dimethylsulfoniopropionate (DMSP) in the euphotic layer of oceans. DMSP is thought to play a role in the regulation of the osmotic pressure in phytoplankton cells, and DMS has been found to be mainly produced during senescence phases of phytoplankton and grazing by zooplankton [1, 2]. However, surface sea water is always supersaturated in DMS, so that a fraction of DMS is emitted to the atmosphere at the ocean-atmosphere interface, the remaining is cycled within the water column through bacterial transformation/consumption and oxidation to DMSO. Once in the atmosphere, DMS oxidises in species such as sulfur dioxide (SO_2), methanesulfonic acid (CH_3SO_3H, MSA) and sulfuric acid (H_2SO_4), which contribute to the acidity of aerosols and serve as precursors of tropospheric cloud condensation nuclei [3]. DMS is the only sulfur containing biogenic gas whose flux is estimated to be of the same order of magnitude as the anthropogenic flux of sulfur dioxide. However, large uncertainties remain about its global flux since its concentrations can vary by as much as two orders of magnitude between summer levels in coastal areas and winter levels in open sea [4–6]. For instance, the continental margins and coastal waters could be important sources of volatile biogases due to their high biological activity, but the DMS flux emitted from these areas has not been accurately taken into account till now due to the limited data set.

This paper presents the DMS concentrations in surface sea water and its flux in the atmosphere from the open ocean and continental European margins of the North Atlantic Ocean during three OMEX cruises and distinct seasons.

Sampling and analysis

Sampling was carried out aboard the R/V Charles Darwin between Las Palmas, Canaries, and Barry, Wales, from 13th December 1993 to 14th January 1994; aboard the R/V Belgica between Bordeaux, France, Cork, Ireland, and Zeebrugge, Belgium, from 19th April 1994 to 5th May 1994; and aboard the R/V Poseidon between Reykjavik, Iceland, and Lisbon, Portugal, from 31st August to 11th September 1995 This sampling was performed simultaneously with biological sampling using the same Niskin bottles both when stationary and during transit, from the ship's sea water pumping system, a non-toxic supply, the inlet of which is situated at the bows 5 meters below sea level. About 60 to 180 mL of surface sea water were sampled, generally from the ship's sea water pumping system,. DMS was extracted from sea water by helium bubbling at a flow rate of 100 mL/min and concentrated in a cryogenic trap at –90 °C before being analysed by GC/FPD [5]. The precision of the analysis is estimated to be about 10 %. When stationary, some sea water samples were collected at 5 m depth from a Niskin bottle on a rosette sampler attached to a CTD system. A comparison between 7 pairs of samples collected simultaneously from the non-toxic supply of the boat and a Niskin bottle shows a DMS ratio of 0.96 ± 0.15, which indicates no significant difference in DMS concentration from the two sampling methods.

Results and discussion

Vertical profiles of DMS and DMSP concentrations

DMS and DMSP profiles during the Belgica cruise are presented in Fig. 3.5.1. For both profiles, the temperature remains constant within the first 100 m. DMS has higher concentration in the euphotic layer, decreasing sharply below 60m.

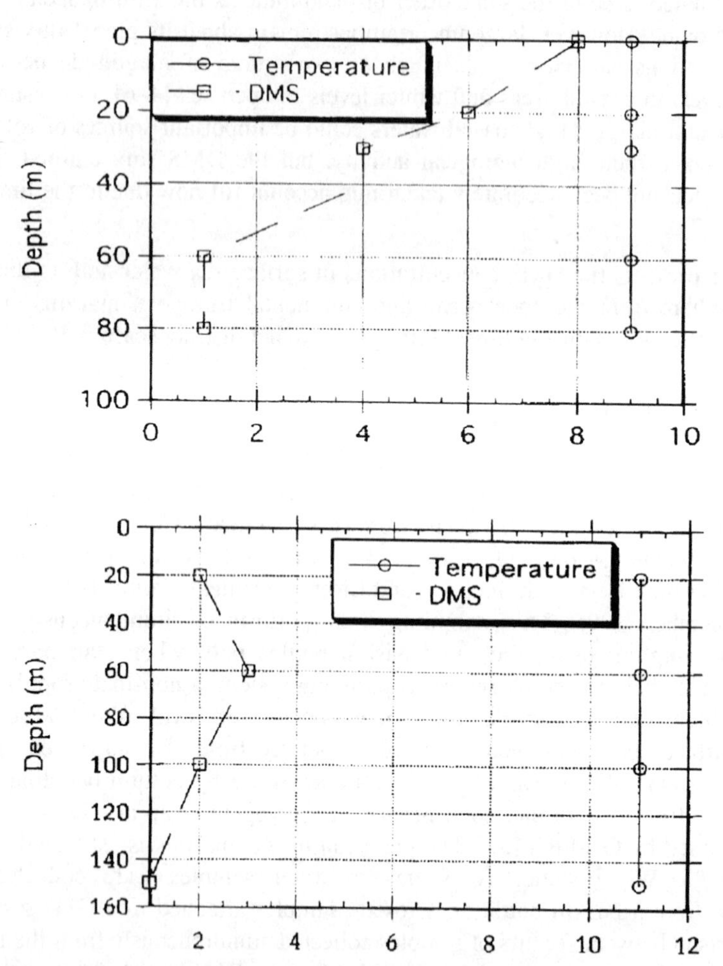

Fig. 3.5.1: DMS (nmol/L) and temperature (°C) during the Belgica cruise. Station CS, 51°33' N/8°15' W (above); Station 9, 47°23' N 7°16' W (below).

Sea water temperature during the Darwin cruise was always constant in the first 40m and sometimes down to 100 m, suggesting a quite deep mixed layer. DMS and DMSP profiles presented the same feature with a nearly constant concentration down to 40–100 m depth (range 1.5–2 and 15–17 nmol/L for DMS and DMSP, respectively) and a sharp decrease in their concentrations to the detection limit below (Fig. 3.5.2). No marked sub-superficial maximum in DMS or DMSP was observed and the DMS/DMSP ratio was generally constant (11.5 % ± 0.5) throughout the first 100 m, as was the chlorophyll *a* concentration which was always smaller than 0.1 µg/L.

These observations indicate that DMS and DMSP are produced principally in the euphotic layer (0 to 100 m) .

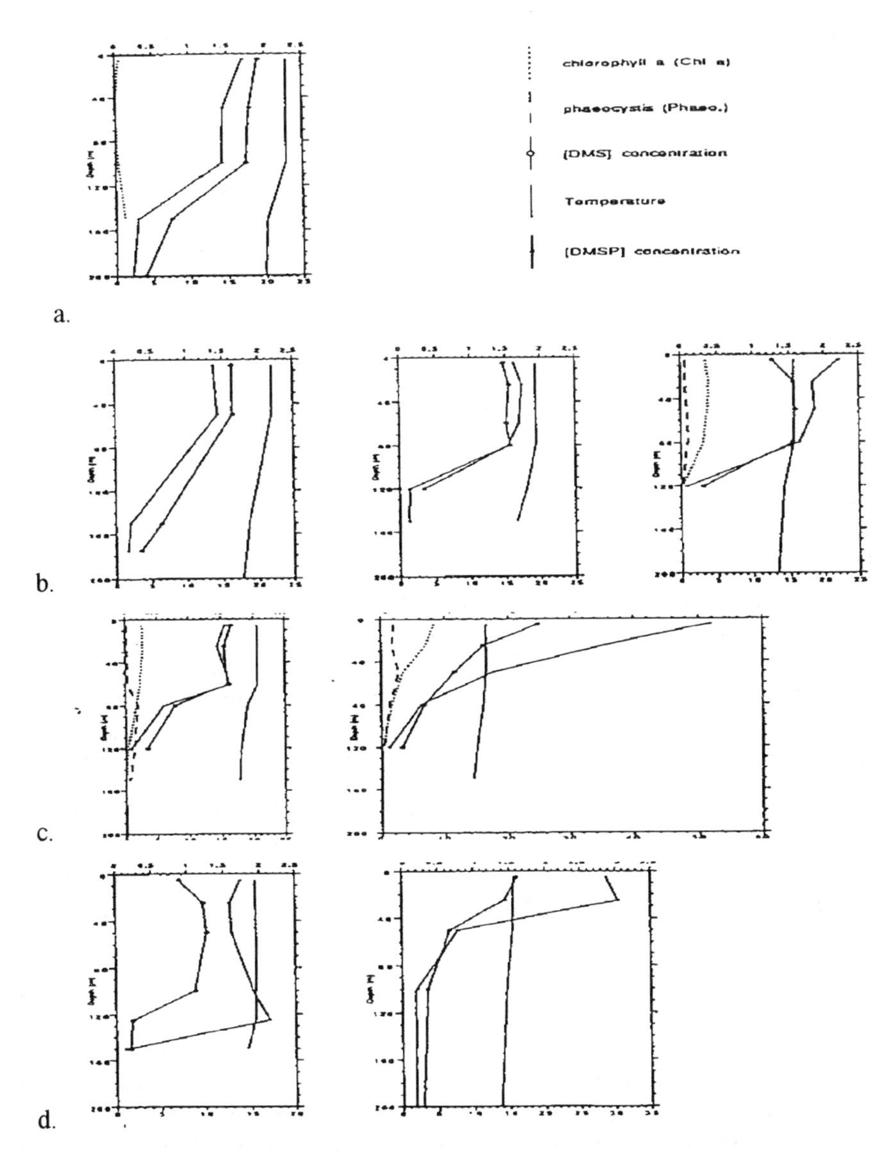

Fig. 3.5.2: Profiles observed: (a) in the open ocean around 25° N; (b) in the open ocean between 26° N and 39° N; (c) over the slopes of tablemounts; and (d) in the vicinity of the Portugal margin.

Variation of DMS concentrations in sea surface water and ocean-atmosphere flux between shelf, shelf-edge and open ocean.

Fig. 3.5.3a indicates the ship track of the Belgica cruise in April–May 1994 and the location of the OMEX stations. This cruise crossed different areas, shelf, shelf-edge and open ocean. DMS concentrations in sea surface-water during the spring cruise are presented in conjunction with meteorological parameters in Fig. 3.5.4.

DMS concentrations varied from 0.3 to 21 nmol/L (mean 5.14 ± 4.1) (Fig. 3.5.4a) Average DMS concentrations between shelf, shelf-edge and open ocean during this cruise are:

3.47 nmol/L (depth 0–100 m, 18 values);
5.26 nmol/L (depth 100–500 m, 10 values);
13.11 nmol/L (depth 500–1000 m, 6 values);
3.30 nmol/L (depth 1500–3000 m, 3 values) see Fig. 3.5.5.

Surface sea water has been reported to be supersaturated in DMS with respect to the atmosphere, and although a large fraction of the dissolved DMS is recycled within the water column, there is a positive DMS flux (F_{DMS}) from the ocean to the atmosphere. F_{DMS} can be estimated using a gas exchange model [7], and the relationship

$$F_{DMS} = k_w[DMS]_{sea\ water}$$

where k_w is the piston velocity depending on the local wind speed and SST [5, 8]. During the Belgica cruise, large variations in wind speed (Fig. 3.5.4c) resulted in k_w values ranging from 0.1 to 25.0 cm hr^{-1} (mean 9.0). Excluding the 2nd–3rd May period, DMS sea-air flux was estimated to range between 0.1 and 40 µmol m^{-2} d^{-1} (mean 11.7, Fig. 3.5.4d). This average is very close to the estimates by Malin *et al.* [9] for the northeast Atlantic in June and July 1987.

However, during the 2nd–3rd May period when a combination of high wind speed and high DMS concentration in sea water occurred, the DMS flux increased to 80 µmol m^{-2} d^{-1} (Fig. 3.5.4d). Taking into account these 2 days, a mean sea-air DMS flux of 12.9 µmol m^{-2} d^{-1} is calculated for the whole period.

Fig. 3.5.5 illustrates the whole dimethylsulfide concentrations in sea surface-water and sea-air fluxes from the Atlantic European continental margins and shows a maximum at the shelf-edge with DMS concentrations varying from 6.57 to 21.03 nmol/L with an average 13.11 nmol/L and a flux varying from 7.80 to 31.21 µmol m^{-2} d^{-1} with an average 18.28 µmol m^{-2} d^{-1}. This maximum of DMS concentrations in the region 500–1000 m deep can be explained by the maximum of microzooplankton biomass (10.00 µg C/L) observed by Edwards *et al.* [10] during the same period April/May 1994 and in the same area (depth 500–1000 m) aboard the R/V Charles Darwin OMEX across the shelf break from 49°05' N–13°18' W to 49°30' N–10°59' W. This maximum of DMS concentrations is probably due to the grazing of phytoplankton by zooplankton [1].

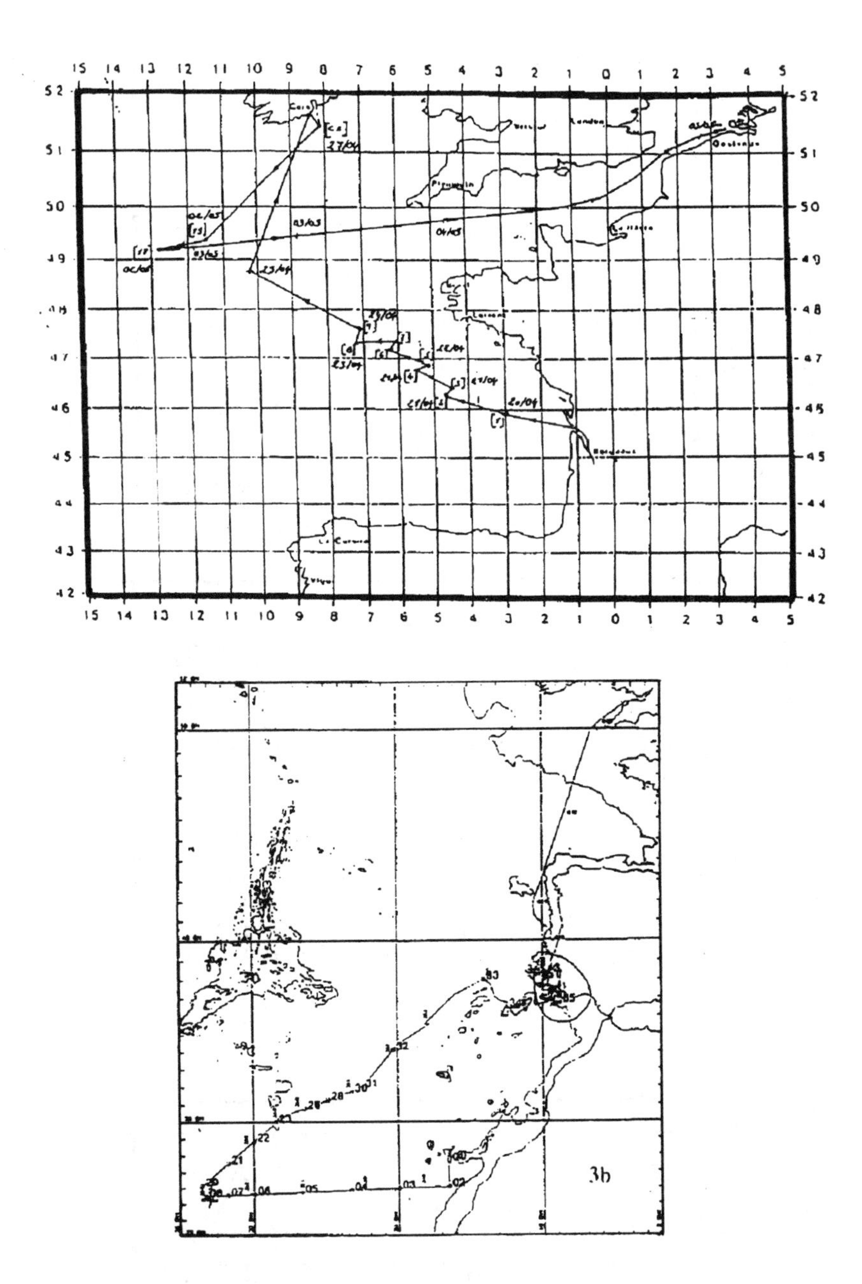

Fig. 3.5.3: Maps showing the ships' tracks during (a) the Belgica cruise (20th April 1994 to 5th May 1994), and (b) the Charles Darwin cruise (14th Dec. 1994 to 14th Jan. 1995).

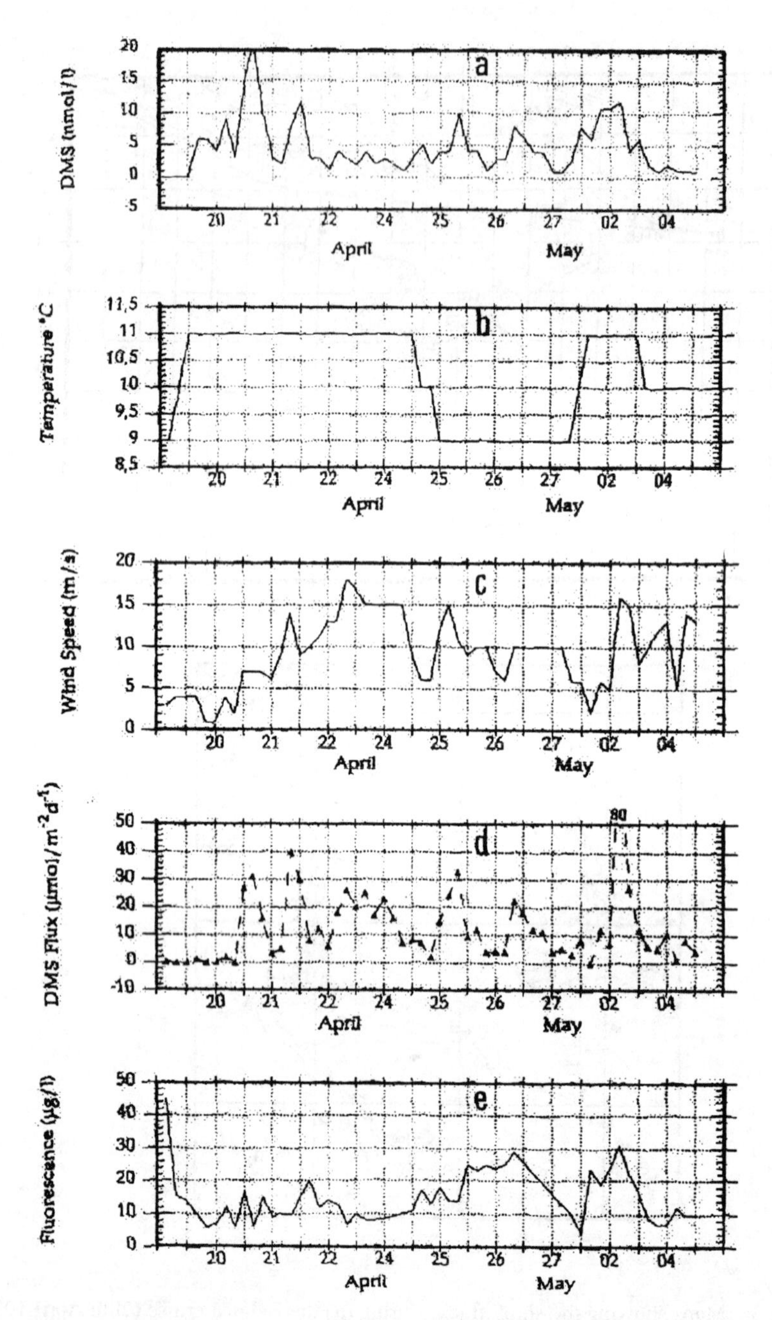

Fig. 3.5.4: Variation of DMS flux and concentration in surface sea water, temperature, wind speed and fluorescence during the Belgica cruise.

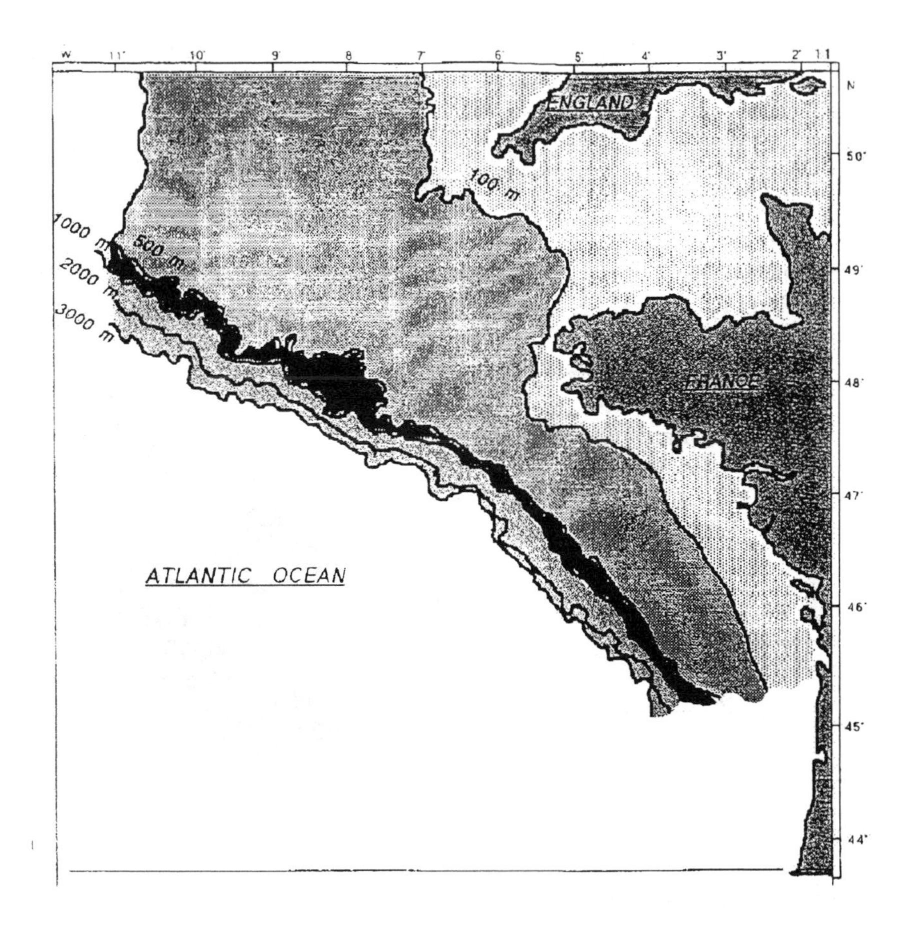

Fig. 3.5.5: Dimethylsulfide concentrations in surface sea water, and air-sea fluxes from Atlantic European Continental Margins.

	Depth	Average Concentration (nmol L^{-1})	Range	Average Fluxes (Nmol $m^{-2}d^{-1}$)	Range
	10–100	3.47	0.30–9.45	4.67	0.018–12.76
	100–500	5.26	2.39–10.09	17.20	0.12–33.60
	500–1000	13.11	6.57–21.03	18.28	7.80–31.21
	1500–3000	3.30	2.68–3.97	9.07	5.86–12.91

Latitudinal variation of DMS concentrations in sea water and ocean-atmosphere flux

In addition to the above Belgica results, data obtained aboard the R/V Poseidon and Darwin allow us to study the latitudinal variation of DMS concentrations in sea water and its flux from the North Atlantic Ocean.

Poseidon cruise.

DMS concentrations in sea water during the Poseidon late summer cruise are presented in Fig. 3.5.6. DMS concentrations vary from 0.71 to 11.81 nmol/L.

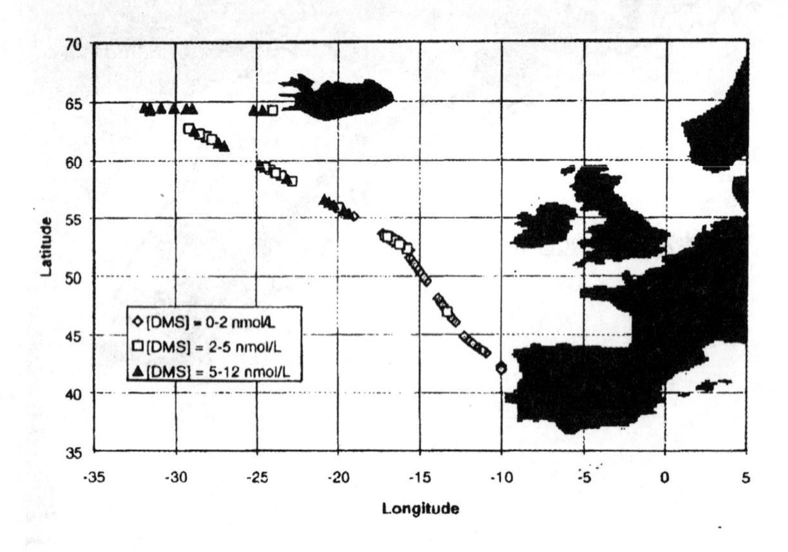

Fig. 3.5.6: Concentration of DMS, Poseidon cruise.

A very clear latitudinal variation of DMS concentrations in sea water can be observed with a maximum from 5 to 11 nmol/L between 65° N to 55° N and decrease to the south. The maximum DMS concentration of 11 nmol/L observed in the high latitude 55° N–65° N during this cruise are very close to those observed by Malin *et al.* [9].

According to the formulas of Liss and Merlivat [7], we obtained the estimates of DMS sea-air flux, plotted in Fig. 3.5.7, ranging from 0.5 to 86.3 μmol m^{-2} day^{-1}, with an average of 23.30 μmol m^{-2} day^{-1} in the 55° N–65° N band, 9.13 μmol m^{-2} day^{-1} in the 45° N–55° N band and 8.24 μmol m^{-2} day^{-1} in the 41° N–45° N band.

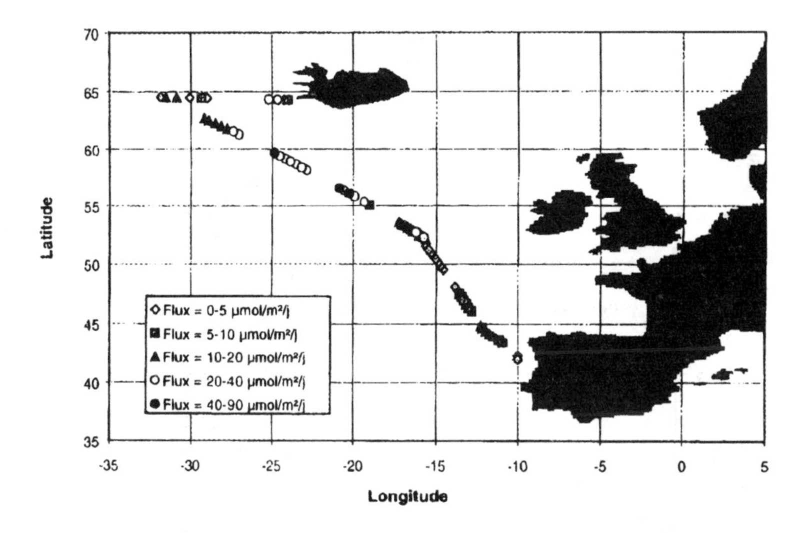

Fig. 3.5.7: Flux of DMS, Poseidon cruise.

Darwin cruise

The Charles Darwin cruise No. 83 departed from Las Palmas (27°45' N–18°02' W, Canaries) on 13th December 1993 and arrived at Barry in Wales on 14th January 1994 (Fig. 3.5.1b).

Sea surface concentrations of DMS during the Darwin winter cruise are presented in Fig. 3.5.8a in conjunction with meteorological parameters (Fig. 3.5.8b). Sampling was performed along two transects, one crossing the subtropical front (37° N–38° N) and the other steaming back to Barry (38° N–50° N). DMS concentrations varied from 0.2 to 2.5 nmol/L (1.3 ± 0.5). No significant difference was observed between continental margins and the open ocean. DMS concentrations were 1.1 and 1.3 nmol/L respectively. It is worth noting that only 6 samples were collected during the winter cruise at the continental margins and in coastal areas.

Dimethylsulfide ocean-atmosphere flux was estimated to range from < 0.1 to > 10 µmol m^{-2} d^{-1} (Fig. 3.5.8d). Over the whole period of the experiment, the mean DMS sea-air flux was 3.4 µmol m^{-2} d^{-1}.

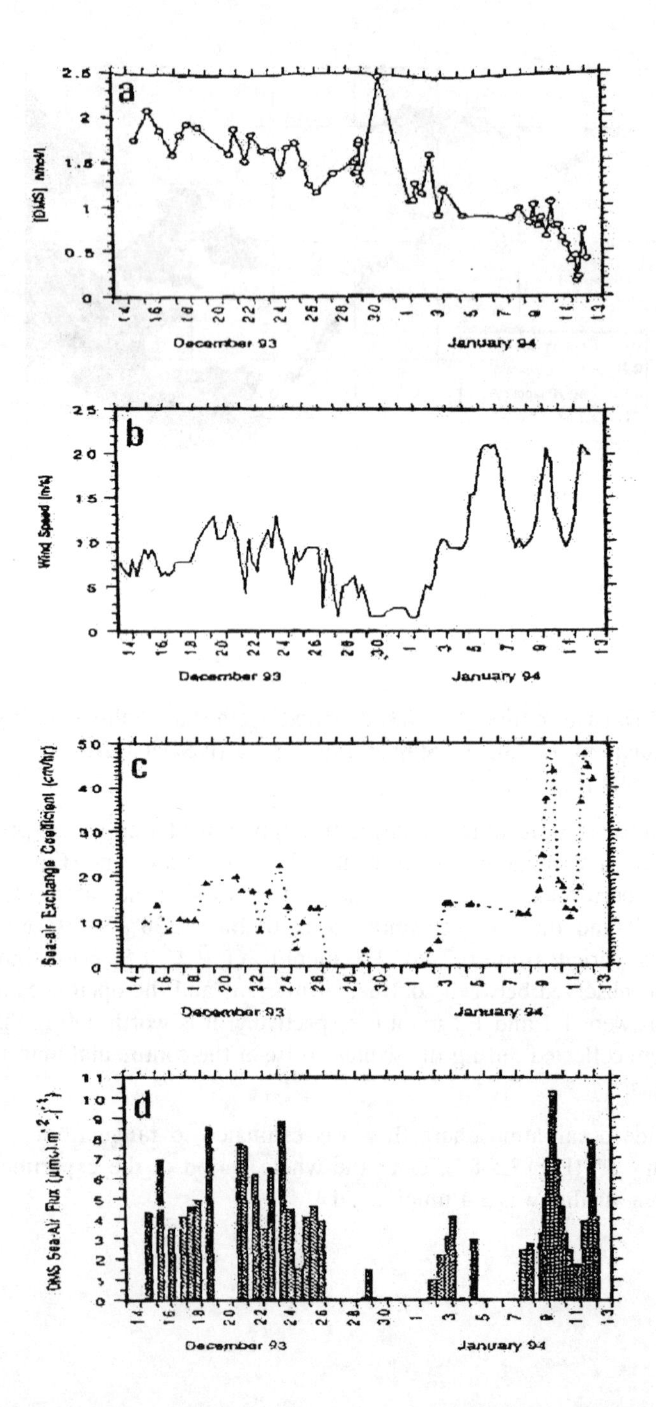

Fig. 3.5.8. Variation of DMS flux and concentration in surface sea water, wind speed and air-sea exchange coefficient during the Charles Darwin cruise.

Seasonal variation of DMS concentrations in sea water and ocean-atmosphere flux

Tables 3.5.1 and 3.5.2 show the seasonal variation of DMS concentrations in sea water and ocean-atmosphere flux respectively.. As well we can observe an important seasonal variation of DMS concentration and its flux in high latitudes (higher than 45 °N) and no significant seasonal variations in low latitudes (lower than 45 °N). In the band of latitude 45 °N to 55 °N we observed a variation of DMS concentration in sea water and its flux between spring, summer and winter from 5.5 nmol/L to 0.42 nmol/L and from 10.6 to 3.4 μmol $m^{-2}.d^{-1}$ respectively.

Table 3.5.1: Seasonal and Latitudinal variations of DMS concentration in seawater (nmol/L)

Latitude °N	Mean	Spring	Summer	Auumn	Winter
55–65	Mean range		6.10 (30) 3.6–11.2		
45–55	Mean range	5.50 (37) 0.3–21	1.77 (30) 1–3.18		0.42 (7) 0.17–0.73
35–45	Mean range		0.84 (75) 0.5–1.6		1.16 (32) 0.75–1.57
25–35	Mean range		1.22 (7) 0.77–1.67	3.1 (7) 2.25–3.95 (upwelling)	1.65 (20) 1.41–1.89

Table 3.5.2: Seasonal and latitudinal variations of DMS ocean-atmosphere flux (μmol m^{-2} d^{-1})

Latitude °N	Mean	Spring	Summer	Auumn	Winter
55–65	Mean range		23.30 (30) 0.4–62		
45–55	Mean range	10.64 (37) 0.3–32	9.12 (30) 0 1–35		3.4 (7) 1.1–5.6
35–45	Mean range		8.24 (10) 2.9–15.7		2.7 (32) 0.3–5.1
25–35	Mean range				4.5 (20) 2.4–6.6

Average DMS concentrations are 0.42 nmol/L (range 0.17–0.73) in winter, 5.5 nmol/L (range 0.3 to 21) in spring and 1.77 nmol/L (range 1–3.18) in summer.

A first observation which can be drawn from these data sets is that the spring values present a greater variability than the winter values. The great variability in chlorophyll *a* concentrations (more than an order of magnitude) deduced from fluorescence data during the Belgica cruise (Fig. 3.5.4), could in some extent explain the higher DMS concentrations and variability observed in spring, relative to winter.

Secondly, a clear seasonal signal was observed for DMS, with higher concentrations during spring than in winter time. These results can be, to some extent, explained by the variations of chlorophyll *a* simultaneously measured. Our results give seasonal spring/winter signal ratios up to 8.

Table 3.5.3 compares the DMS concentrations measured during five cruises with that observed in other marine regions. Our winter mean for the open sea of 1.3 nmol/L is higher to those measured by Bates *et al.* [11] and Nguyen *et al.* [5] in the Pacific and Indian Oceans, but one has to take into account that during these experiments we crossed regions with large sea temperature gradients (from 10 to 240 °C).

Table 3.5.3: Comparison between DMS concentrations measured during the 2 cruises with those reported for the other temperate marine regions.

Region and season	Conc (nmol/L)	Reference
North Atlantic Ocean		
spring	5.1	This work
winter	1.3	This work
North Pacific Ocean		
summer	2.2	[11]
winter	0.7	[11]
Indian Ocean	0.4	[5]
UK coastal and shelf waters		
summer	7.0	[4]
winter	0.1	[4]
Peru coastal waters		
summer	6.9	[12]
winter	1.3	[12]
Ligurian Sea (spring)	4.6	[13]

For the shelf and coastal areas our winter value is higher than that of Turner *et al.* [4] but in good agreement with that measured at the coastal areas of Peru [12]. Our spring values for open sea, shelf and coastal areas are also in agreement with those measured by Turner *et al.* [4] and Belviso *et al.* [13].

Table 3.5.4 compares our estimated fluxes with some calculated from other marine regions. Our winter DMS flux average for the open sea is higher than those measured by Bates *et al.* [11] and Nguyen *et al.* [5] in the Pacific and Indian Oceans, and falls within the lower limit of the mean value proposed by Andreae [14] for the world's temperate seas. In addition for the shelf areas, our winter value is higher than that of Turner *et al.* [15] which, to our knowledge, is the only calculation of winter time flux of DMS from European margins. It is not possible to compare our spring values since no values exist in the bibliography for that period, with the exception of Turner *et al.* [4] during the spring bloom at the shelf area around England. However, our mean spring values fall within the range proposed by Andreae [14] for the world's temperate coasts and shelf.

Table 3.5.4: Comparison between DMS fluxes measured during the 2 cruises with those reported for the other temperate marine regions.

Region and season	Flux (μmol.m^{-2} d^{-1})	Reference
North Atlantic Ocean		
spring	12.9	This work
winter	3.4	This work
North Pacific Ocean		
summer	5.0	[11]
winter	2.1	[11]
Indian Ocean		
summer	3.0	[5]
winter	1.3	[5]
U.K. coastal and shelf waters		
summer	29.2	[4]
winter	0.50	[4]
World's temperate seas	3.3-9.9	[14]
World's shelf and coastal	5.6-11.2	[14]

Conclusions

From the results obtained aboard the five R/V cruises, Belgica, Poseidon, Darwin, Le Suroit and Atalante, performed in the north-east Atlantic Ocean during different seasons both at the European continental margins and the open ocean, we can draw the following conclusions regarding the spatial and temporal variability of DMS concentrations:

* DMS and DMSP are produced principally in the euphotic layer (0 to 100 m)

* An important DMS flux at European Atlantic margins compared to open ocean, particularly in spring has been observed. DMS fluxes at shelf, shelf-edge and open ocean were 4.1, 13.1 and 3.3 μmol m^{-2} d^{-1} respectively with a factor from 3 to 4 between the shelf-edge and the shelf or open ocean.

* A clear seasonal signal of DMS concentration in sea water was observed, with higher a DMS concentration during spring than in winter time. Spring/winter ratios can be very high, up to 8. These results could, to some extent, be explained by the phytoplankton bloom particularly in spring.

* Clear latitudinal gradients of DMS concentration in sea water and its flux were observed with highest DMS concentrations in the high latitude between 55 °N–65 °N.

References

1. Dacey, J.W.H., Wakeham, S.G., Oceanic dimethylsulfide: production during zooplankton grazing on phytoplankton, *Science* **233** (1986) 1314–1316.
2. Nguyen, B.C., S. Belviso, N. Mihalopoulos, J. Gostan, R Nival, Dimethylsulflde production during natural phytoplanktonic blooms, *Marine Chem.* **24** (1988) 133–141.
3. Charlson R.J., J.E. Lovelock, M.O. Andreae, S.G. Warren, Oceanic phytoplankton, atmospheric sulphur, cloud albedo and climate, *Nature* **326** (1987) 655–661.
4. Turner S.M., G Malin., P. Liss, The seasonal variation of dimethylsulfide and dimethylsulfoniumpropionate concentrations in nearshore waters, *Limnol. and Oceanography* **33** (1988) 364–375.
5. Nguyen, B.C., N. Mihalopoulos, S. Belviso, Seasonal variation of atmospheric dimethyl sulflde at Amsterdam Island in the southern Indian Ocean, *J. Atmos. Chem.* **11** (1990) 123–143.
6. Nguyen, B.C., N. Mihalopoulos, J.P. Putaud, A. Gaudry, L.Gallet, W.C.Keene, J.N. Galloway, Covariation in oceanic dimethylsulfide, its oxidation products and rain acidity at Amsterdam Island in the southern Indian Ocean. *J. Atmos. Chem.* **15** (1992) 39–53.
7. Liss, P.S., L Merlivat, Sea-air gas exchange rates: Introduction and synthesis; in: P. Buat-Ménard (ed), *The Role of sea-air Exchange in Geochemical Cycling,* D. Reidel, Horwell, Mass. 1986, pp. 113–128.
8. Monfray, P., Echanges océan-atmosphère du gaz carbonique: variability avec l'état de la mer, Ph. D. Thesis, Université de Picardie, 1987.
9. Malin, G., S. Turner, P. Liss, P. Holligan, Dimethylsulphide and dimethylsulphoniopropionate in the Northeast Atlantic during the summer coccolithophore bloom, *Deep sea Research* **40** (1993) 1487–1508.
10. Edwards, E.S., P.H. Burkill, C.E. Stelfox, Microzooplankton abundance and biomass in the Goban Spur, Celtic Sea, *OMEX 2nd Ann. Rep.* 1995, D.83–D 89.
11. Bates, T.M., J.D. Cline, R.H. Gammon, S.R. Kelly-Hansen, Regional and seasonal variation of the flux of oceanic dimethylsulflde to the atmosphere, *J. Geophys. Res.* **92** (1987) 2930–2938.
12. Coopper W.J., Matrai P.A., Distribution of dimethylsulfide in the oceans; in: E.Saltzman, W.J. Cooper (eds), *Biogenic Sulfur in the Enviromnent*, Am. Chem. Soc. Washington, DC 1989, pp 140–151.
13. Belviso, S., P. Buat-Ménard, J.P. Putaud, B.C. Nguyen, H. Claustre, J. Neveux, Size distribution of dimethylsulfoniopropionate in areas of the tropical northeastern Atlantic ocean and the Meditenanean sea, *Marine Chem.* **44** (1993) 55–71.
14. Andreae M.O., Ocean-atmosphere interactions in the global biogeochemical sulfur cycle, *Marine Chem.* **30** (1990) 1–29.
15. Turner S.M., G Malin, P.Liss Dimethylsulphide and dimethylsulfoniumpropionate in European coastal and shelf waters; in: E.S. Saltzman, W.J. Cooper (eds), *Biogenic Sulfur in the Environment.* 1989, pp. 183–200.

3.6 The Biogeochemical Cycling of Carbonyl Sulfide and Dimethyl Sulfide at the European Continental Margin

G. Uher, V.S. Ulshöfer, O.R. Flöck, S. Rapsomanikis and M.O. Andreae

Max Planck Institute for Chemistry, Biogeochemistry Dept., P.O.Box 3060, 55020 Mainz, Germany

Carbonyl sulfide (COS) is formed photochemically from chromophoric dissolved organic matter (CDOM) in surface sea water. Its inertness against photo-oxidation allows COS to be transported into the stratosphere, where it is photodissociated and oxidised to sustain the background concentration of the stratospheric sulfate aerosol. This sulfuric acid aerosol is believed to affect the earth's radiation balance as well as heterogeneous reactions linked to ozone depletion.

Dimethyl sulfide DMS is formed mainly by the enzymatical cleavage of dimethylsulfonium propionate (DMSP) which is a metabolic product of marine phytoplankton. The taxonomic position of the phytoplankton populations, cell lysis during the senescence of blooms, zooplankton grazing, and bacterial production and consumption are believed to be the main factors controlling the DMS concentration in sea water. Air-sea exchange processes result in the emission of dissolved DMS into the atmosphere. Here it is oxidised into sulfate particles, which act as cloud condensation nuclei (CCN) and thus affect the reflectance of clouds and therefore the earth's radiation budget and climate. The Charlson-Lovelock-Andreae-Warren hypothesis proposes that phytoplankton vary their production of DMS in response to changes in surface temperature and sunlight and modulate the planet's climate in a feedback loop.

Activities within ASE

Both COS and DMS exhibit a pronounced seasonality in sea water. In addition, the sea surface concentrations are characterised by a high degree of patchiness in the case of DMS and by a pronounced diurnal cycle in case of COS (caused by a combination of photoproduction, hydrolysis, and downward mixing). An understanding of the processes controlling the sea surface concentrations of these trace gases requires a better spatial and temporal data coverage. The aims of this work have been:

1. to develop analytical systems for the semi-continuous measurements of atmospheric and dissolved COS and DMS;
2. to improve the seasonal and geographical data coverage;

3. to relate sulfur gas production to UV absorbance and fluorescence properties of CDOM in the case of COS, and to phytoplankton pigment concentration in the case of DMS; and

4. to investigate the cycling of COS and DMS in the marine environment at the European continental margin.

Our group participated in three oceanographic cruises in the northeast Atlantic:

1. R/V Meteor cruise M27/1: Hamburg, Germany, to La Coruña, Spain, Jan. 1994. (atmospheric and dissolved COS (n = 120), COS depth profiles and sunlight irradiations, ancillary parameters: chl a, UV absorbance and fluorescence properties of CDOM)

2. R/V Meteor cruise M30/1: Las Palmas, Gran Canaria, to Bremen, Germany, Sept. 1994. (dissolved DMS (n = 208); condensation nuclei (continuously); atmospheric and dissolved COS (n = 235); COS depth profiles and sunlight irradiations; ancillary parameters: chl a, UV absorbance and fluorescence properties of CDOM; black carbon and ^{222}Radon activity)

3. R/V Valdivia cruise V154: Cork; Ireland, to Hamburg (via Cascais, Portugal), July 1995. (dissolved DMS (n = 500); condensation nuclei (continuously); atmospheric and dissolved carbon dioxide (n = 250); ancillary parameters: atmospheric ^{222}Radon activity)

Principal results

Carbonyl sulfide

For the September 1994 cruise an automated sampling system was developed for alternating air and water measurements at 30 minute intervals. This system was tested successfully under field conditions and allowed semicontinuous day and night sampling with high time resolution.

Atmospheric and dissolved COS concentrations were measured on 335 samples during two cruises in the northeast Atlantic Ocean. Together with the results from our JGOFS cruise (aboard R/V Meteor, from Dublin, Ireland, to Funchal, Madeira, April/May 1992) these data covered three seasons. In January 1994, persistent undersaturation of COS in sea water with respect to the atmosphere was observed. This is the first data set to show a strong and persistent undersaturation with the mean saturation ration (SR) being 46 % and the standard deviation 13 %. In April 1992, the sea water was slightly supersaturated, with a SR of 126 ± 58 %. Only in September 1994, strong supersaturation of 214 ± 86% was observed. The measured air concentrations were relatively uniform, averaging 410 ± 67 ppt in January 1994, 466 ± 42 ppt in April/May 1992, and 396±18 ppt in September. Sea-to-air fluxes were estimated using three different exchange models. We obtained moderate to low COS emissions in September (19 to 33 nmol m^{-2} d^{-1}) and April/May (5 to 10 nmol m^{-2} d^{-1}), in contrast to a significant flux from atmosphere into the ocean in January (–76 to –31 nmol m^{-2} d^{-1}). The strong seasonal variation of COS emissions with the possibility of reversed fluxes into the ocean must be considered in future oceanic source estimates. The possible effect of an open ocean winter sink on global marine emissions of COS could be a reduction by some 10–15 %.

Depth profiles of dissolved COS were taken using non-contaminating, gas tight, GoFlo water samplers. In general, the vertical profiles showed maxima at the sea surface and an approximately exponential decay to a background level beneath the mixed layer. Although the COS concentration in deep water samples was very low, some transport or non-photochemical production mechanisms are required to maintain this background level and compensate losses due to hydrolysis.

We analyse our data using a simple kinetic box model, which includes the following parameters: a zeroth order COS photoproduction constant, surface UV light intensity, light independent COS production, hydrolysis removal, and air-sea exchange. This model satisfactorily describes time series from both sunlight irradiations and sea water measurements. First results indicate that the COS photoproduction is closely related to the occurrence of CDOM in surface sea water.

Dimethyl sulfide

During the September 1994 and July 1995 cruises, more than 700 surface water samples were analysed for DMS. We applied a newly developed automated GC-FPD system which allowed the semicontinuous sampling of dissolved DMS at 20 to 60 minutes intervals (Fig. 3.6.1). This system was tested successfully under field conditions. It provides a new tool for studying the link between air-sea exchange of DMS and the formation of sulfate aerosol particles at an improved time resolution.

In September 1994, dissolved DMS ranged from 1 nmol l^{-1} up to 12 nmol l^{-1} with an average of 2.8 nmol l^{-1} for the entire cruise track. The DMS concentration increased slowly from 1 nmol l^{-1} north of the Canary Islands up to a maximum of 4 nmol l^{-1} in the open ocean near the Celtic Sea margin. During the transect across the shelf edge, dissolved DMS varied from 1.5 nmol l^{-1} to 4 nmol l^{-1} but no concentration gradient could be identified. Highest DMS concentration (up to 33 nmol l^{-1}) were observed while passing the Ushant front which separates the stratified North Atlantic water from the well mixed English Channel water. Dissolved DMS and chlorophyll a were well correlated in the open ocean and across the Ushant front. However, no such correlation could be observed across the continental margin and on the Celtic Sea shelf, indicating that the phytoplankton populations in these regions were dominated by minor DMS/DMSP producers.

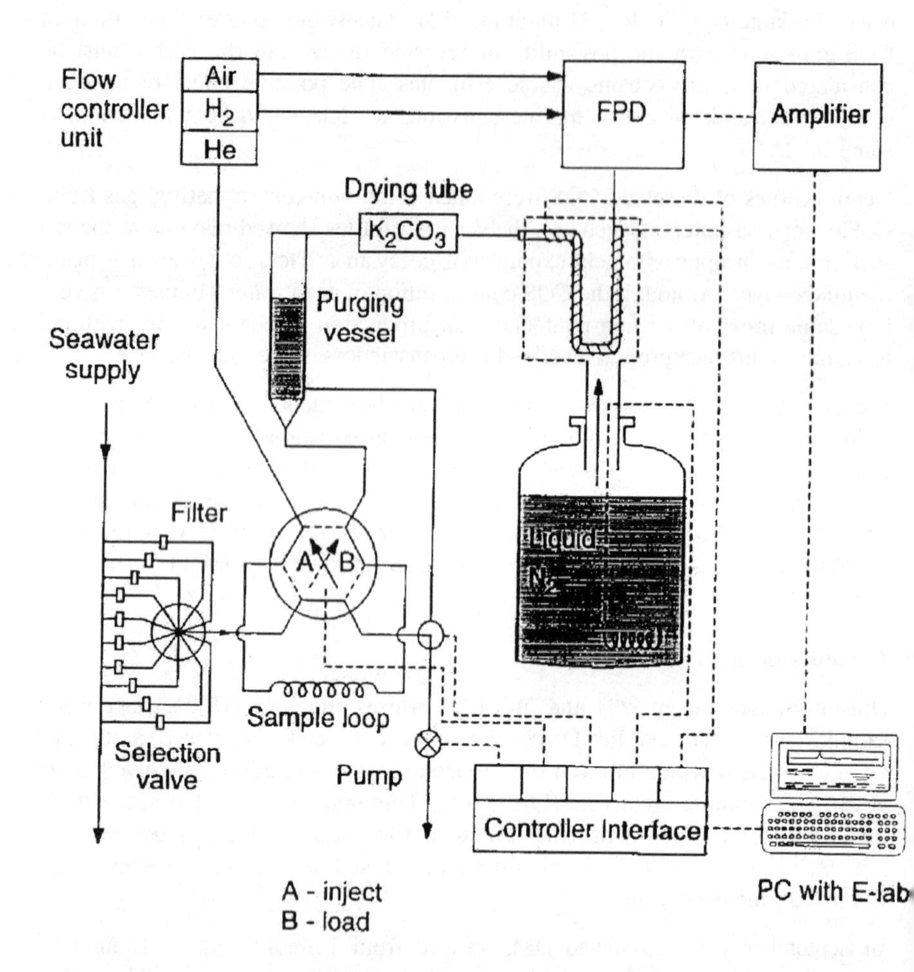

Fig. 3.6.1: Analytical system for automated semi-continuous measurements of dissolved DMS.

The sea-to air flux of DMS was estimated using two different air-sea exchange models. Preliminary results indicate, that condensation nuclei number density and DMS sea-to air flux were significantly correlated in air masses free from continental influence (Fig. 3.6.2). These results, however, must be viewed with precaution since several assumptions have to be made when comparing such DMS flux estimates to simultaneous measurements of the atmospheric aerosol number concentration.

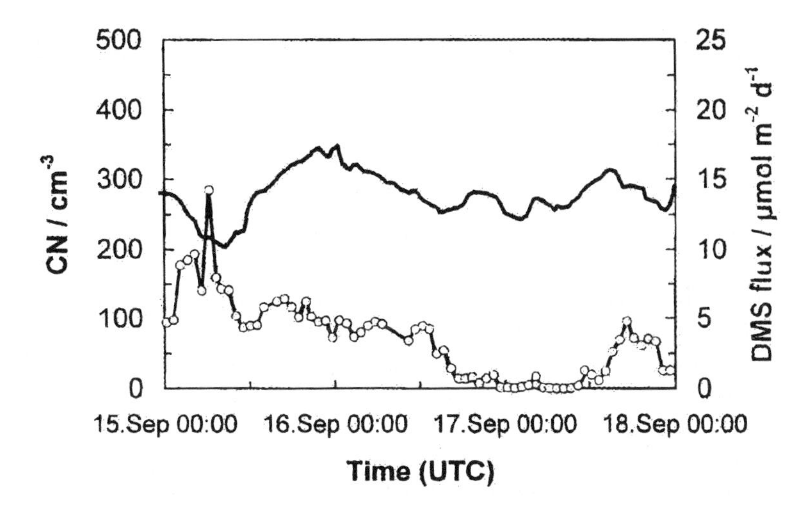

Fig. 3.6.2: Condensation nuclei (CN) number density and DMS sea-to air flux during the R/V Meteor cruise from the Canary Islands to Bremen, Germany, September 1994. The "clean air" period presented here was characterised by apparent black carbon concentrations equal or lower than 20 ng m^{-3}. The DMS flux was estimated using the air-sea exchange model of Wanninkhof.

Two of these assumptions are:

1. Sea surface DMS concentration and hence DMS flux are representative for the oceanic region in contact with the analysed air masses.
2. The air masses analysed are homogeneous with respect to there chemical composition.

Data processing is still in progress and new results will be presented later.

3.7 Air-Sea Exchanges of Non-Methane Hydrocarbons

B. Bonsang

Centre des Faibles Radioactivités, Laboratoire mixte CNRS/CEA,
91198 Gif-sur-Yvette, France

Short lived NMHC have been detected in the remote atmosphere, and their origin was ascribed to local emissions from the ocean surface [1–4]. Superficial sea water in the ocean appears supersaturated in light non methane hydrocarbons by two or three orders of magnitude and acts as a net source for the atmosphere. Despite the great variability of one order of magnitude in hydrocarbons concentrations, its relative composition remains fairly constant and is characterised by a high proportion of alkenes, mainly ethene and propene. In order to determine the origin of these light hydrocarbons, systematic vertical profiles were undertaken in the Mediterranean in order to follow the seasonal variability of light unsaturated and mono-unsaturated hydrocarbons in connection with biological parameters such as chlorophyll and photosynthetic pigments. Also vertical profiles were performed in the Pacific Ocean near hydrothermal sources in order to investigate the possible origin of acetylene.

The vertical distribution of NMHC with depth shows the existence of a concentration maximum in the euphotic layer for most of the light C_2–C_6 NMHC. Three different kinds of species present typical distributions: acetylene concentration is constant with depth, and presents a low seasonal variability with a constant mixing ratio. Light C_2–C_5 alkenes and alkanes present a smooth maximum of concentration at the thermocline level. Isoprene has been identified for the first time in sea water, it is supersaturated at the surface; its distribution is typical and similar to that of fluorescence, with a clear maximum at 40 m depth.

Global estimates of the marine production of non methane hydrocarbons leads to a figure of 35 Tg C yr^{-1} with a very significant contribution of alkenes (70 %)

Aims of this research

Non methane hydrocarbons (NMHC) and particularly alkenes oxidised either by ozone, OH and NO_3 radicals can have a significant impact on the photochemistry of the remote background atmosphere [5] and are able in turn to influence the budget of these oxidants. Strong evidence of a residual tropospheric background in hydrocarbons has been reported by several authors working in remote areas [2, 3, 6]. Particularly alkenes, despite their high reactivity, are found at significant levels from 10 ppt to 100 ppt in the southern hemisphere an even in the Antarctic continent far from usual anthropogenic sources [7]. The existence of a marine production by outgasing of superficial sea water was first reported by Lamontagne *et al.* [1] working in the Pacific Ocean, and the origin of light hydrocarbons was ascribed to local emissions from the ocean surface by photochemical degradation of dissolved organic carbon [8].

In order to understand the role of the ocean in the budget of NMHC and eventually in the tropospheric photochemistry, the fluxes of NMHC by superficial sea water have been evaluated by different authors, [2, 4] and discussed by Donahue and Prinn [5]. A large discrepancy, mainly due to the different method of estimation used, was pointed out with a significant consequence on the uncertainty of the impact of oceanic NMHC emissions on the tropospheric background chemistry. A major aspect is the variability of this oceanic source which appears to be an important factor, and must be parametrised by knowing the production mechanisms of NMHC in sea water.

In the context of the EUROTRAC ASE subproject, coordinated research was undertaken in order to study the fluxes, exchange constants and production mechanisms of reactive species at the ocean/atmosphere interface. Particularly, the geographical variability of this oceanic source was investigated and appears to be an important factor. Vertical profiles of the dissolved NMHC content in sea water were undertaken in order to understand the production in the sea water column. The two experimental sites were the Mediterranean for the hydrocarbon seasonal variability in connection with biological parameters (Dyfamed/JGOFS programme), and the Gulf of California for the investigation of a possible hydrothermal origin of acetylene (Guaynaut cruise in the Pacific Ocean, in collaboration with Ifremer). Global budgets in the water column and the atmosphere are also compared in order to validate the different methods of flux determination at the air-sea interface.

Principal experimental results

Methodology

In order to study the sea water content in hydrocarbons an extraction procedure methodology has been developed. Concentrations of dissolved NMHC in superficial sea water are determined from stripping with ultrapure helium prepared in the laboratory by removing hydrocarbons on charcoal/tenax adsorbent traps. Sea water samples are collected by mean of oceanographic bottles or stainless steel devices checked against contamination, or continuously pumped through stainless steel lines extending beyond the hull of the ship. Sea water samples are transferred into a glass chamber where volatile NMHC are stripped from 1 to 1.5 L of water with a flow rate of 100 mL/mn and collected in 6 L stainless steel canisters for further analysis. Theoretical calculations or experimental determinations of the efficiency of this procedure agree for a quantitative extraction of the light NMHC. Concentrations in sea water are expressed usually in pico moles/litre or nL of gas per litre of water (nL L^{-1}) under standard conditions. A typical chromatogram is shown in Fig. 3.7.1.

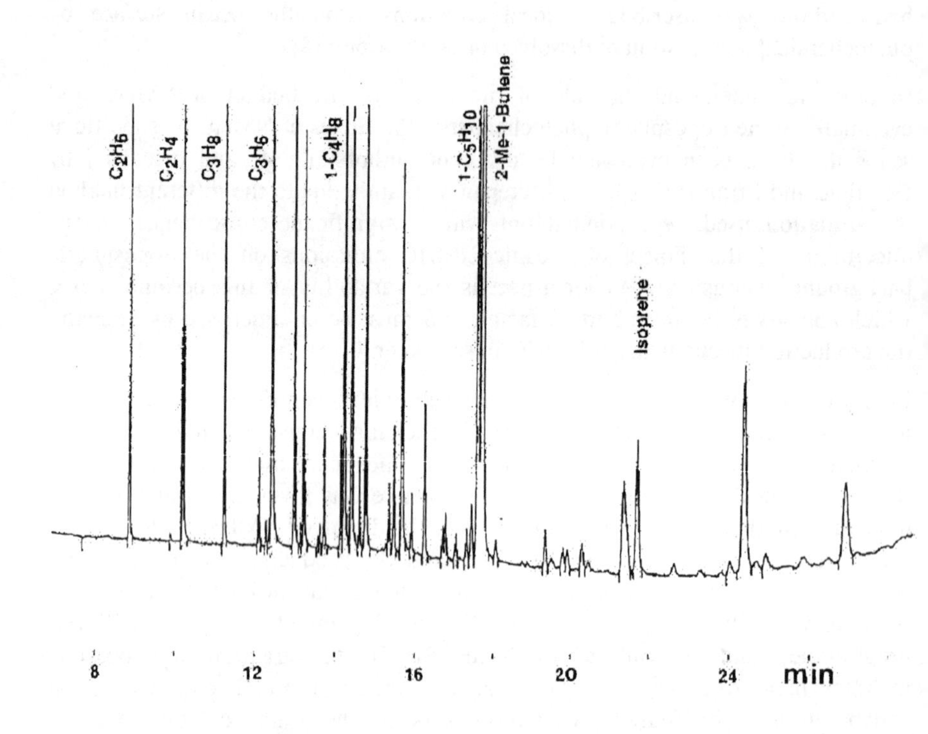

Fig. 3.7.1: Chromatogram of dissolved hydrocarbons in sea water.

Geographical variability of dissolved NMHC in sea water

During the past years, measurements of dissolved NMHC in superficial sea water were performed by different authors: in the Pacific Ocean by Lamontagne *et al.* [1] and Donahue and Prinn [9]; in the Indian Ocean and in the Pacific near Hawaii and Polynesia by Bonsang *et al.* [4, 10]; and in the Atlantic Ocean by Plass *et al.* [11]. All these authors agree that the dominant contribution is of dissolved alkenes (mainly ethene and propene) in superficial sea water, also with a significant contribution of C_2–C_3 alkanes. For comparison, averaged concentrations obtained for the different oceanic areas that we have investigated are presented in Fig. 3.7.2. It shows the geographical variability for typical species; a variability of about one order of magnitude is observed for the hydrocarbons, depending on the area. It can be observed that ethene concentration range from 2–5.3 nL L^{-1} in the Pacific Ocean to 36 n L L^{-1} in the south Indian Ocean, propene varies from 0.6–1 nL L^{-1} for the Pacific Ocean to 15 nL L^{-1} in the south Indian Ocean. Very high values are usually obtained in the continental shelf as observed by Bonsang *et al.* [12] close to the atoll of Polynesia. In the Atlantic Ocean an extended set of values reported by Plass *et al* [11] range from 1.5 to 3.5 nL L^{-1} for ethene, and 0.74 to 1.75 nL L¯1 for propene. Significant latitudinal gradients are also observed. A difference, by a factor 4 to 5, was observed for the hydrocarbon content in sea water between the north and south Indian Ocean [4]. Identically, Plass *et al.* [11] have shown the occurrence of the very clear and regular north to south gradient of dissolved NMHC for the Atlantic Ocean.

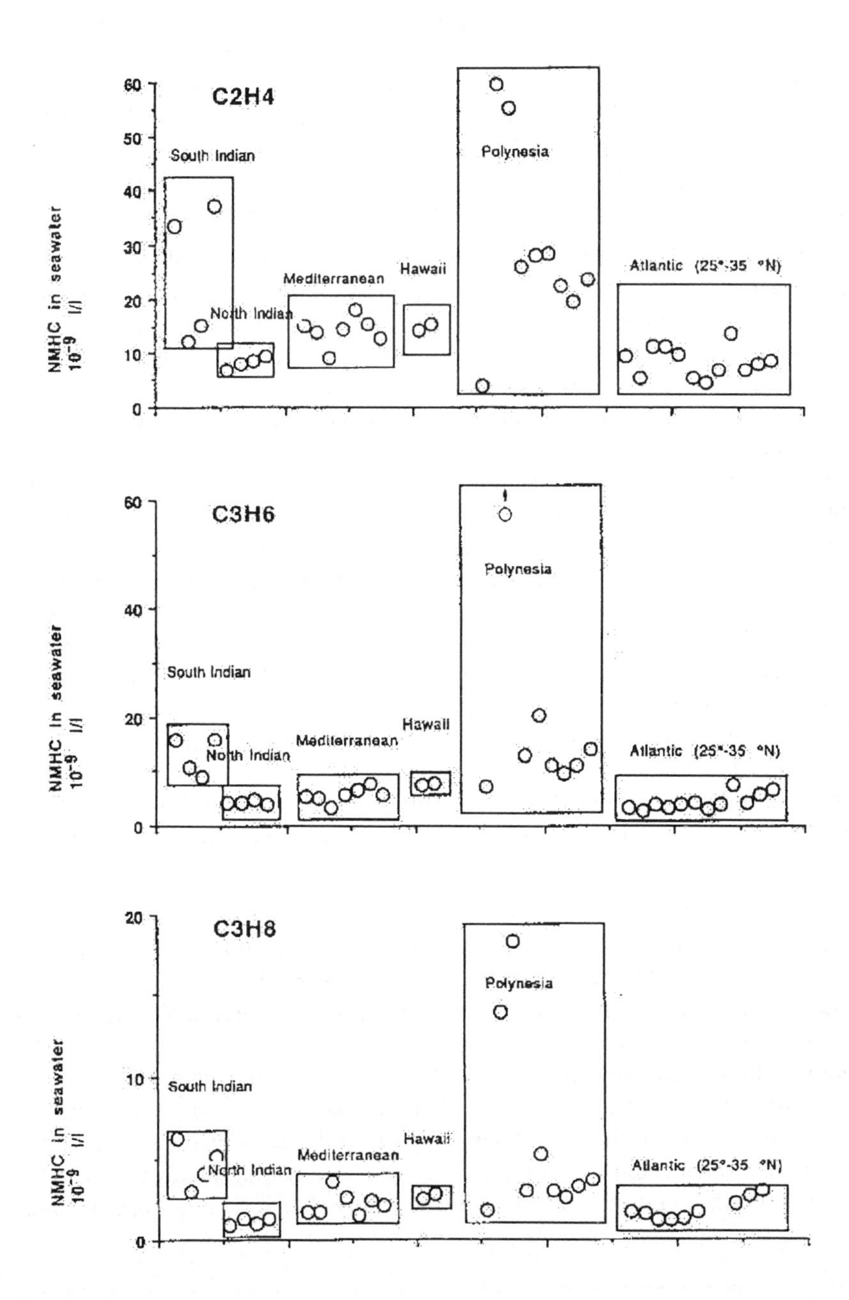

Fig. 3.7.2: Geographical variability of some selected hydrocarbons in surface sea water, (units: nL L^{-1})

Besides light C_2–C_6 hydrocarbons, other species have been detected in surface sea water, including alkanes and alkenes up to C_6 [4]. Acetylene was also detected in superficial sea water at concentration of the order of 0.5–1 nL L^{-1} [13] and also observed by Plass et al. [11] in the Atlantic Ocean at concentrations ranging from 0.08 to 0.1 nL L^{-1}. One main finding of our research is the existence of dienes and particularly isoprene (2 methyl 1,3 butadiene) in superficial sea water of the Mediterranean and the Pacific Ocean at concentrations of the order of 0.2 nL L^{-1} in surface waters [14].

Seasonal variation of hydrocarbons distribution in sea water

We have shown the occurrence of an hydrocarbon maximum in the euphotic layer near the thermocline. Systematic vertical profiles were undertaken in order to follow the seasonal variability of hydrocarbons in sea water, particularly the distribution in winter and spring. The seasonal variability is shown in Fig. 3.7.3 for January, February, May and October and for three typical species, ethene, propene and acetylene. For ethene and propene, which are the two major hydrocarbons present in sea water, a minimum is observed in winter. Concentrations in the water column in spring and in autumn are significantly greater by a factor two to five. This observation confirms the hypothesis of a production mechanism linked to the photodegradation of dissolved organic mater [8, 15]. The relationship between hydrocarbons and biological parameters such as photosynthetic pigments have also been studied. However no clear relationship was established, only isoprene presents a vertical distribution linked to fluorescence and possibly chlorophyll *b*. Looking at the seasonal variability in depth below the euphotic layer, it is observed that concentrations do not show any clear seasonal variability. One hypothesis could be the existence of a different production mechanism in depth. Concerning acetylene (Fig. 3.7.3) a constant vertical profile is observed with no seasonal variability. From this observation it seems that the production of acetylene could result from a mechanism independent of biological processes.

In order to investigate this possibility, the composition of dissolved hydrocarbons were measured in the Gulf of California near hydrothermal sources down to a depth of 2600 m. Dilution of hydrothermal waters in the ocean were measured through their content in magnesium. Correlations were established between hydrocarbons and magnesium. Two different sites were investigated. At one site a relatively high acetylene content of 31 nL L^{-1} was measured, compared to the 7.8 nL L^{-1} in surface and the 1 nL L^{-1} level in the open ocean or the Mediterranean. The hydrothermal origin of acetylene appears to be consequently a reasonable hypothesis, which could explain, providing a relatively long lifetime of acetylene in sea water, the constant vertical profile generally observed in the water column.

On the other hand, isoprene is clearly produced in the euphotic zone, the vertical profiles display a clear maximum near the thermocline level, with concentrations of the order of 1.5 to 3 nL L^{-1} compared to 0.2 nL L^{-1} at the surface.

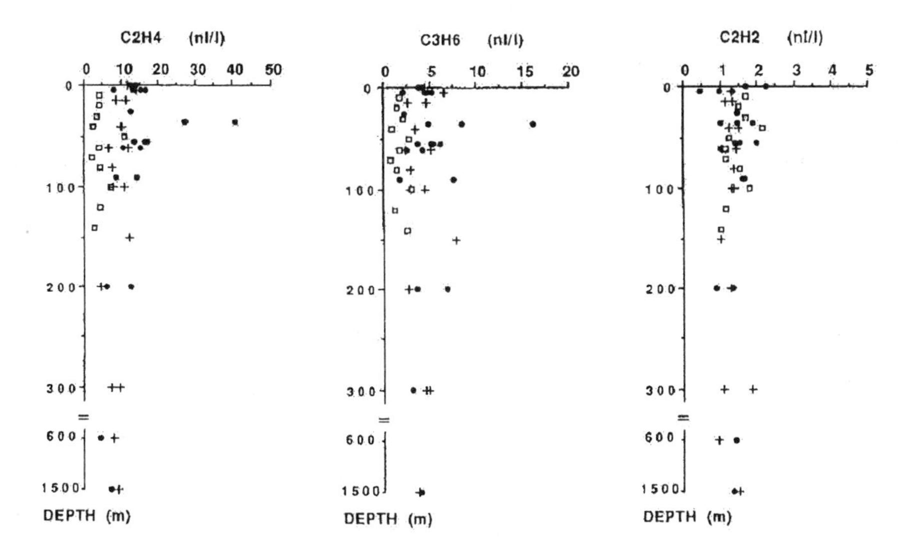

Fig. 3.7.3: Seasonal variability of hydrocarbons in the sea water column (Mediterranean 43°24' N–07°52' E) Squares: January Crosses: May Circles: October

The isoprene maximum appears below the thermocline and slightly above the fluorescence peak located near 60 m depth. (Fig. 3.7.4). A direct link between isoprene and biological parameters is not obvious, however some profiles reveal a significant correlation between isoprene and chlorophyll *b* (Fig. 3.7.5) which ascertain the hypothesis of an origin by living organisms. On the other hand the variability of isoprene concentration does not present any clear diurnal trend which could establish a link with a possible photodegradation of organic carbon. Further investigation in order to understand its production mechanisms. Moore *et al.* [16] have shown that isoprene could be directly produced by diatom cultures and the biological origin of this compound in sea water is clearly established.

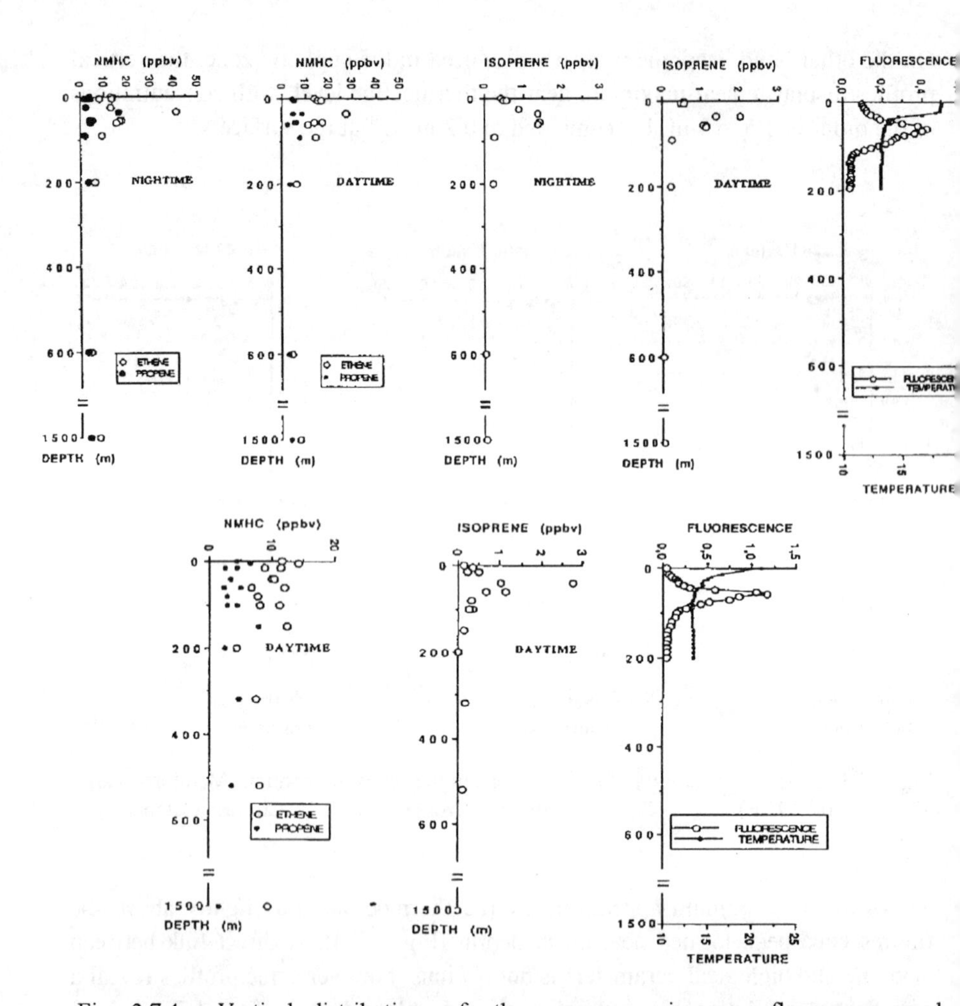

Fig. 3.7.4: Vertical distributions of ethene, propene, isoprene, fluorescence and temperature in Mediterranean sea water. (top) for October and (bottom) for May.

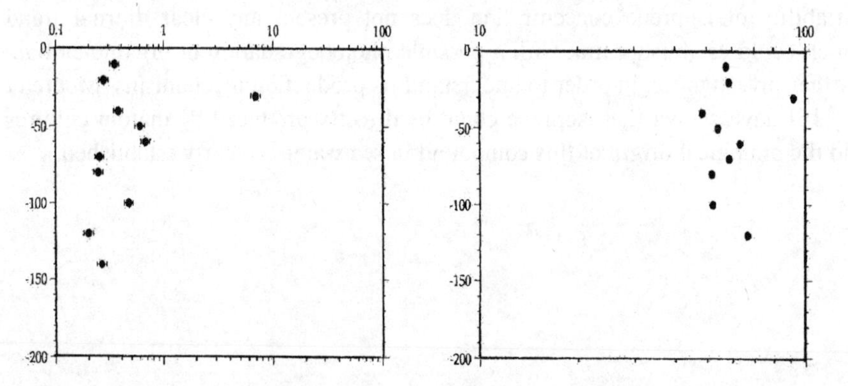

Fig. 3.7.5: Isoprene (nL L^{-1}; left) and chlorophyll *b* (µg L^{-1}; right) versus depth (m) in the euphotic layer (Mediterranean 43°24' N–07°52' E)·

Composition of dissolved NMHC in sea water

The existence of a typical composition of dissolved NMHC in sea water was first pointed out by Bonsang *et al.* [4, 12]. The histogram reported in Fig. 3.7.6 represents a typical sea water composition that we have measured in the Atlantic Ocean (28° to 32° N; 20° to 40° W). As this composition seems to be a characteristic of superficial sea water, an averaged composition normalised to propane has been established for the Atlantic, Pacific and Indian Oceans, it shows a dominant contribution of alkenes whose concentrations exceed those of their saturated homologues. For each group of alkenes or alkanes, the concentration decreases with increasing carbon number. One major fact is a very good agreement between several authors for surface sea water composition in C_2–C_3 alkenes. For the different areas investigated and previously presented in Fig. 3.7.1, the ethene/propene ratio appears to be constant with an average value of 2.2 ± 0.6 (Fig. 3.7.7a). It can be compared to the values reported by Plass *et al.* [11] who reported an ethene/propene ratio of 2.1 ± 0.4 for the Atlantic Ocean waters, and Donahue and Prinn [9] gave 2.2 ± 0.5 for the Pacific Ocean, whereas Lamontagne *et al.* [1] reported a slightly higher figure of 3.6 ±1.5. A relatively good agreement is also obtained for the ethane/propane ratio (Fig. 3.7.7b) which seems however to be more variable: for the Atlantic ocean, Plass *et al.* [11] establish the existence of a quasi constant figure of 3.2 greater than the average of 2.1 deduced from our data which includes different oceanic areas.

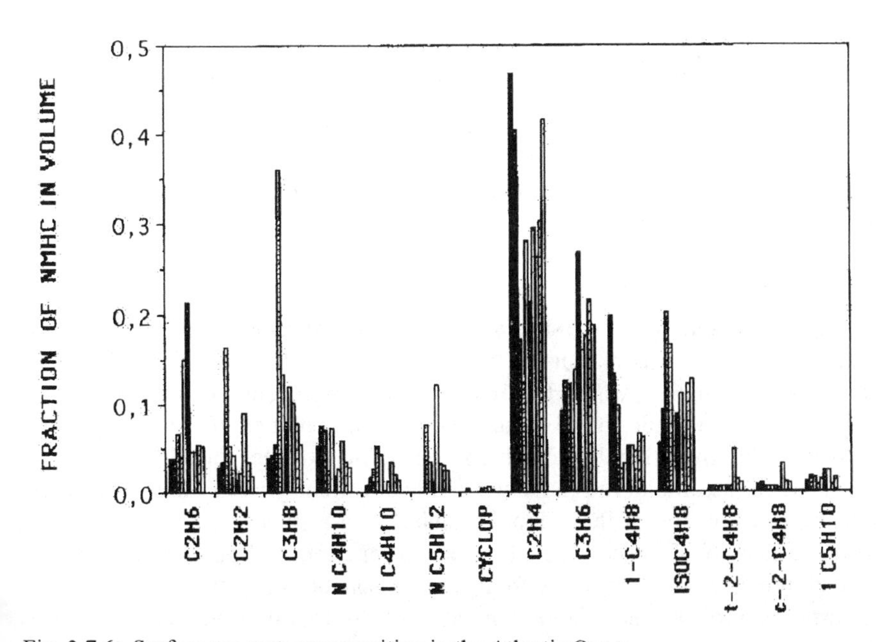

Fig. 3.7.6: Surface sea water composition in the Atlantic Ocean.

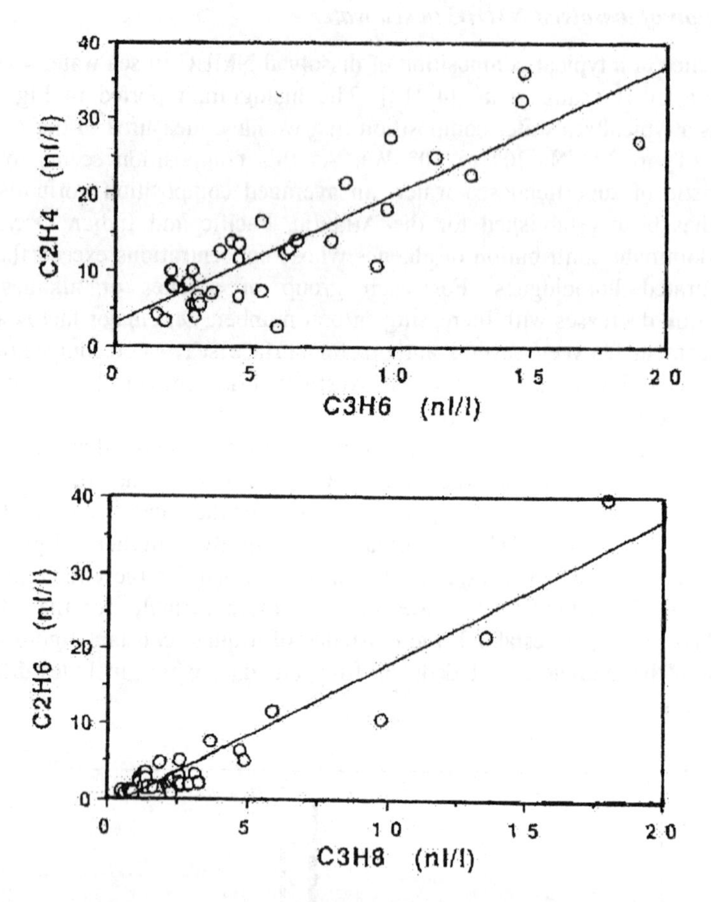

Fig. 3.7.7: Ethene to propene relationship and ethane propane relationship for dissolved hydrocarbons in surface sea water.

Global budget for NMHC marine production

Taking into account the large supersaturation of the different NMHC with respect to the atmosphere, it is reasonable to assume that the flux from the ocean to the atmosphere is simply given by the product of the piston velocity times and the sea water concentration. From the estimation of the flux of a single species, a global estimation for the other hydrocarbons can be undertaken considering that the fluxes are in the ratio of the abundances of hydrocarbons in sea water. This piston velocity is a function of the Schmidt number, and a small correction depending on the diffusivity of the considered species in water must be applied. The oceanic flux of propane was estimated from comparison of the global distribution of atmospheric propane and ^{222}Radon mainly emitted from the continents and partly from the ocean [3, 12]. Using an experimental linear Log-Log relationship between propane and ^{222}Radon, and knowing the global continental flux of ^{222}Radon, the global continental flux of propane is estimated at 20 TgC yr^{-1}. From this value a global oceanic flux of propane of 1.9 TgC yr^{-1} is obtained.

Taking into account the average surface sea water composition, the fluxes of other species are deduced and presented in Fig. 3.7.8; global estimates of oceanic fluxes are of the order of 7.5 and 8.3 Tg C yr^{-1} for propene and ethene respectively, 3 to 5 Tg yr^{-1} for 1-pentene 1-butene and *iso*-butene, and lie in the range of 1 to 2 Tg C yr^{-1} for C_2 to C_6 alkanes

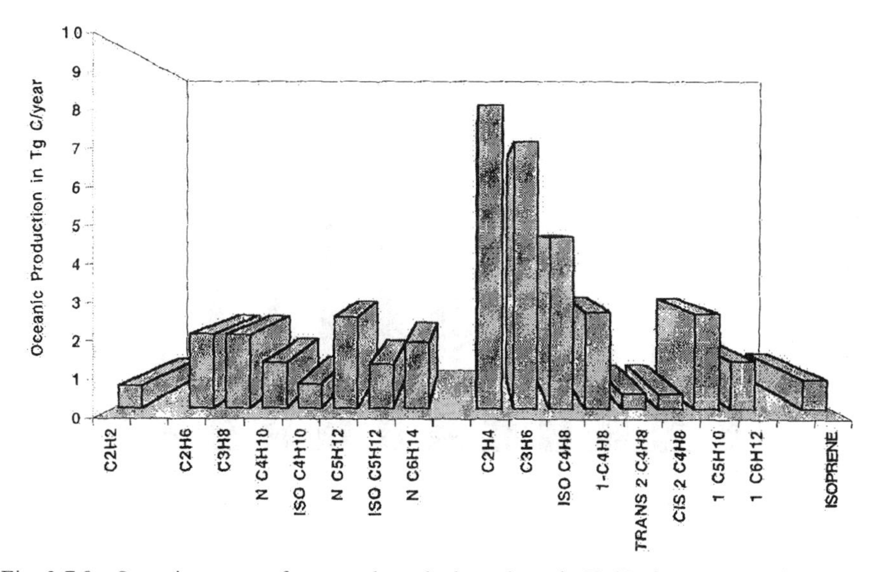

Fig. 3.7.8: Oceanic source of non methane hydrocarbons in Tg/Carbon per year based on an oceanic flux of propane of 2 Tg C/year, and an averaged surface sea water composition.

The global budget of the oceanic NMHC sources were estimated recently using different approaches: an estimation by Plass *et al.* [11] based on their intensive set of data in the Atlantic Ocean leads to an oceanic source of 1.0 Tg C_2H_6 yr^{-1}, 0.4 Tg C_3H_8 yr^{-1}, 1.7 Tg C_2H_4 yr^{-1} and 1.3 Tg C_3H_6 yr^{-1}, which is significantly lower than our estimates, particularly for alkenes. Plass-Dülmer *et al.* [17] have estimated global marine NMHC productions based on their own data and a study of the available data sets of surface sea water hydrocarbons. They estimated a global production of 2.1 Tg C_2–C_4 NMHC yr^{-1} scaled to their estimate of ethene production which represents 40 % of the total. A different approach based on the computation of the atmospheric removal in an atmospheric column was used by Donahue and Prinn [5] and, as pointed out by these authors, this determination leads to high values, which can be due either to an overestimation of column removal rates, or advection from continental or remote sources

A validation of our oceanic flux of ethane and propane was made by using the monitoring of this species in the oceanic atmosphere (Amsterdam Island) compared to the modelling of their atmospheric concentrations with the three dimensional climatological model, Moguntia [18]. In Fig. 3.7.9 is presented the ethane computed seasonal variability based on its sources, distribution and emissions and the corresponding measurements obtained on two years monitoring (1986 and 1994) A good agreement is observed between experimental results and modelling for an oceanic source of 1.9 Tg C_2H_6 yr^{-1} (1.5 Tg C–C_2H_6 yr^{-1}) which validates the order of magnitude of our estimate. Validation of other NMHC marine sources particularly alkenes are currently in progress.

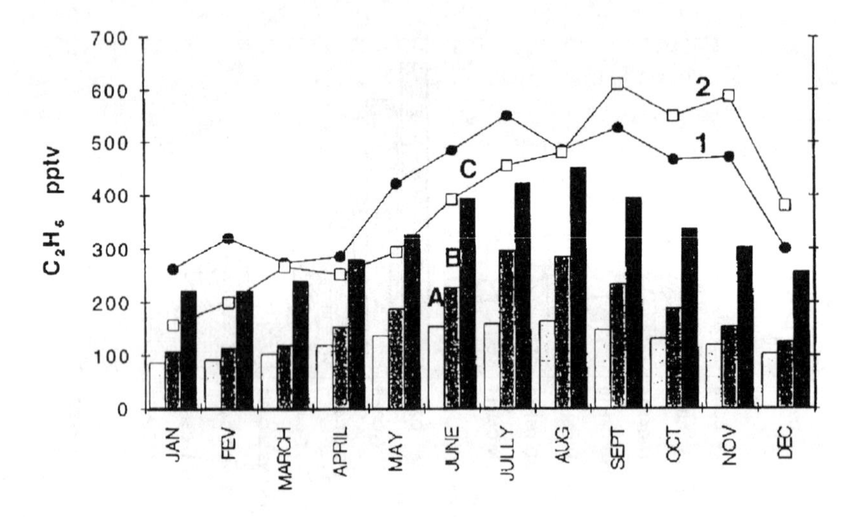

Fig. 3.7.9: Comparison of computed and measured seasonal variability of atmospheric ethane at Amsterdam Island (37°47' S–77°31' E).

Histograms: Modelling results:

A: Without biomass burning and without oceanic source

B: With biomass burning and without oceanic source

C: Basic run with biomass burning and an oceanic source of 1.9 Tg C_2H_6/yr.

Curves Measurements: 1: year 1986, 2: year 1994.

Main conclusions

In surface sea water and in depth, the main species found are ethene and propene. A typical composition, particularly the ratios ethene/propene and ethane/propane ratios remain fairly constant in the euphotic zone

The superficial sea water is supersaturated in light non-methane hydrocarbons with respect to the atmosphere by two or three orders of magnitude and acts as a net source for the atmosphere. NMHC concentrations in surface sea water of the order of nL of gas per litre of water present a great geographical variability of more than one order of magnitude, however, its seems that it is possible to define a typical composition of this marine source, characterised by a high proportion of alkenes (mainly ethene and propene). The origin of hydrocarbons in sea water is not clearly established, but seems to be related, at least for alkanes and alkenes, to the photodegradation of dissolved organic carbon. Isoprene, which displays a very clear and reproducible vertical distribution with a maximum in depth linked to fluorescence appears as a product of the metabolism of living organisms; acetylene, in contrast, is relatively constant in depth and season and a non-biological origin is highly probable. This hypothesis is confirmed by the vertical distributions of NMHC with depth and their seasonal variability. On a global scale, the oceanic source of hydrocarbons could reach a figure of 35 Tg carbon per year with a contribution of reactive alkenes reaching 70 %.

Acknowledgements

This work was supported by the "Centre National de la Recherche Scientifique", "Ministère de la Recherche", and French committees EUROTRAC and PACB. We thank the Station Zoologique de Villefranche-sur-mer, and IFREMER Brest for helpful collaboration and seawater samplings.

References

1. Lamontagne R.A., J.W. Swinnerton, W.J. Linnenbom, Hydrocarbons in the North and South pacific, *Tellus* **26** (1973) 71–77.
2. Rudolph J., D.H. Ehhalt, Measurements of C_2–C_5 hydrocarbons over the North Atlantic, *J. Geophys. Res.* **86** (1981) 11959–11964.
3. Bonsang B., G. Lambert, Nonmethane hydrocarbons in an oceanic atmosphere, *J. Atmos.Chem.* **2** (1985) 257–271.
4. Bonsang B., M. Kanakidou, G. Lambert, P. Monfray, The marine source of C_2–C_6 aliphatic hydrocarbons. *J. Atmos.Chem.* **6** (1988) 3–20.
5. Donahue N.M., R.G. Prinn, Non methane hydrocarbons chemistry in the remote marine bounsdary layer, *J. Geophys. Res.* **95** (1990) 18387–18411.
6. Greenberg J.P., P.R. Zimmerman, Non methane hydrocarbons in remote troposphere, *J. Geophys. Res.* **89** (1984) 4767–4778.
7. Rudolph J., A. Khedim, D. Wegenbach, The seasonal variations of light nonmethane hydocarbons in the antarctic troposphere, *J. Geophys. Res.* **94** (1989) 13039–13044.
8. Wilson D.F., J.W. Swinnerton, R.A. Lamontagne, Production of carbon monoxyde and gaseous hydrocarbons in seawater: relation to dissolved organic carbon. *Science* **168** (1970) 1577–1579.
9. Donahue N.M., R.G. Prinn, *In-situ* nonmethane hydrocarbon measurements on SAGA 3, *J. Geophys. Res.* **98** (1993) 16915–16932.
10. Bonsang B., M. Kanakidou, G. Lambert, NMHC in marine atmosphere: preliminary results of monitoring at Amsterdam Island., *J. Atmos. Chem.* **11** (1990) 169–178.
11. Plass DülmerC., R. Koppmann, J. Rudolph, Light hydrocarbons in the surface water of the Mid-Atlantic, *J. Atmos. Chem.* **15** (1992) 235–231.
12. Bonsang B., D. Martin, G. lambert, M. Kanakidou, J.C. Le Roulley, G. Sennequier, Vertical distribution of non methane hydrocarbons in the remote marine boundary layer. *J. Geophys. Res.* **96** (1991) 7313–7324.
13. Kanakidou M., B. Bonsang , J.C. Le Roulley, G. Lambert, D. Martin, G. Sennequier, Marine source of atmospheric acetylene, *Nature* **333** (1988) 51–52.
14. Bonsang B., C. Polle, G. lambert, Evidence for marine production of isoprene, *Geophys. Res. Lett.* **11** (1992) 1129–1132.

15. Ratte M., Plass DülmerC., Koppmann R., Rudolph J., Denga J. Production mechanism of C_2–C_4 hydrocarbons in sea water: field measurements and experiments *Global Biogeochemical Cycles* **7** (1993) 369–378.
16. Moore R.M., Oram D.E., Penkett S.A., Production of isoprene by marine phytoplankton cultures. *Gephys. Res. Lett.* **21** (1994) 2507–2510.
17. Plass Dülmer C., Koppmann R., Rudolph J., Light non methane hydrocarbons in seawater, *Global Biogeochemical Cycles* **9** (1995) 79–100.
18. Kanakidou M., Crutzen P.J., Zimmermann P.H., Bonsang B., A 3 dimensional global study of the photochemistry of ethane and propane in the troposphere: production and transport of organic nitrogen compounds. in: H. Van Dop, G. Kallos (eds), *Air Pollution Modeling and its Applications* IX, Plenum Press, New York 1992, pp. 415–426.

3.8 The Oceanic Source of Light Non-methane Hydrocarbons

Jochen Rudolph, M. Ratte, Ch. Plass-Dülmer and R. Koppmann

Institut für Atmosphärische Chemie, Forschungszentrum Jülich, 52425 Jülich, Germany

The oceanic concentrations of several light non-methane hydrocarbons (NMHC) were studied during several ship cruises. Simultaneous measurements of the atmospheric mixing ratios of these compounds were made. The results clearly demonstrate that the oceans are, compared to the atmosphere, supersaturated with light alkanes and alkenes. The supersaturation of ethyne is only small, in a few cases an undersaturation was observed.

As a consequence the ocean emits light NMHC into the atmosphere. The emission rates were calculated from the NMHC concentrations and ocean/atmosphere transfer velocities. The best estimate for the total world-wide annual emission of the C_2–C_4 hydrocarbons is 2.1 Tg yr^{-1}, the upper limit (90 % confidence limit) is 5.5 Tg yr^{-1}. These results are at the low side of previous estimates. In spite of the still existing uncertainties this strongly suggests that oceanic emissions of light NMHC are of minor importance for the global budget of NMHC and the chemistry of the remote marine atmosphere.

The alkenes contribute about 80 % to the total oceanic NMHC emissions. They are formed in the euphotic zone of the oceans photochemically from DOC (dissolved organic compounds). This process is most effective in the visible light at short wavelength and in the near UV. The emission into the atmosphere is the most important loss process for alkenes from the ocean. Case studies of the alkene budget in the ocean surface layer support this conclusion. Within a factor of less than two production and emission rates balance.

The mixing ratios of light NMHC in the remote marine atmosphere are generally low, the turnover rates of light NMHC are very low compared to methane. The light alkenes found in the remote marine boundary layer are predominantly of oceanic origin whereas for the light alkanes advection from the continent dominates.

Aims of the research

Oceanic emissions of non-methane hydrocarbons (NMHC) may be important both for their impact on the photochemistry of the marine atmosphere and their contribution to the global NMHC budget. As a consequence of the large area of the oceans, even small fluxes of NMHC per unit area could result in substantial global emission rates. Based on a comparison of their own measurements of light hydrocarbons in the atmosphere and the concentrations of NMHC in the ocean reported by Swinnerton and Lamontagne [1] and Lamontagne *et al.* [2], Rudolph and Ehhalt [3] concluded that

the oceans are supersaturated with light NMHC. They estimated NMHC fluxes limited ocean areas and suggested that the alkenes found in the remote mar atmosphere may be of oceanic origin. They did not give global emission estima Later Ehhalt and Rudolph [3] published an estimate of the global oceanic emiss rates for light NMHC based on these results. The estimate of about 20 Tg y indicated that the ocean may be a substantial NMHC source, but due to the v limited data base these estimates were highly uncertain. Based on measurements light NMHC made by Rasmussen and Khalil [4] in and above the marine bound layer, Penkett [5] estimated that the oceanic emissions of light alkenes may be as hi as
70 Tg yr^{-1}. The first simultaneous measurements of light NMHC in the ocean and the atmosphere were published by Bonsang *et al.* [6]. They estimated that the glol source strength of light NMHC is around 50 Tg yr^{-1}. These estimates indicated t NMHC emissions from the oceans may be substantial, but the uncertainties were v high, the existing data sets were very limited and the factors and mechanis determining the oceanic NMHC emissions poorly understood.

The purpose of this project was to determine the impact of oceanic emissions of lig NMHC on the chemistry of the marine atmosphere. For this purpose the followi investigations were made:

* measurements of the mixing ratios of NMHC in the remote marine tropospher
* measurements of the concentrations of light NMHC in the ocean surface water;
* studies of the processes determining the production of NMHC in the ocean;
* estimates of the emission rates of NMHC into the atmosphere; and
* estimates of the budgets of light NMHC in the marine boundary layer.

Results

Three series of shipborne measurements of the concentrations of light hydrocarbo both in the atmosphere and in the ocean were made: Two ship cruises over t Atlantic in September/October 1988 and August/September 1989 and one cruise ov the Mediterranean Sea, the mid and North Atlantic and the North Sea in April a May 1991. Furthermore during one ship cruise in March/April 1987 over the Atlan measurements of the atmospheric mixing ratios of light NMHC were made.

The atmospheric measurements were made by gas chromatography in combinati with a cryogenic preconcentration step. The analysis of NMHC in sea water was ma by a purge and trap method which allowed us to use a sample of nearly 1 L of s water. The experimental methods for the measurements of the light NMHC in t atmosphere and in the ocean are described by Rudolph and Johnen [7], Rudolph *et* [8], Plass *et al.* [9], and Koppmann et al [10]. The measurements were always ma directly after sampling on board the ships, in most cases directly on-line without a storage of the samples.

addition to the field measurements several series of experiments were made to study : production mechanisms of light NMHC, especially alkenes, in sea water. This :luded both studies in the laboratory and experiments during a ship cruise.

e results of most of the investigations are already published, in this paper we will :refore only present a brief summary of the results.

e distribution of non-methane hydrocarbons in the marine tropospheric undary layer

:tails of the latitudinal distribution of light NMHC and some halocarbons in the ırine boundary layer have been published in Rudolph and Johnen [7] and ɔppmann *et al.* [10–13]. Measurements of the seasonal cycles of NMHC can be ınd in Rudolph *et al.* [8, 14].

ı overview of the mixing ratios of some selected organic trace gases in the marine nosphere is shown in Fig. 3.8.1. For a comparison the mixing ratios over rural ntinental regions are also given.

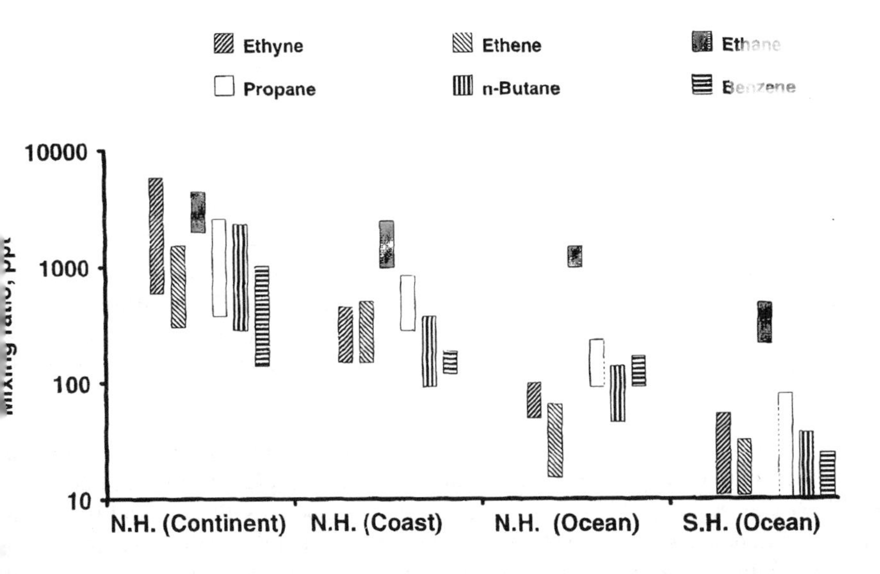

ig. 3.8.1: Range of mixing ratios of selected NMHC observed over different regions.

xamples for the latitudinal distributions of the mixing ratios of some alkanes are iven in Fig. 3.8.2. The distribution of an organic trace gas in the marine atmosphere epends strongly on its lifetime. An overview of the atmospheric lifetimes of a number f organic substances is given in Table 3.8.1.

Table 3.8.1: Atmospheric life times of selected NMHC

Compound	Life time*	Compound	Life time*
methane	10 years	benzene	2 weeks
ethane	2,5 months	toluene	2.5 days
propane	2.5 weeks	ethylbenzene	2.,5 days
n-butane	7 days	*m*-xylene	1.5 days
i-butane	7 days	1,3,5 trimethylbenzene	7.5 hours
n-pentane	4 days		
i-pentane	4.5 days	cyclopentane	4 days
n-hexane	3 days	cyclohexane	4 days
n-heptane	2.5 days	methylcyclohexane	2 days
n-octane	2 days	cyclohexene	3 hours
ethene	15 days	isoprene	3 hours
propene	11 hours	α-pinene	4 hours
1-butene	10 hours	limonene	1 hour
ethyne	35 weeks	tetrachloroethene	45 months

* The lifetimes are calculated for an average OH-radical concentration of 6.5×10^5 cm^{-3} [25], average ozone concentration of 7×10^{11} cm^{-3} (corresponding to 30 ppb at standard temperatu and pressure) and a temperature of 280 K (as recommended by Warneck [26]). The ra constants are taken from studies of Atkinson *et al.* [27–32], Cohen [33], Droege and Tully [34 Greene and Atkinson [35], Harris and Kerr [36] and Treacy *et al.* [37].

For hydrocarbons and halocarbons with atmospheric life times of months, reasonab well defined latitudinal and seasonal variations (see examples in Figs. 3.8.2a, 3.8.2 and 3.8.3) can be given [15]. For more reactive compounds the variability strong increases with decreasing atmospheric life times [8]. In the remote marine atmosphe the mixing ratios of reactive, short lived NMHC are generally at the lower end of tl ranges given in Fig. 3.8.1. Over the remote ocean the mixing ratios of alkene alkylbenzenes and higher alkanes ($\geq C_4$) are a few ten ppt or less. As a consequen of these low mixing ratios the turnover of NMHC in the remote marine atmosphere low compared to continental or coastal regions (Fig. 3.8.4.).

Higher mixing ratios of alkanes or alkylbenzenes can in most cases be explained b the impact of continental air masses. In these continentally influenced air masses th changes in the relative pattern can often be explained by the different rates atmospheric removal[7, 16]. An example is shown in Fig. 3.8.5. The ratios of th NMHC mixing ratios follow within their uncertainty the behaviour predicted fro their rate constants for the reaction with OH radicals.

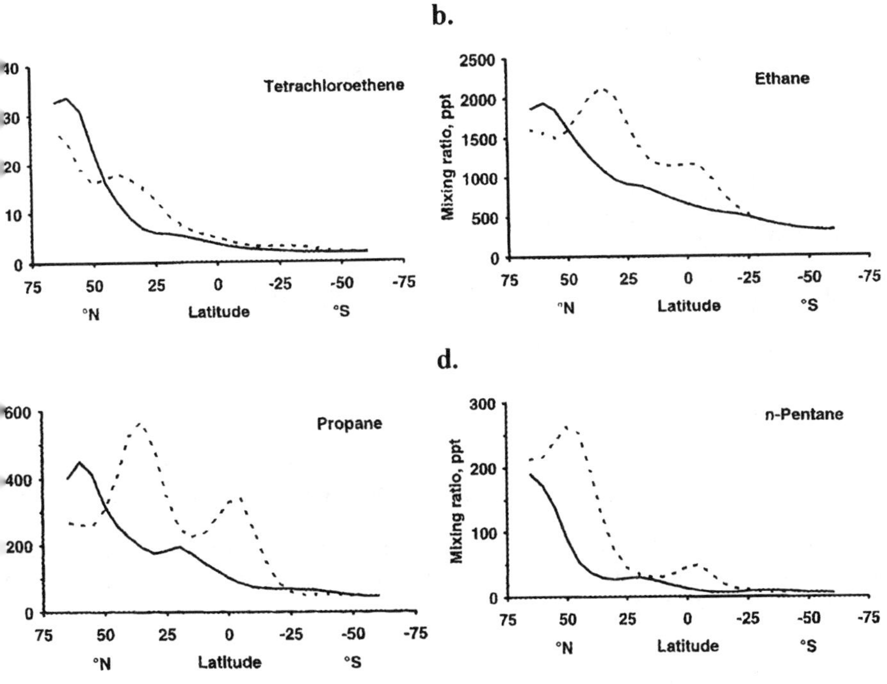

ig. 3.8.2: Latitudinal distribution some selected organic trace gases. The measurements were ade over the east coast of North America or the west coast of South America (solid line) and e east coast of South America and the west coast of Africa and Europe (broken line).

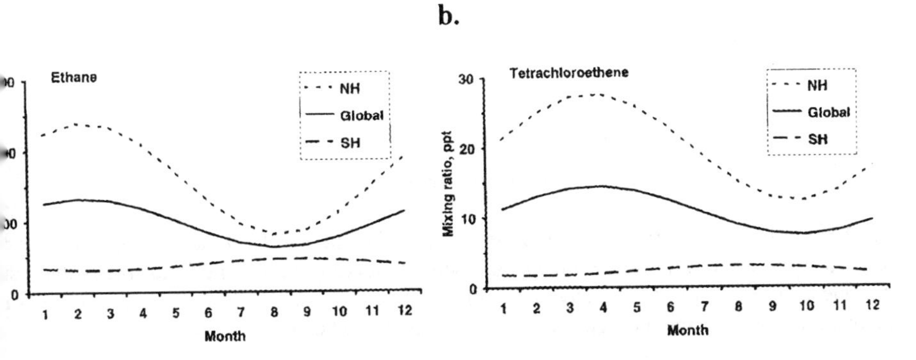

'ig. 3.8.3: Average seasonal variation of the mixing ratios of ethane and tetrachloroethene.

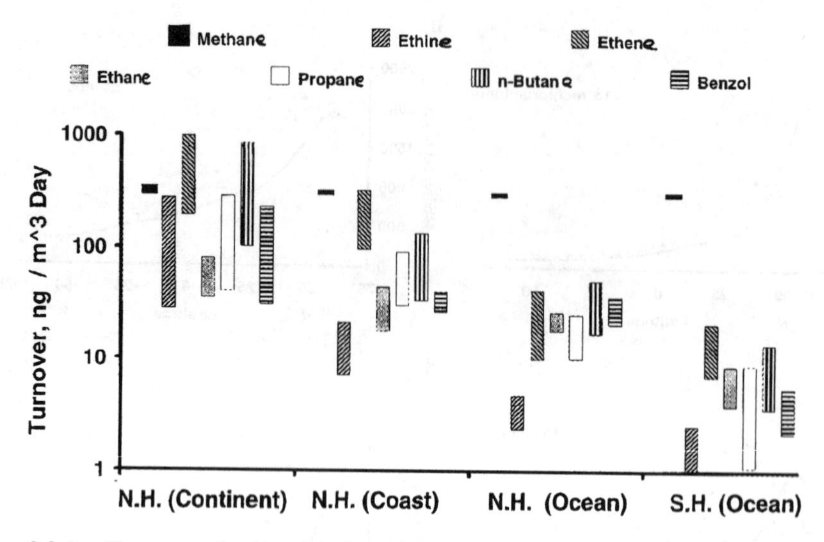

Fig. 3.8.4: Turnover of selected hydrocarbons over different regions. The turnover rates based on the same rate constants and OH radical and ozone concentrations as the atmosphe lifetimes given in Table 3.8.1.

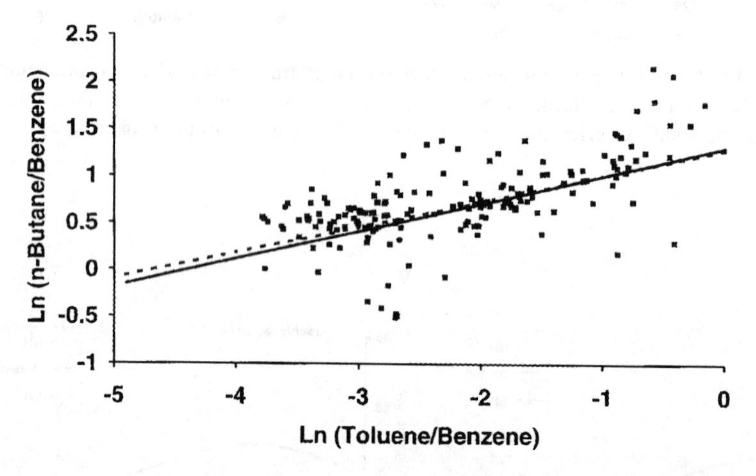

Fig. 3.8.5: Plot of logarithms of the ratios of the *n*-butane and benzene mixing ratios vers the logarithm of the ratio of the toluene and benzene mixing ratios. The solid line is the res of a linear least square fit, the slope of the broken line is calculated from the rate constants the reactions of *n*-butane, benzene and toluene with OH radicals.

The concentration of non-methane hydrocarbons in the ocean surface water

The present state of the art does not allow to measure the emission rates of lig NMHC from the oceans directly. However, reasonable emission rates can determined from the difference of the concentrations between ocean and atmosphe and the ocean-atmosphere transfer velocities. Thus the key to a quantitati ermination of oceanic emission rates of light NMHC is the knowledge of their centrations in the oceans.

e measurements of the concentrations of light NMHC made as part of this project published by Plass *et al.* [17] and Plass-Dülmer *et al.* [18, 19] and Ratte *et al.*]. An overview of the existing measurements of the concentrations of light rocarbons in sea water is given by Plass-Dülmer *et al.* [21].

e results of the measurements made in 1988 and 1989 in the Atlantic Ocean surface ter are summarised in Fig. 3.8.6. Global mean values derived from the more than) measurements made as part of this project and about 500 measurements published the literature are shown in Fig. 3.8.7. The details of the individual data sets and ir evaluation are described by Plass-Dülmer *et al.* [210].

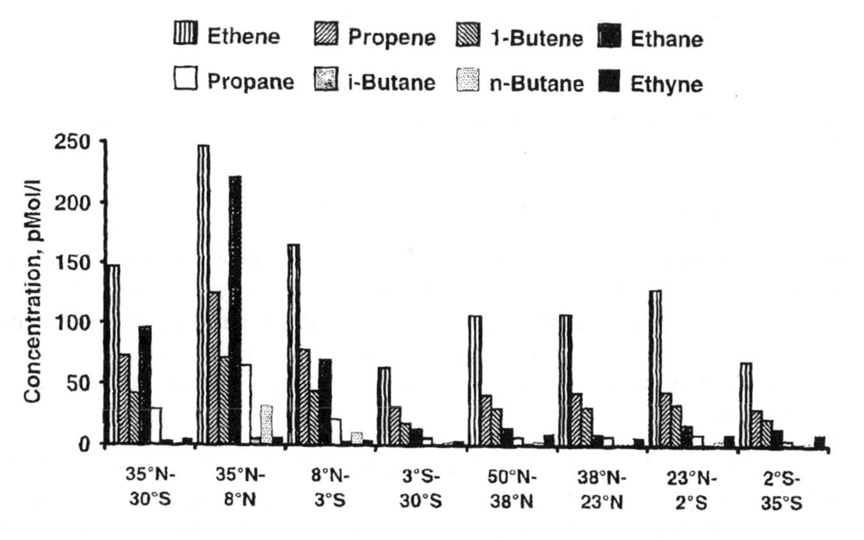

. 3.8.6: Concentrations of light hydrocarbons in the ocean for several latitude ranges. The asurements were made 1988 (left part of the graph) and 1989 (right part of the graph) in the face water of the Atlantic.

e concentrations of light NMHC in sea water range mostly from a few ol L^{-1} to several hundred pmol L^{-1}. In general the alkenes are more abundant than alkanes and the concentrations decrease with increasing number of carbon atoms. yne is the only detectable compound with a triple bond. Its concentrations range und 10 pmol L^{-1}.

e depth profiles of the alkane concentrations showed often an increase at depths ich correspond to a light intensity of 10 % of the surface intensities. The alkene files showed little systematic variability in the mixed surface layer of the ocean. ly under conditions of very low wind velocities significantly elevated ethene, pene and 1-butene concentrations were found close to the surface.

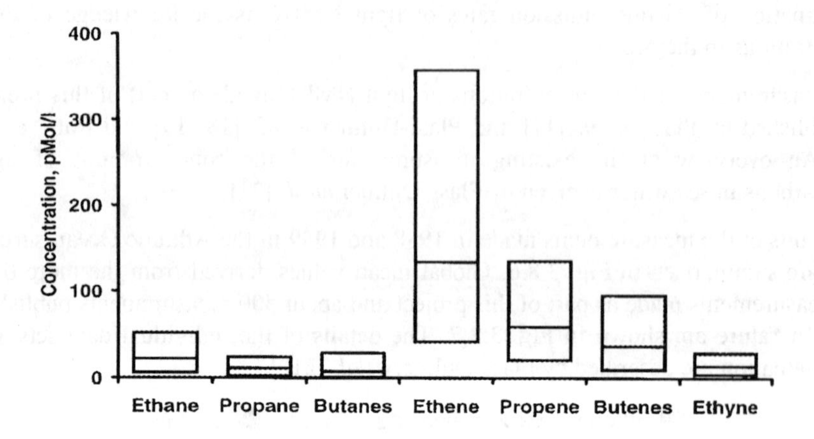

Fig. 3.8.7: Average concentration of light hydrocarbons in ocean surface water. The vert bars indicate the 90 % confidence interval of the measurements. The results are based on m than 800 *in-situ* measurements.

Formation of light NMHC in the ocean surface water

Studies of the production mechanisms of light NMHC in sea water were made both the laboratory and in the field. The results of these investigations are published Ratte *et al.* [20, 22]. Several of the factors which determine the production rate ethene and propene in sea water are reasonably well understood. However, parameters which influence the production of light alkanes in the ocean surface wa remain unknown.

It is established that the production of ethene and propene in the ocean surface wa depends on the light intensity and the concentration of DOC. However, not only concentration of DOC but also the photochemical reactivity of the DOC is import for the formation rates of light alkenes. The efficiency of the alkene formation depe on the wavelength. Irradiation with light in the wavelength range of 360–420 nm W m^{-2}) gave alkene formation rates which were more than an order of magnit higher than those obtained from the irradiation with light of longer wavelength similar intensity.

Experiments with sea water made during a ship cruise (NATAC 91) under ambi conditions gave formation rates around 50 pmol L^{-1} d^{-1} for ethe 30 pmol L^{-1} d^{-1} for propene, and 10–15 pmol L^{-1} d^{-1} for 1-butene. The formation ra of alkanes were considerably lower than for the homologues alkenes. The format rates of the alkenes agreed well with the rates of increase observed in the surface wa during a period of very low wind velocities. Under these conditions the emissions fr the ocean into the atmosphere is negligible and the rate of the increase is equivalen the net production. The observed increase rates of 36 pmol L^{-1} d^{-1} for ethene, 16 pm L^{-1} d^{-1} for propene and 8 pmol L^{-1} d^{-1} are comparable with the formation ra observed in the experiments and support the assumption that the results of experiments are applicable to the ocean.

ceanic emissions of light non-methane hydrocarbons

s mentioned above, the emission rate of a given NMHC (F_{NMHC}) can be calculated om the concentrations of the NMHC in the ocean (${}^{O}c_{NMHC}$) and in the atmosphere c_{NMHC}), the transfer velocities for the ocean/atmosphere ($k_{O/A}$)exchange and the lubility of the NMHC (H_{NMHC}).

$$F_{NMHC} = k_{O/A} * ({}^{O}c_{NMHC} - {}^{A}c_{NMHC} / H_{NMHC}) \quad (1)$$

r most of the NMHC, except ethyne, the ocean is highly supersaturated $c_{NMHC} >> {}^{A}c_{NMHC} / H_{NMHC}$) and Eq. 1 can be simplified:

$$F_{NMHC} = k_{O/A} * {}^{O}c_{NMHC} \quad (2)$$

is important to note that Eq. 2 strongly overestimates the emission rates if the ndition ${}^{O}c_{NMHC} >> {}^{A}c_{NMHC} / H_{NMHC}$ is not fulfilled. Thus the use of Eq. 2 for the timation of the ethyne emission rates would result in a much too high oceanic hyne source strength, since for ethyne the ocean is only marginally supersaturated, casionally even undersaturated.

r the calculations of emission rates the transfer velocities were calculated according the ocean/atmosphere transfer formulation of Liss and Merlivat [23]. The Schmidt ımber for CO_2, which is used in this formulation, was transformed for individual MHC similar to the method described to Kanakidou [24]. The wind velocities eded to calculate the transfer velocities were either taken from observations made nultaneously with the NMHC measurements or from climatological data. Details of procedures used to derive emission rates from the NMHC concentration are given Plass *et al.* [9] and Plass-Dülmer *et al.* [19, 21]. These papers also contain details the oceanic emission rates determined during this project.

overview of the emission rates determined for the Atlantic Ocean in 1988 and 89 is given in Fig. 3.8.8. Global mean annual emission rates derived from the more an 300 measurements of NMHC in the ocean made as part of this project and about 0 measurements published in the literature are shown in Fig. 3.8.9. The details of individual data sets and their evaluation are described by Plass-Dülmer *et al.* [19,]. The largest emission rates are found for the alkenes, especially ethene. With out 0.9 Tg yr^{-1}. the ethene emissions contribute about 40 % of the total emissions of ht NMHC. Propene and the butenes together have emission rates comparable to ene and the light alkanes in total contribute 20 % to the oceanic emissions of light MHC of 2.1 Tg yr^{-1}. The emission rates for ethyne (not shown in Fig, 3.8.9) are very v, around 0.02 Tg yr^{-1}. (range 0–0.05 Tg yr^{-1}.). The 90 % confidence interval for total annual emissions of light NMHC is 5.5 Tg yr^{-1}.

significant anticorrelation between the wind velocities and the concentrations of MHC was found for several of the measurement series. Roughly 50 % of the served variability could be ascribed to variations in the transfer velocities. From a tailed data analysis and a simple conceptual model describing the oceanic NMHC ncentrations as a function of the transfer velocity it was concluded that the issions into the atmosphere are the major loss process for NMHC in the ocean face water [19]. For the alkenes this result was supported by a case study.

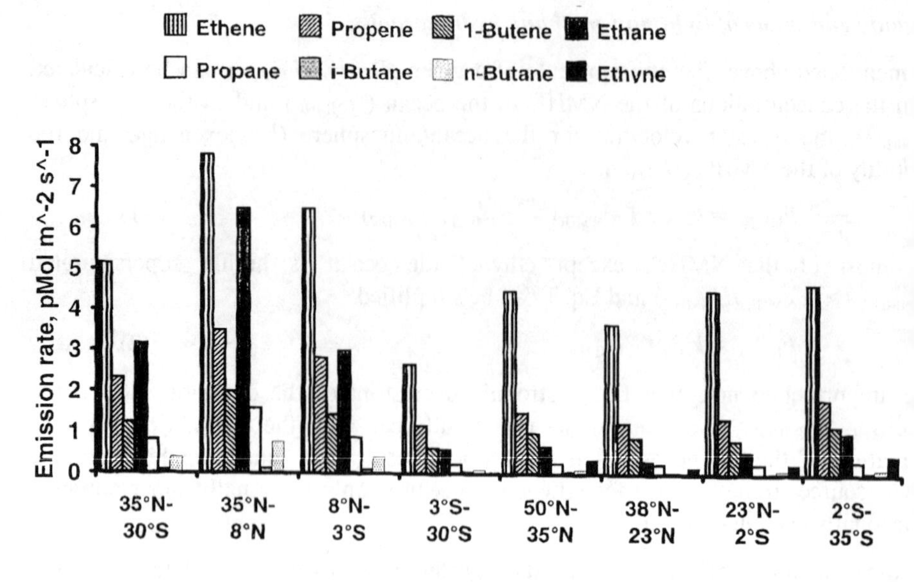

Fig. 3.8.8: Emission rates of light hydrocarbons from the ocean for several latitude ranges. emission rates are based on measurements made 1988 (left part of the graph) and 1989 (r part of the graph) in the surface water of the Atlantic.

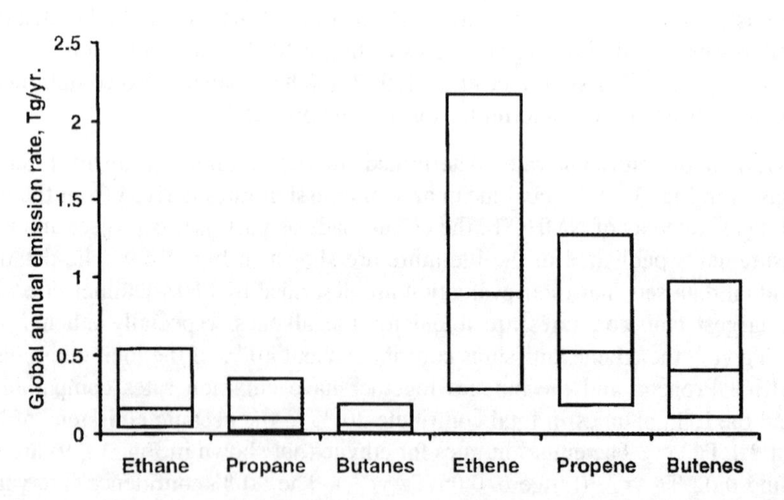

Fig. 3.8.9: Global annual emission rates of light NMHC. The vertical bars indicate 90 % confidence interval of the measurements. The results are based on more than *in situ* measurements.

A comparison of the emission rates with the production rates derived fr experiments and extrapolated to the euphotic zone of the ocean showed a g agreement (Fig. 3.8.10). Also the observed increase of the alkene concentrations in ocean surface water under conditions of very low wind velocities (neglig emissions) supports this conclusion (see previous section).

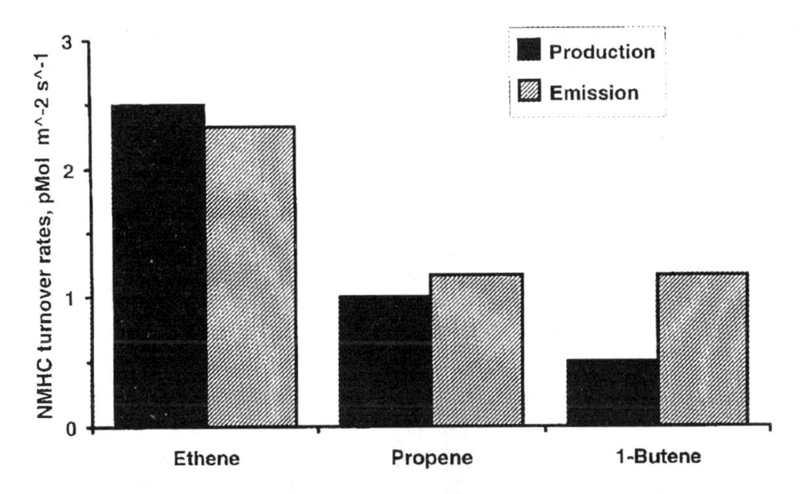

ȝ. 3.8.10: Comparison of production rates in the surface ocean water and the emission rates ɒo the atmosphere for light alkenes.

e budget of light hydrocarbons in the remote marine boundary layer

om the oceanic emission rates of NMHC and the removal rates of these compounds the marine boundary layer the steady state concentrations of NMHC in the remote arine boundary layer can be estimated. This was done with a simple one dimensional odel [10, 19]. The results are shown in Fig. 3.8.11. For the sake of comparison also e measured NMHC concentrations are included. It can be seen that generally the easured mixing ratios exceed the calculated values substantially. Only for ethene the lues are, within their uncertainty, of comparable magnitude. It should be noted that r the Polarstern cruise in 1989 the propene concentrations were nearly always below e detection limit of 5–10 ppt. This agrees qualitatively with the calculated low lues around 2–3 ppt. However a detailed comparison is not possible.

r the alkanes and ethyne the oceanic emissions can explain only a small fraction of e observed mixing ratios. These findings strongly suggest that the mixing ratios of ;ht alkanes and ethyne observed in the marine troposphere are mainly due to vection from the continents.

onclusions

ıe oceans are supersaturated with light alkanes and alkenes and thus act for the mosphere as a source of these compounds. However, the emission rates are quite w, around 2 Tg yr^{-1}. Although this estimate has a rather high uncertainty the upper nit of the oceanic NMHC source strength of 5.5 Tg yr^{-1} (90 % confidence limit) early indicates that the oceanic NMHC emissions contribute only marginally to the obal budget of NMHC (Table 3.8.2). The oceanic NMHC emission rates obtained in is study are at the low side of all previous estimates. The light alkenes observed in e remote marine atmosphere are mainly due to oceanic emissions, for alkanes and hyne transport from the continents generally dominates. The turnover of NMHC in e remote marine atmosphere is quite low, well below that of methane. In spite of their low concentrations the light alkenes dominate in the remote marine atmosph the turnover of NMHC. For the alkenes the oceanic source has a substantial influe on their mixing ratios in the remote marine atmosphere.

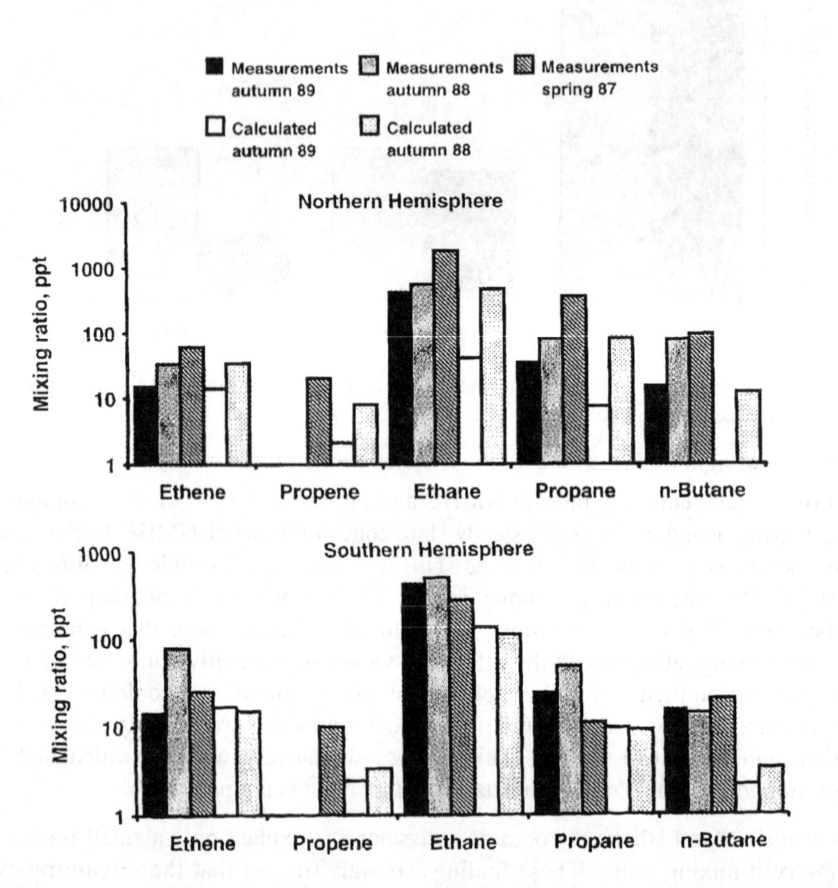

Fig. 3.8.11: Comparison of the measured mixing ratios of light NMHC in the mar atmosphere with calculations based on a steady state between oceanic emissions atmospheric removal rates.

The alkenes contribute about 80 % to the total oceanic NMHC emissions. They formed in the euphotic zone of the oceans photochemically from DOC. This process most effective in the visible light at short wavelengths and in the near UV. Emissi into the atmosphere is the most important loss process for alkenes from the oce Case studies of the alkene budget in the ocean surface layer support this conclusi Within a factor of less than two production and emissions balances.

ble 3.8.2: Global annual emission rates of some NMHC (Tg yr^{-1})

Compound	Industrial sources*	Vegetation §	Biomass burning	Oceans^{+}	Total
ethyne	2.5	0	$4.7^{\&}$	< 0.1	7.3
ethene	10	5.4	$13.1^{\&}$	0.9	29.4
ethane	6	1.2	$6.1^{\&}$	0.2	13.5
propene	10	0.6	$3.6^{\&}$	0.5	14.7
propane	6	0	$4.5^{\&}$	0.1	10.6
butane	10	0	$0.2^{\$}$	0.1	10.3
pentane	10	0	$0.3^{\$}$	< 0.1	10.3
hexane	15.5	0	$0^{\$}$	< 0.1	15.6
benzene	5	0	$2.4^{\$}$	< 0.1	7.4
toluene	15	0	$1.5^{\$}$	< 0.1	16.6
sum	90	7.2	36.5	2.1	135.9

Hough [38]; § Ehhalt and Rudolph [39];$^{+}$ Plass-Dülmer *et al.* [21]; $^{\&}$ based on the emission tios from Rudolph *et al.* [40];$^{\$}$ Lobert *et al.* [41]

eferences

. J. W. Swinnerton, R. A. Lamontagne, The oceanic distribution of low-molecular-weight hydrocarbons. *Environ. Sci. and Technol.* **8** (1974) 657–663.

R.A. Lamontagne, J. W. Swinnerton, W. J. Linnenboom, C_1–C_4 hydrocarbons in the north and south Pacific, *Tellus* **26** (1974) .

. J. Rudolph, D.H. Ehhalt, Measurements of C_2–C_5 hydrocarbons over the North Atlantic, *J. Geophys. Res.* **86** (1981) 11.959–11964.

. R.A. Rasmussen, M.A.K. Khalil, Latitudinal distributions of trace gases in and above the boundary layer, *Chemosphere*, **11** (1982) 227–235.

. S. A. Penkett, Non-methane organics in the remote troposphere, in: E.D. Goldberg (ed), *Atmospheric Chemistry*, Springer Verlag, New York 1982, pp. 329–355.

. B. Bonsang, M. Kanakidou, G. Lambert, P. Monfray, The marine source of C_2–C_5 aliphatic hydrocarbons, *J. Atmos. Chem.* **6** (1988) 3–20.

. J. Rudolph, F.J. Johnen, Measurements of light atmospheric hydrocarbons over the Atlantic in regions of low biological activity, *J. Geophys. Res.* **95** (1990) 20 583.

. J. Rudolph, A. Khedim, T. Clarkson, D. Wagenbach, Long Term Measurements of Light Alkanes and Acetylene in the Antarctic Troposphere, *Tellus* **44B** (1992) 252–261.

. Ch. Plass, R. Koppmann, J. Rudolph, Measurements of dissolved non-methane hydrocarbons in sea water, *Fresenius J. Anal. Chem.* **339** (1991) 746–749.

10. R. Koppmann, Bauer, F.J. Johnen, C. Plass-Dülmer, J. Rudolph, The distribution of li non-methane hydrocarbons over the mid - Atlantic: Results of the Polarstern cruise A VII/1. *J. Atmos. Chem.* **15** (1992) 215–234.

11. R. Koppmann, F.J. Johnen, C. Plass, J. Rudolph, The latitudinal distribution of light nc methane hydrocarbons over the mid Atlantic between 40 °N and 30 °S, G. Restelli, G. Angeletti (eds), *Physico - Chemical Behaviour of Atmospheric Pollutar Proc. 5th European Symp. Varese,* Kluwer Academic Publishers, Dordrecht 1990, pp. 6: 662

12. R. Koppmann, F.J. Johnen, C. Plass-Dülmer, J. Rudolph, The Distribution of Met Chloride, Dichloromethane, Trichloroethene, and Tetrachloroethene over the Atlantic. *Geophys. Res.* **97** (1993) 20517–20526.

13. R. Koppmann, F.J. Johnen, C. Plass-Dülmer, and J. Rudolph, The Distribution Dichloromethane, Trichloroethene and Tetrachloroethene over the Atlantic, G. Restelli, G. Angeletti (eds), *Proc. 6th European Symp. on the Physico-Chemi Behaviour of Atmospheric Pollutants*, Report EUR 15609/1 EN, Brussels-Luxemburg 19 pp. 417–423.

14. J. Rudolph, A. Khedim, D. Wagenbach, The seasonal variation of light non-metha hydrocarbons in the Antarctic Troposphere, *J. Geophys. Res.* **94** (1989) 13039–13044.

15. J. Rudolph, The Tropospheric Distribution and Budget of Ethane. *J. Geophys. Res.* **1** (1995) 11383–11391.

16. J. Rudolph, R. Koppmann, Sources and atmospheric distribution of light hydrocarbons, P.J. Crutzen, J.-C. Gerard, R. Zander (eds), *Proc. 28th Liege Astrophysical Colloqui "Our Changing Atmosphere", 1989,* Liege, Belgium 1990, pp. 435–447.

17. Ch. Plass, F.J. Johnen, R. Koppmann, J. Rudolph, The latitudinal distribution of NMHC the Atlantic and their fluxes into the atmosphere, in: G. Restelli, G. Angeletti (ed *Physico - Chemical Behaviour of Atmospheric Pollutants, Proc. 5th European Symposiu Varese/Italy 1989*, Kluwer Academic Publishers, Dordrecht, 1990, pp. 663–668.

18. Ch. Plass-Dülmer, R. Koppmann, J. Rudolph, Light hydrocarbons in the surface water the mid-Atlantic, *J. Atmos. Chem.* **15** (1992) 235–251.

19. Ch. Plass-Dülmer, R. Koppmann, F.J. Johnen, J. Rudolph, H. Kuosa, Emissions of Lig Non-methane Hydrocarbons from the Atlantic into the Atmosphere, *Global Biogeochemic Cycles* **7** (1993) 211–228.

20. M. Ratte, Ch. Plass-Dülmer, R. Koppmann, J. Rudolph, Horizontal and vertical profiles light hydrocarbons in sea water related to biological, chemical and physical parameter *Tellus B* **47** (1995) 607–623.

21. Ch. Plass-Dülmer, R. Koppmann, M. Ratte, and J. Rudolph, Light Non-methar Hydrocarbons in Seawater, *Global Biogeochemical Cycles* **9** (1995) 79–100.

22. M. Ratte, Ch. Plass-Dülmer, R. Koppmann, J. Rudolph, Production Mechanism of C_2–(Hydrocarbons in Sea Water: Field Measurements and Experiments, *Global Biogeochemic Cycles* **7** (1993) 369–378.

23. P. S. Liss, and L. Merlivat, Air-sea gas exchange rates: introduction and synthesi in:P. Buat-Menard (ed), *The role of Air-Sea Exchange in Geochemical Cyclin* D. Reidel, Hingham, Mass. 1986, pp. 113–127.

24. M. Kanakidou, Contribution à l`étude des sources des hydrocarbures légers nc méthaniques dans l´atmosphère, Ph. D. thesis, Univ. Paris VII, 1988.

A. Volz, D. H. Ehhalt, and R. D. Derwent, Seasonal and Latitudinal Variation of ^{14}CO and the tropospheric concentration of OH radicals, *J. Geophys. Res.* **86** (1981) 5163–5171.

P. Warneck, *Chemistry of the natural atmosphere*, Academic Press, Inc., New York 1988.

R. Atkinson, Kinetics and mechanisms of the gas phase reactions of the hydroxyl radical with organic compounds under atmospheric conditions, *Chem. Rev.* **86** (1986) 69.

R. Atkinson, S. M. Aschmann, Kinetics of the gas-phase reactions of Cl-atoms with chloroethenes at 299±2 K and atmospheric pressure, *Int. J. Chem. Kinet.* **19** (1987) 1097.

R. Atkinson, D.L. Baulch, R.A. Cox, R.F. Hampson, Jr, J.A. Kerr, J. Troe, Evaluated kinetic and photochemical data for atmospheric chemistry. Supplement III, JUPAC subcommittee on gas kinetic data evaluation for atmospheric chemistry, *J. Phys. Chem. Ref. Data* **19** (1989) 881–1095.

R. Atkinson, Gas-phase Tropospheric Chemistry of Organic Compounds: a Review, *Atmos. Environ.* **24A** (1990) 1–41.

R. Atkinson, D. Hasegawa, S.M. Aschmann, Rate constants for the gas phase reactions of O_3 with a series of monoterpenes and related compounds at 296 ± 2 K, *Int. J. Chem. Kinetics* **22** (1990) 871.

R. Atkinson, D.L. Baulch, R.A. Cox, R.F. Hampson, Jr, J.A. Kerr, J. Troe, Evaluated kinetic and photochemical data for atmospheric chemistry. Supplement IV, IUPAC subcommittee on gas kinetic data evaluation for atmospheric chemistry, *J. Phys. Chem. Ref. Data* **21** (1992) 1125–1568.

N. Cohen, Are rate coefficients additive? Revised transition state theorie calculations for OH + alkane reactions, *Int. J. Chem. Kinetics* **23** (1991) 397–417.

A. T. Droege, F.P. Tully, Hydrogen abstraction from alkanes by OH. 6. Cyclopentane and cyclohexane, *J. Phys. Chem.* **91** (1987) 1222.

C. R. Greene, R. Atkinson, Rate constants for the gas-phase reactions of O_3 with a series of Alkenes at 296 ± 2 K, *Int. J. Chem. Kinetics* **24** (1992) 803–811.

J. Harris, J.A. Kerr, Relative rate measurements of some reactions of hydroxyl radicals with alkanes studied under atmospheric conditions, *Int. J. Chem. Kinetics* **20**, (1988) 939.

J.M. Treacy, El Hag, D. O'Farrell, H. Sidebottom, Reaction of ozone with unsaturated organic compounds, *Ber. Bunsenges. Phys. Chem.* **96** (1992) 422–427.

A.M. Hough, Development of a two-dimensional global tropospheric model: Model chemistry, *J. Geophys. Res.* **96** (1991) 7325–7362.

D.H. Ehhalt, J. Rudolph, On the importance of light hydrocarbons in multiphase atmospheric systems, *Berichte der Kernforschungsanlage Jülich,* 1984.

J. Rudolph, A. Khedim, R. Koppmann, B. Bonsang, Field Study of the Emissions of Methyl Chloride and other Halocarbons from Biomass Burning in Western Africa. *J. Atmos. Chem.* **22** (1995) 67–80.

J.M. Lobert, D. H. Scharffe, W.-M. Hao, T.A. Kuhlbusch, R. Seuwen, P. Warneck, P. Crutzen, Experimental evaluation of biomass burning emissions: Nitrogen and carbon containing compounds, in: J. S. Levine (ed), *Global biomass burning: Atmospheric, climatic, and biospheric implications*, MIT Press, Cambridge 1991, pp. 289–304.

Chapter 4

ASE Publications: 1988–1995

1988

Allen, A.G., R.M. Harrison, M.T. Wake; A meso-scale study of the behaviour of atmospheric ammonia and ammonium, *Atmos. Environ.* **22** (1988) 1347–1353.

Bonsang, B., M. Kanakidou, G. Lambert, P. Monfray; The marine source of C_2–C_6 atliphatic hydrocarbons, *J. Atmos. Chem.* **6** (1988) 3–20.

Kersten, M., M. Dicke, M. Kriews, K. Naumann, D. Schmidt, M. Schulz, M. Schwikowski, M. Steiger.; Distribution and fate of heavy metals in the North Sea, in: W. Salomons, B.L. Bayne, E.K. Duursma, U. Förstner (eds), *Pollution of the North Sea - An Assessment,* Springer Verlag 1988, pp. 300–347.

Losno, R., G. Bergametti, P. Buat-Ménard; Zinc partioning in Mediterranean rainwater, *Geophys. Res. Lett.* **15** (1988) 1389–1392.

Nguyen, B.C., S. Belviso, N. Mihalopoulos, J. Gostan, P. Nival; Dimethysulfide production during natural phytoplanktonic blooms, *Marine Chem.* **24** (1988) 133–141.

Schulz, M., M. Steiger, M. Schwikowski, M. Kriews, K. Naumann, W. Dannecker; Variability of ambient trace element concentrations at the North Sea with respect to air mass history, *J. Aerosol Sci.* **19** (1988) 1171–1174.

Schwikowski, M., M. Schulz, M. Steiger, W. Dannecker; Contribution of airborne nitrate and ammonium to the eutrophication of the North Sea, *ICES CM. 1988/C:43* (1988).

1989

Allen, A.G., R.M. Harrison; Field Measurements of the Dissociation of Ammonium Nitrate and Ammonium Chloride Aerosols, *Atmos. Environ.* **23** (1989) 1591–1599.

Bergametti, G., A.L. Dutot, P. Buat–Ménard, R. Losno, E. Remoudaki; Seasonal variability of the elemental composition of atmospheric aerosol particles over the northwestern Mediterranean, *Tellus* **41**B (1989) 353–361.

Bergametti, G., L. Gomes, E. Remoudaki, M. Desbois, D. Martin, P. Buat-Ménard; Present transport and deposition patterns of African dusts to the north-western Mediterranean, in: M. Leinen, M. Sarnthein (eds), *Palaeoclimatology and Paleometeorology: Modern and Past Patterns of Global Atmospheric Transport, NATO/ASI Series* No. 282, Kluwer Academic Publishers, Dordrecht 1989, pp. 227–252.

Bonsang, B., M. Kanakidou, G. Lambert; Sur la faible variabilite de composition relative des hydrocarbures legers non methaniques dissous dans l'eau de mer superficielle, *C. R. Acad. Sci. Serie II* (1989) 495–500.

Cornille, P., W. Maenhaut, J.M. Pacyna; Size distribution of atmospheric trace elements in Birkenes, Norway, during spring 1987, in. J.P. Vernet (ed) *Int. Conf. on Heavy Metals in the Environment*, CEP Consultants Ltd., Edinburgh 1989, pp. 156–159.

Davidson, C.I., L.A. Barrie, G. Bergametti, C.F. Boutron, R.M. Harrison, K. Kemp, U. Krell, W. Maenhaut, J. Müller, W.H. Schröder, H. Ross; Research needs in understanding processes of transformation, and dry and wet deposition of atmospheric metals, in: J.M. Pacyna, B.Ottar (eds), *Control and Fate of Atmospheric Trace Metals, NATO ASI Series*, Kluwer Academic Publ., Dordrecht 1989, pp. 355–364.

Edson, J.B., C.W. Fairall, S.E. Larsen, P.G. Mestayer; The HEXIST Lagrangian simulation of the transport of evaporation jet drops, in: W.A. Oost, S.D. Smith, K.B. Katsaros (eds), *Proc. NATO Advanced Workshop on Humidity Exchange over the Sea*, University of Washington, AK-40, Seattle 1989, pp. 164–177.

Larsen, S.E.; Identification of critical research topics, in: E.C. Monahan, M.A. van Patten (eds), *Climate and Health Implications of Bubble-Mediated Sea-Air Exchange*, Connecticut Sea Grant College Program. University of Connecticut, Groton 1989, pp. 165–170.

Leeuw, G. de; Investigations on turbulent fluctuations of particle concentrations and relative humidity in the marine atmospheric surface layer, *J. Geophys. Res.* **94** (1989) 3261–3269.

Leeuw, G. de; Investigations on turbulent fluctuations of particle concentrations and relative humidity in the marine atmospheric surface layer, *J. Geophys. Res.* **94** (1989) 3261–3269.

Leeuw, G. de; The occurrence of large salt water droplets at low elevations over the open ocean, in: E.C. Monahan, M.A. van Patten (eds), *Climate and Health Implications of Bubble-Mediated Sea-Air Exchange*, Connecticut Sea Grant College Program. University of Connecticut, Groton 1989, pp. 65–82.

Maenhaut, W., P. Cornille, J.M. Pacyna, V. Vitols; Trace element composition and origin of the atmospheric aerosol in the Norwegian Arctic, *Atmos. Environ.* **23** (1989) 2551–2569.

Maenhaut, W.; Analytical techniques for atmospheric trace elements, in: J.M. Pacyna, B.Ottar (eds), *Control and Fate of Atmospheric Trace Metals, NATO ASI Series*, Kluwer Academic Publ., Dordrecht 1989, pp. 259–301.

Mestayer, P.G., J.B. Edson, C.W. Fairall, S.E. Larsen, D.E. Spiel; Turbulent transport and evaporation of droplets generated at an air-water interface, *Turbulent Shear Flows* 6, Springer, Berlin/Heidelberg 1989, pp. 129–147.

Mestayer, P.G.; Enhancement of sea-to-air moisture flux by spray droplets associated with breaking waves and bubble entrainment. Simulations in the IMST wind tunnel, in: E.C. Monahan, M.A. van Patten (eds), *Climate and Health Implications of Bubble-Mediated Sea-Air Exchange*, Connecticut Sea Grant College Program. University of Connecticut, Groton 1989, pp. 101–119.

Mihalopoulos, N., B. Bonsang, B.C. Nguyen, M. Kanakidou, S. Belviso; Field observations of carbonyl sulfide deficit near the ground. Possible implication on vegetation, *Atmos. Environ.* **23** (1989) 2159–2166.

Pacyna, J.M., A. Bartonova, P. Cornille, W. Maenhaut; Modelling of long-range transport of trace elements. A case study; *Atmos. Environ.* **23** (1989) 107–114.

Rouault, M., S.E. Larsen; Spray droplets under turbulent conditions, *Risø-M-2845*, Risø National Laboratory, Roskilde 1989, pp. 1–60.

Rudolph, J., F.J. Johnen, A. Khedim, G. Pilwat; The use of automated on line gas chromatography for the monitoring of organic trace gases in the atmosphere at low levels, *Int. J. Environ. Anal. Chem.* **38** (1989) 143.

Schulz M.,M. Schwikowski, M. Steiger, W. Dannecker; Land-sea-transformations of polluted air masses investigated with optical particle counters and short time resolved aerosol sampling, *J. Aerosol Sci.* **20** (1989) 1217–1220.

Steiger M., M. Schulz, M. Schwikowski, K. Naumann, W. Dannecker; Variability of aerosol size distributions above the North Sea and its implication to dry deposition estimates, *J. Aerosol Sci.* **20** (1989) 1229–1232.

Sturges, W.T., R.M. Harrison; The use of nylon filters to collect HCl: Efficiencies, interferences and ambient concentrations, *Atmos. Environ.* **23** (1989) 1987–1996.

Vawda, Y., I. Colbeck, R.M. Harrison, K.W. Nicholson; The effects of particle size on deposition rates, *J. Aerosol Sci.* **20** (1989) 1155–1158.

1990

Artaxo, P., F. Andrade, W. Maenhaut; Trace elements and receptor modelling of aerosols in the Antartic peninsular, *Nuclear Instruments and Methods in Phys. Res.* **B49** (1990) 383–387.

Artaxo, P., F. Andrade, W. Maenhaut; Trace elements in aerosols from the Antartica peninsular, in: S. Masuda, K. Takahashi (eds), *Proc. 3rd Int. Aerosol Conf.on Aerosols in Science, Industry Health and Environment,* Pergamon Press 1990, pp. 1136–1139.

Asman, W.A.H., J.A. Van Jaarsveld; A variable-resolution statistical transport model applied for ammonia and ammonium, *Report* 228471007, National Institute of Public Health and Environmental Protection (RIVM), Bilthoven 1990.

Asman, W.A.H.; A detailed ammonia emission inventory for Denmark, *Report DMU LUFT*-A133, National Environmental Research Institute (DMU), Roskilde 1990.

Asman, W.A.H.; Ammonia emission in Europe: updated emission and emission variations, *Report DMU LUFT*-A132, National Environmental Research Institute (DMU), Roskilde 1990. *Report* 228471008 National Institute of Public Health and Environmental Protection (RIVM), Bilthoven 1990.

Asman, W.A.H.; Atmosfaerisk ammoniak og ammonium i Danmark (Atmospheric ammonia and ammonium in Denmark), *NPO Report* A18, Danish National Agency of of Environmental Protection, Copenhagen 1990, 96 pp.

Baeyens, W., H. Dedeurwaerder, F. Dehairs; *Atmos. Environ.* **24**A (1990) 1693–1703.

Belviso, S., B.C. Nguyen, P. Buat-Ménard, S.K. Kim, F. Rassoulzadegan; Production of dimethylsulfonium propionate and dimethylsulfide by the microbial food web in the Ligurian Sea, Ocean Sciences Meeting, New Orleans, 1990, *EOS* **71** (1990) 104.

Belviso, S., S.K. Kim, F. Rassoulzadegan, B. Krajka, B.C. Nguyen, N. Mihalopoulos, P. Buat-Ménard; Production of dimethylsulfonium propionate (DMSP) and dimethylsulphide (DMS) by a microbial foodweb, *Limnol. Oceanogr.* **85** (1990) 1810–1821.

Bonsang, B., E. Fontaine, D. Martin, G. Lambert, M. Kanakidou, J.C. Le Roulley; Vertical distribution of none methane hydrocarbons in the remote marine boundary layer, in: *Proc. 7th Int. Symp. of the Commission on Atmospheric Chemistry and Global Pollution*, Chamrousse, 1990, p. 28.

Bonsang, B., M. Kanakidou, G. Lambert; NMHC in marine atmosphere: preliminary results of monitoring at Amsterdam Island, *J. Atmos. Chem.* **11** (1990) 169–178.

Cornille, P., W. Maenhaut, J.M. Pacyna; PIXE analysis of size-fractionated aerosol samples collected at Birkenes, Norway, during spring 1987, *Nuclear Instruments and Methods in Phys. Res.* **B49** (1990) 376–382.

Edson, J.B., P.G. Mestayer, M.P. Rouault, S.E. Larsen; Simulations of evaporating droplet diffusion in oceanic conditions using the HEXIST/CLUSE numerical models, in: *Proc. 9th AMS Symp. on Turbulence and Diffusion*, Roskilde, 1990, Amer. Met. Soc., Boston 1990, pp. 184–187.

Edson, J.B., S. Levi Alvares, J.-F. Sini, P.G. Mestayer; GWAIHIR, A combined Eulerian/Lagrangian model of the transport of heavy particles in turbulent flows, *Note interne*, LMTTD-H.90.03, 1990.

Edson, J.B.; Simulating droplet motion above a moving surface, in: P.G. Mestayer, E.C. Monahan, P.A. Beetham (eds), *Proc. Workshop on Modelling the Fate and Influence of Marine Spray*, Whitecap Rep. No. 7, Marine Sciences Institute, University of Connecticut 1990, pp. 84–94.

Fairall, C.W.; Modeling the fate and influence of marine spray: A Review, in: P.G. Mestayer, E.C. Monahan, P.A. Beetham (eds*), Proc. Workshop on Modelling the Fate and Influence of Marine Spray*, Whitecap Rep. No. 7, Marine Sciences Institute, University of Connecticut 1990, pp. 1–16.

Hillamo, R., W. Maenhaut; Aerosol size distribution measurements at Dye 3, Greenland, *EOS Trans. Amer. Geophys. Union* **71 No.43** (1990) 1261.

Huebert, B.J., T.S. Bates, A. Bandy, S.E. Larsen, R.A. Duce; IGAC/MAGE: International planning of chemical air/sea exchange research, *EOS* **71**.(35) (1990) 1051–1057.

Hummelshøj, P., N.O. Jensen, S.E. Larsen, C. Hansen; Modelling dry deposition of particles to the ocean, in: *Proc. 9th AMS Symposium on Turbulence and Diffusion*, Amer. Met. Soc., Boston 1990, pp. 321–324.

Larsen, S.E, J.B. Edson, P.G. Mestayer, C.W. Fairall, G. de Leeuw; Deposition velocity of submicron particles determined in the IMST tunnel as a function of the surface aerosol production rate and its relation to over-ocean conditions, in: P.G. Mestayer, E.C. Monahan, P.A. Beetham (eds), *Proc. Workshop on Modelling the Fate and Influence of Marine Spray*, Whitecap Rep. No. 7, Marine Sciences Institute, University of Connecticut 1990, pp. 155–167.

Leeuw, G. de, J.B. Edson, C.W. Fairall, S.E. Larsen, M.P. Rouault, P.G. Mestayer; CLUSE numerical models: fate and influence of aerosol droplets near water surfaces, in: G. Trusty, P. Roney (eds), *Proc. Workshop on infrared propagation in the maritime aerosol layer*, Naval Research Laboratory, Washington 1990.

Leeuw, G. de; *J. Geophys. Res.* **95** (1990) 9779–9782.

Leeuw, G. de; Profiling of aerosol concentrations, particle size distribution and relative humidity in the atmospheric surface layer over the North Sea, *Tellus* **42**B (1990) 342–354.

Leeuw, G. de; Spray droplet source function: from laboratory to open ocean, in: P.G. Mestayer, E.C. Monahan, P.A. Beetham (eds.), *Proc. Workshop on Modelling the Fate and Influence of Marine Spray*, Whitecap Report No. 7, Marine Sciences Institute, University of Connecticut 1990, pp. 17–28.

Losno, R., J.L. Colin, N. Le Bris, G. Bergametti, T. Jickells, B. Lim; Chemistry of the dissolution of atmospheric Al and Fe in marine rainwaters, in: *Proc. 7th Int. Symp. of the Commission of Atmospheric Chemistry and Global Pollution* (CACGP), Chamrousse 1990.

Maenhaut, W.; Recent advances in nuclear and atomic spectrometric techniques for trace element analysis. A new look at the position of PIXE, *Nuclear Instruments and Methods in Phys. Res.* **B49** (1990) 518–532.

Mestayer, P.G., E.C. Monahan, P.A. Beetham (eds); in: *Proc. Workshop on Modelling the Fate and Influence of Marine Spray*, Whitecap Rep. No. 7, Marine Sciences Institute, University of Connecticut 1990.

Mestayer, P.G., J.B. Edson, M.P. Rouault, C.W. Fairall, S.E. Larsen, G. de Leeuw, D.E. Spiel, J. DeCosmo, K.B. Katsaros, E.C. Monahan, R. Schiestel; CLUSE simulations of vapor flux transformations by droplet evaporation, in: P.G. Mestayer, E.C. Monahan, P.A. Beetham (eds), *Proc. Workshop on Modelling the Fate and Influence of Marine Spray*, Whitecap Rep. No. 7, Marine Sciences Institute, University of Connecticut 1990, pp. 100–105.

Mestayer, P.G.; Sea water droplet evaporation in CLUSE model, in: P.G. Mestayer, E.C. Monahan, P.A. Beetham (eds), *Proc. Workshop on Modelling the Fate and Influence of Marine Spray*, Whitecap Rep. No. 7, Marine Sciences Institute, University of Connecticut 1990, pp. 65–76.

Nguyen, B.S., N. Mihalopoulos, S. Belviso; Seasonal variation of atmospheric dimethyl sulfide at Amsterdam Island in the Southern Indian Ocean, *J. Atmos. Chem.* **11** (1990) 123–141.

Rouault, M., S.E. Larsen; Spray droplets under turbulent conditions, Risø-M-2847, 1990, 59 pp.

Rouault, M.P.; The return of living droplets: Transformation of the CLUSE model for sea–salt droplets over the ocean. in: P. G. Mestayer, E. C. Monahan, P. A. Beetham (eds), *Proc. Workshop on Modelling the Fate and Influence of Marine Spray*, Whitecap Rep. No. 7, Marine Sciences Institute, University of Connecticut 1990, pp. 77–83.

Rudolph, J., F.J. Johnen; Measurements of light atmospheric hydrocarbons over the Atlantic in regions of low biological activity, *J. Geophys. Res.* **95** (1990) 20583.

Schulz, M., M. Schwikowski, M. Steiger, K. Naumann, W. Dannecker; Aircraft measurements of aerosol transformation processes and vertical mixing above the North Sea. in: *Proc. 7th Int Symp. of the Commission of Atmospheric Chemistry and Global Pollution* (CACGP), Chamrousse 1990.

Schwikowski, M., M. Schulz, M. Steiger, K. Naumann, W. Dannecker; Transformation and transport of nitrogen compounds above the North Sea investigated by aircraft measurements, *J. Aerosol Sci.* **S21** (1990) S311-S314.

Smith, S.D., K.B. Katsaros, W.A. Oost, P.G. Mestayer; Two major experiments in the humidity exchange over the sea program, *Bull. Amer. Met. Soc.* **71** (1990) 161–172.

Steiger M., K. Naumann, M. Schulz, M. Schwikowski, W. Dannecker; Quantitative determination of sources in urban aerosols using chemical receptor models, *J. Aerosol Sci.* **21** (1990) 271–274.

Turner, S.M., G. Malin, L.E. Bagander, C. Leck; Interlaboratory calibration and sample analysis of dimethyl sulphide in water, *Marine Chem.* **29** (1990) 47–62.

Vawda, Y., I. Colbeck, R.M. Harrison, K.W. Nicholson; The effects of particle size on deposition rates, *J. Aerosol Sci.* **20** (1989) 1155–1158

Xianren, Q., W. Baeyens; *J. Chromatogr.* **514** (1990) 362–370.

1991

Baeyens, W., H. Dedeurwaerder; Particulate trace metals above the southern bight of the North Sea, I. Analytical procedures and average aerosol concentrations, *Atmos. Environ.* **25**A (1991) 293–304.

Baeyens, W., H. Dedeurwaerder; Particulate trace metals above the southern bight of the North Sea, II. Origin and behaviour of the trace metals. *Atmos. Environ.* **25**A (1991) 1077–1092.

Baeyens W., Lansens P., C. Casais, C. Meuleman; Evaluation of some chromatographic columns for the determination of methyl- ethyl- and phenylmercury in polar and non-polar solvents, *J. Chromatography* **58** (1991) 329–340.

Baeyens, W., M. Leermakers, H. Dedeurwaerder; P. Lansens modelizations of the mercury fluxes and the air-sea interface, *Water Air Soil Pollut.* **56** (1991) 731–744.

Batchvarova, E., S. E. Gryning; An applied model for the growth of the daytime mixed layer, *Boundary-Layer Meteorol.* **56** (1991) 261–274.

Belviso, S., B. C. Nguyen, P. Buat-Ménard, S.K. Kim, F. Rassoulzadegan; Production of dimethylsulfonium propionate and dimethylsulfide by the microbial food web in the Ligurian sea, in: P. Borrell, P.M. Borrell, W. Seiler (eds), *Proc. EUROTRAC Symp. '90*, SPB Academic Publishing bv, The Hague 1991, pp. 79–80.

Bonsang, B., D. Martin, G. Lambert, M. Kanakidou, J.C. Le Roulley, G. Sennequier; Vertical distribution of non-methane hydrocarbons in the remote marine boundary layer, *J. Geophys. Res.* **96** (1991) 7313–7324.

Bonsang, B., G. Lambert; Air Sea Exchanges of non methane hydrocarbons, in: P. Borrell, P.M. Borrell, W. Seiler (eds), *Proc. EUROTRAC Symp. '90*, SPB Academic Publishing bv, The Hague 1991, pp. 81–82.

Buijsman, E, P.J. Jonker, W.A.H. Asman, T.B. Ridder; Chemical composition of precipitation collection on a weathership in the North Atlantic, *Atmos. Environ.* **25**A (1991) 873–883.

Donard, O.F.X., M. Wells, L.M. Mayer, S. Ackelson , M.M. Souza Sierra; Contribution of photoreduction of marine waters to the speciation of trace metal fluxes to the atmosphere: preliminary results, in: P. Borrell, P.M. Borrell, W. Seiler (eds), *Proc. EUROTRAC Symp. '90*, SPB Academic Publishing bv, The Hague 1991, pp. 83–84.

Ducastel, G., W. Maenhaut, R. Hillamo, T.A. Pakkanen, J.M. Pacyna; Detailed size distributions of trace elements in southern Norway during spring 1988 and 1989, in: J.G. Farmer (ed), *Int. Conf. on Heavy Metals in the Environment*, CEP consultants, Edinburgh 1991, pp. 270–273.

Edson, J.B., C.W. Fairall, P.G. Mestayer, S.E. Larsen; A study of the inertial-dissipation method for computing air-sea fluxes, *J. Geophys. Res.* **96**, C6 (1991) 10689–10711.

Flossman, A.I; The scavenging of two different types of marine aerosol particles calulated using a two-dimensional detailed cloud model, *Tellus* **43**B (1991) 301–321.

Hummelshøj, P., N.O. Jensen, S.E. Larsen; Particles dry deposition to a sea surface, in: S.E. Schwartz, W.G.N. Slinn (eds), *Precipitation Scavenging and Atmosphere Surface Exchange Processes*, Hemisphere Publishing Co., London 1991, pp. 829–840.

Jickells, T., R.R. Yaaqub, M-M Kane, A. Rendell, T. Davies, R. Harrison, C. Otley, R. Chester, G. Bradshaw, J. Miller, M. Schultz; Atmospheric Inputs into the North Sea, A Progress Report, in: P. Borrell, P.M. Borrell, W. Seiler (eds), *Proc. EUROTRAC Symp. '90*, SPB Academic Publishing bv, The Hague 1991, pp. 85–86.

Larsen, S.E., J.B. Edson, P.G. Mestayer, C.W. Fairall , G. de Leeuw; Sea spray and particle deposition, air/water tunnel experiment and its relation to over-ocean conditions, in: P. Borrell, P.M. Borrell, W. Seiler (eds), *Proc. EUROTRAC Symp. '90*, SPB Academic Publishing bv, The Hague 1991, pp. 87–91.

Leeuw, G. de, S.E. Larsen, P.G. Mestayer, J.B. Edson, D. Spiel; Deposition, dynamics and influences of particles in the marine atmospheric boundary layer, laboratory study, *TNO Report* 1991, TNO Physics and Electronics Laboratory, The Hague 1991.

Lim, B. T.D. Jickells, T.D. Davies; Sequential sampling of particulate, major ions and total trace metals in wet deposition, *Atmos. Environ.* **25** (1991) 745–762.

Lim, B., T.D. Jickells; Dissolved, particulate and acid leachable trace metal concentrations in North Atlanctic precipitation collected on the Global Change Expedition, *Global Biogeochem. Cycles* **4** (1991) 445–458.

Liss, P.S.; Introduction to ASE, in: P. Borrell, P.M. Borrell, W. Seiler (eds), *Proc. EUROTRAC Symp. '90*, SPB Academic Publishing bv, The Hague 1991, p. 77.

Liss, P.S.; The sulphur cycle, in: J. C. Duplessy, A. Pons, R. Fantechi (eds), *Climate and Global Changes*, CEC, Bruxelles 1991, pp. 75–89.

Mestayer, P.G., S.E. Larsen, C.W. Fairall, J.B. Edson; Turbulence sensor dynamic calibration using real-time spectral computations, *J. Atmos. Oceanic. Technol.* **7** (1991) 841–851.

Mihalopoulos, N., J.P. Putaud, B.C. Nguyen, S. Belviso; Annual variation of atmospheric carbonyl sulphide in the marine atmosphere in the southern Indian Ocean, *J. Atmos. Chem.* **13** (1991) 73–82.

Ottley, C.J., R.M. Harrison; The atmospheric input flux of trace metals to the North Sea: A review and recommendations for research, *Sci. Tot. Environ.* **100** (1991) 301–318.

Pakkanen, T.A., R.E. Hillamo, W. Maenhaut, G. Ducastel, J.M. Pacyna; Atmospheris size distrubitions of pollutant elements in southern Norway, *Nordic Symp. on Atmospheric Chemistry*, Stockholm - Tallinn 1991.

Plass, C., R. Koppmann, J. Rudolph; Measurements of dissolved nonmethane hydrocarbons in sea water, *Fresenius J. Anal. Chem.* **339** (1991) 746.

Rojas, C.M., P.M. Otten, R.E. Van Grieken, R. Laane; Dry aerosol deposition over the North Sea estimated from aircraft measurements in: H. van Dop, D.G. Steyn (eds) *Air Pollution Modeling and Its Application* - VIII, Plenum Press, New York-London 1991, pp. 419–425

Rouault, M.P., P.G. Mestayer, R. Schiestel; A model of evaporating spray droplet dispersion, *J. Geophys. Res.* **96** (1991) 7181–7200.

Wells, M., L.M. Mayer, O.F.X. Donard, M.M. Souza Sierra, S. Ackelson; *Nature* **353** (1991) 240–250.

Winkler, P.,M. Schulz, W. Dannecker; Problems with the determination of heavy metals in precipitation, *Fresenius J. Anal. Chem.* **340** (1991) 575–579.

Xhoffer, C., P. Bernard, R. Van Grieken, L. Van der Auwera; Chemical characterization and source apportionment of individual aerosol particles over the North Sea and the English Channel using multivariate techniques, *Environ. Sci. Technol.* **25** (1991) 1470–1478.

Yaaqub, R.R., T.D. Davies, T.D. Jickells, J.M. Miller; Tracer metals in daily collected aerosols at a site in southeast England, *Atmos. Environ.* **25** (1991) 966–985.

1992

Allerup, P., J. Joergensen, H. Madsen, S. Overgaard, F. Vejen, W.A.H. Asman; Precipitation regime in the Kattegat sea area, *Report Havforskning fra Miljøstyrelsen No.* 13, Danish Environmental Protection Agency, Copenhagen 1992, 110 pp. (in Danish).

Amouroux, D., O.F.X. Donard; H_2O_2 variability in estuarine and oceanic environments of the Gulf of Gascogne, in: J.-C. Sorbe and J.M. Jouanneau (eds), *Actes du IIleme Colloque International d'Oceanographie du Golfe de Gascogne*, Session: Chimie et Polluant, 1992, pp. 93–96.

Artaxo, P., M.L.C. Rabello, W. Maenhaut, R. van Grieken; Trace elements and individual particle analysis of aerosol particles from the Antarctic peninsula, *Tellus* **44**B (1992) 318–334.

Asman, W.A.H., J.A. van Jaarsveld; A variable-resolution transport model applied for NH_x in Europe, *Atmos. Environ.* **26**A (1992) 445–464.

Asman, W.A.H.; A detailed ammonia emission inventory for Denmark and some deposition calculations, in: G. Klaassen (ed), *Ammonia Emissions in Europe: Emission Coefficients and Abatement Costs*, IIASA, Laxenburg, Austria 1992, pp. 159–168.

Baeyens, W.; Speciation of mercury in different compartments of the environment, *Trends in Analytical Chemistry* **11** (1992) 245–254.

Bonsang, B., C. Polle, G. Lambert; Evidence for marine production of isoprene, *Geophys. Res. Lett.* **19** (1992) 1129–1132.

Colin, J.-L., T. Jickells; Convenors report, EGS Seventeenth General Assembly, OA8 session: gas-particle-water interactions in the marine rainwater, *EGS Newletter* **45** (1992) 14.

Dierck, I., D. Michaud, L. Wouters, R. Van Grieken; Laser microprobe mass analysis of individual North Sea aerosol particles, *Environ. Sci. Technol.* **26** (1992) 802–808.

Donard, O.F.X., F.M. Martin; Hyphenated technics applied to environmental studies, *T.R.A.C.* **11**(1) (1992) 17–26.

Eijk, A.M.J van, G. de Leeuw; Modeling aerosol particle size distributions over the North Sea from HEXMAX data, TNO Physics and Electronics Laboratory, *Report* FEL-92-A119 (1992).

Eijk, A.M.J. van, G. de Leeuw; Modeling aerosol extinction in a coastal environment, in: A. Kohnle, W.B. Miller (eds), *Atmospheric Propagation and Remote Sensing*, *Proc. SPIE* **1688** (1992) 28–36.

Eijk, A.M.J. van, G. de Leeuw; Modeling aerosol particle size distributions over the North Sea, *J. Geophys. Res.* **97C** (1992) 14417–14429.

François, F., J. Cafmeyer, C. Gilot, W. Maenhaut; Chemical composition of size-fractionated marine aerosols, *Annales Geophysicae, Suppl. II*, **10** (1992) C220.

Gaudry,A., M. Kanakidou, N. Mihalopoulos, B. Bonsang, G. Bosang P. Monfray, R. Tymen, B.C. Nguyen; Atmospheric trace compounds at a european coastal site: Application to CO_2, CH_4 and COS flux determination, *Atmos. Environ.* **26A** (1992) 145–157.

Geernaert, G., S.E. Larsen, 1992. On the role of humidity in estimating radar cross-sections, wind stress, heat flux, and stratification, *Annales Geophysicae, Suppl. II*, **10**, (1992) C210.

Geernaert, G., S.E. Larsen; On the role of humidity in estimating marine surface layer stratification and scatterometer cross section, *J. Geophys. Res.* **98,** C1 (1992) 927–932.

Haster, T., M. Schulz, B. Schneider, W. Dannecker; The use of short-term measurements of trace element size-distributions to investigate aerosol dynamics in a marine Lagrangian-type experiment, *J. Aerosol Sci.* **23** (1992) 703–706.

Hummelshøj, P., N.O. Jensen and S.E. Larsen; Particle dry deposition to the seasurface, in: S.E. Schwartz, W.G.N. Slinn (eds), *Proc. 5th Int. Conf. on Precipitation Scavenging and Atmosphere-Surface Exchange Processes*, Hemisphere Publishing Corporation, Washington 1992, pp. 829–840.

Injuk, J., Ph. Otten, R. Laane, W. Maenhaut, R. van Grieken; Atmospheric concentrations and size distributions of aircraft-sampled Cd, Cu, Pb and Zn over the Southern Bight of the North Sea, *Atmos. Environ.* **26**A (1992) 2499–2508.

Jickells, T., M.M. Kane, A. Rendell, T. Davies, M. Tranter, K.E. Jarvis; Applications of inductively coupled plasma techniques to the analysis of atmospheric precipitation, *Anal. Proc.* **29** (1992) 288–291.

Jickells, T.D., T.D. Davies, M. Tranter, S. Landsberger, K. Jarvis, P. Abrahams; Trace elements in snow samples from the Scotish Highlands: sources and dissolved/particulate distributions, *Atmos. Environ.* **26** (1992) 393–401.

Jickells, T.D., T.M. Church, J.R. Scudlark, F. Dehairs; Barium in North Atlantic rain water: a reconnaissance, *Atmos. Environ.* **26**A (1992) 2541–2546.

Koppmann, R., R. Bauer, F.J. Johnen, C. Plass, J. Rudolph; Distribution of light nonmethane hydrocarbons over the Mid Atlantic, *J. Atmos. Chem.* **15** (1992) 215–234.

Larsen, S.E., F. Hansen, and M. Schulz; Initial analysis of wind data from a sonic anemometer on FS Alkor in the NOSE 1991 campaign, in: T.D. Jickells, L.J. Spokes (eds), *EUROTRAC Air-Sea Exchange Experiment North Sea, 14--27 September 1991*, School of Environmental Sciences, University of East Anglia, Norwich 1992, pp. 1–12.

Larsen, S.E.; Convenors report, EGS Seventeenth General Assembly, OA7 Session: Dynamics and Bio-Geochemistry of Ocean-Atmosphere Interface, *EGS Newsletter* **45** (1992) 14.

Lassus, P.F.; Suivi saisonnier des hydrocarbures non méthanique légers en Mediterranée, *Report DEA*, Université de Paris VII, 1992.

Leeuw, G. de, G.J. Kunz; NOVAM evaluation from aerosol and lidar measurements in a tropical marine environment, in: A. Kohnle, W.B. Miller (eds), *Atmospheric and Remote Sensing*, *Proc. SPIE* **1688** (1992) 14–27.

Maenhaut, W.; Trace element analysis of environmental samples by nuclear analytical techniques, *Int. J. PIXE* **2** (1992) 609–635.

Malin, G., S.M. Turner, P.S. Liss; Sulfur: The plankton/climate connection, *J. Phycol.* **28** (1992) 590–597.

Mestayer, P.G., S.E. Larsen, G. de Leeuw, J.B. Edson, D.E. Spiel, A. Zoubiri, P. Hummelshøj, A.M.J. van Eijk; Dynamics of submicron aerosol deposition to the coastal sea surface: laboratory and numerical simulations, *Anales Geophysicae, Suppl.* **10** (1992) 219.

Mihalopoulos, N., B.C. Nguyen, C. Boissard, J.M. Campin, J.P. Putaud, S. Belviso, J. Barnes, H.K. Becker; Field study of dimethylsulfide oxidation in the boundary layer: Variation of dimethylsulfide, methanesulfonic acid, sulfur dioxide, non-sea-salt sulfate and Aitken nuclei at a costal site, *J. Atmos. Chem.* **14** (1992) 459–487.

Mihalopoulos, N., B.C. Nguyen, J.P. Putaud, S. Belviso; The oceanic source of carbonyl sulfide (COS), *Atmos. Environ.* **26A** (1992) 1383–1394.

Nguyen, B.C., N. Mihalopoulos, A. Gaudry, J.P. Putaud, L. Gallet, W.C. Keene, N.J. Galloway; Covariations of oceanic dimethylsulfide, its oxidation products and rain acidity in the Southern Indian Ocean, *J. Atmos. Chem.* **15** (1992) 39–53.

Nguyen, B.C., N. Mihalopoulos, C. Boissard, J.M. Campin, J.P. Putaud, S. Belviso; Variations of dimethylsulfide, methanesulfonic acid, sulfur dioxide, non-sea-salt sulfate and Aitken nuclei at a costal site (Penmarc'h) during OCEANO-NO_x experiment, *Air Pollut. Res. Rep.* **35** (1992) 152–161.

Ottley, C.J., R.M. Harrison; The spatial distribution and particle size of some inorganic nitrogen, sulphur and chlorine species over the North Sea, *Atmos. Environ.* **26**A (1992) 1689–1699.

Plass, C., R. Koppmnann, J. Rudolph; The distribution of light nonmethane hydrocarbons in the surface waters of the Mid Atlantic, *J. Atmos. Chem.* **15** (1992) 235–152.

Putard, J.P., N. Mihalopoulos, B.C. Nguyen, J.M. Campin, S. Belviso; Seasonal variation of sulfur dioxide and dimethyl sulfide concentrations at Amsterdam Island in the Southern Indian Ocean, *J. Atmos. Chem.* **15** (1992) 117–131.

Rojas, C.M., R.E. Van Grieken; Electron microprobe characterization of individual aerosol particles collected by aircraft above the Southern Bight of the North Sea, *Atmos. Environ.* **26**A (1992) 1231–1237.

Runge, E.H., W.A.H. Asman, N.A. Kilde; A detailed emission inventory of sulphur dioxide for Denmark, *Report Risø*-M-2937 (1992) 33 pp.

Schulz, M., M. Kriews; Atmospheric input/Box 7a, *Report to the North Sea Task Force*, 1992.

Schulz, M., S.E. Larsen; Overview of the meteorological conditions during the North-Sea experiment 1991, in: T.F. Jickells, L.J. Spokes (eds), *EUROTRAC Air–Sea Exchange Experiment North Sea, 14–27 September 1991*, School of Environmental Sciences, University of East Anglia, Norwich 1992, pp. 1–18.

Schulz, M., W. Baeyens, W. Dannecker, R. van Grieken, R. Harrison, G. Jennings, T. Jickells, S.E. Larsen, W. Maenhaut, G. Petersen, B. Quack, S. Rhapsomanikis, M. Schrader, L. Soerensen; North Sea Experiment 1991 – concept and first results on the change of aerosol size distributions during transport over sea, *Annales Geophysicae, Supp. II*, **10** (1992) C220.

Sempreviva, A.M., S.E. Larsen, N.G. Mortensen; Experimental study of flow modification inland from a coast for nonneutral conditions, *Report Risø*-M-2924 (EN) (1992) 41 pp.

Van Grieken, R. J. Injuk, P. Otten, C. Rojas, H. Van Malderen, R. Laane; Fluxes and sources of heavy metal inputs into the Southern Bight of the North Sea, in. A. Müezzinoglu, M. L. Williams (eds), *Industrial Air Pollution: Assessment and Control, NATO Advanced Study Institute Series G: Ecological Sciences*, Vol. **31**, Springer Verlag, Berlin-Heidelberg 1992, pp. 184–193.

Van Malderen, H., C. Rojas, R. Van Grieken; Characterization of individual giant aerosol particles above the North Sea *Environ. Sci. Technol.* **26** (1992) 750–756.

Xhoffer, C., L Wouters, R. Van Grieken; Characterization of individual particles in the North Sea surface microlayer and underlying seawater: comparison with atmospheric particles, *Environ. Sci. Technol.* **26** (1992) 2151–2162.

1993

Andreae, M.O., R.J. Ferek; Photochemical production of carbonyl sulfide in seawater and its emission to the atmosphere, *Global Biogeochem. Cycles* **6** (1992) 175–183.

Asman, W.A.H. and P.K. Jensen; Wet deposition processes, (in Danish), *The Danish Marine Res. Prog. 90*, Danish Environmental Agency, Rep. No. 26, Copenhagen 1993,

Asman, W.A.H., E.H. Runge, N.A. Kilde; Emission of NH_3, NO_x, SO_2 and NMVOC to the atmosphere in Denmark, (in Danish), *The Danish Marine Res. Prog. 90*, Danish Environmental Agency, Rep. No. 19, Copenhagen 1993.

Asman, W.A.H., R. Berkowicz, O. Hertel, L.-L. Sorensen, K. Granby, H. Nielsen, E.H. Runge, P.K. Jensen, S.-E. Gryning, S.E. Larsen, N. A. Kilde, H. Madsen, P. Allerup, S. Overgaard, J. Jorgensen, F. Vejen, K. Hedegaard; Deposition of nitrogen compounds to Danish coastal waters, in: P.M. Borrell, P. Borrell, T. Cvitaš, W. Seiler (eds), *Proc. EUROTRAC Symp. '92*, SPB Academic Publishing bv, The Hague 1993, pp. 779–782.

Belviso, S., J. P. Putaud, B. C. Nguyen, P. Buat-Ménard; Biogeochemical Cycling of Dimethylsulfide (DMS). Size Fractionation of DMS Precursors in areas of the Mediterranean Sea and the tropical northeastern Atlantic ocean, in: P.M. Borrell, P. Borrell, T. Cvitaš, W. Seiler (eds), *Proc. EUROTRAC Symp. '92*, SPB Academic Publishing bv, The Hague 1993, p. 313.

Belviso, S., P. Buat-Ménard, J.P. Putard, B.C. Nguyen, H. Claustre, J. Neveux; Size distribution of dimethyl-sulfoniopropionate (DMSP) in areas of the tropical northeastern Atlantic Ocean and the Mediterranean Sea, *Mainer. Chem.* **44** (1993) 55–71.

Belviso, S., P. Buat-Ménard, J.P. Putaud, B.C. Nguyen, H. Claustre, J. Neveux; Size distribution of dimethylsulfoniopionate (DMSP) in areas of the tropical northeastern Altantic Ocean and the Mediterranean Sea,, *Marine Chem.* **44** (1993) 55–71.

Bonsang, B., C. Polle, G. Lambert; Production of non-methane hydrocarbons in sea water, *Anales de l'institut oceanographique, Paris* **69** (1) (1993) 125–128.

Bonsang, B.; Hydrocarbon emissions from the ocean, in: *Proc. NATO Workshop Tropospheric Chemistry in the Polar Regions*, Wolfville, Nova Scotia 1993, pp. 251–260.

Chester, R., G.F. Bradshaw, C.J. Ottley, R.M. Harrison, J.L. Merrett, M.R. Preston, A.R. Rendell, M.M. Kane, T.D. Jickells; The atmospheric distribution of trace metals, nutrients and trace organics over the North Sea, *Phil. Trans. R. Soc. London* **A 343** (1993) 543–556.

Chester, R., K.J.T. Murphy, F.J. Lin, A.S. Berry, G.A. Bradshaw, P.A. Corcoran; Factors controlling the solubility of trace metals from non-remote aerosols deposited to the sea surface by the dry deposition mode, *Marine Chem.* **42** (1993) 107–136.

Chin, M., D.D. Davis; Global sources and sinks of OCS and CS_2 and their distribution, *Global Biogeochem. Cycles* **7** (1993) 321–337.

Dannecker, W., H. Hinzpeter, M. Kriews, K. Naumann, M. Schulz, M. Schwikowski, M. Steiger; Atmospheric transport of contaminants, their concentrations and input into the North Sea, in: J. Suendermann (ed), *Circulation and Contaminant Fluxes in the North Sea*, Springer Verlag, Heidelberg 1993, pp. 138–189.

Dekker, H.J., G. de Leeuw; Bubble excitation of surface waves and aerosol droplet production: a simple dynamical model, *J. Geophys. Res.* **98**, C6 (1993) 10223–10232.

Djupstroem, M., J.M. Pacyna, W. Maenhaut, J.W. Winchester, S.-M. Li, G.E. Shaw; Contamination of Arctic air during a haze event in late winter 1986, *Atmos. Environ.* **27**A (1993) 2999–3010.

Donard, O.F.X., M.M. Souza Sierra, M. Ewald, C. Belin, D. Amouroux, H. Etcheber; Fluorescence of continental and marine waters. *Annales de l'Institut Océanographique* **69** (1993) 189–191.

Ducastel, G., W. Maenhaut, K. Beyaert, J.E. Hanssen; On the contribution from biogenic dimethylsulfide to the aerosol non-seasalt sulfate at Ny Aalesund, Spitsbergen, *Annales Geophysycae*, Vol. 11, Suppl II (1993) C236.

Eijk, A.M.J. van and G. de Leeuw; Atmospheric effects on IR propagation, in: B.F. Andresen, F.D. Sheperd (eds), *Infrared Technology XIX. Proc. SPIE* **2020** (1993) 196–206.

Eijk, A.M.J. van, P.G. Mestayer, G. de Leeuw, Conversion of the CLUSE model for application over open ocean, *TNO Physics and Electronics Laboratory, Report FEL-93-A035* (1993).

François, F., J. Cafmeyer, C. Gilot, W. Maenhaut; Chemical composition of size-fractionated marine aerosols, *Annales Geophysicae, Suppl. II,* **10** (1992) C223.

François, F., J. Cafmeyer, C. Gilot, W. Maenhaut; Chemical composition of size-fractionated atmospheric aerosols at some coastal stations and over the North Sea, in: P.M. Borrell, P. Borrell, T. Cvitaš, W. Seiler (eds), *Proc. EUROTRAC Symp. '92*, SPB Academic Publishing bv, The Hague 1993, pp. 788–791.

Gathman, S.G., D.R. Jensen, W.P. Hooper, J.E. James, H.E. Gerber, K.L. Davidson, M.H. Smith, I.E. Consterdine, G. de Leeuw, G.J. Kunz, M. Moerman; NOVAM evaluation utilizing electro-optics and meteorological data from KEY-90, *Technical Report* 1608, Naval Command Control and Ocean Surveillance Center, San Diego 1993.

Hertel, O., R. Berkowicz, J. Christensen, Ø. Hov; Test of two numerical schemes for use in atmospheric transport-chemistry models, *Atmos. Environ.* **27**A (1993) 2591–2611.

Hertel, O., R. Berkowicz, W:A:H: Asman, J. Christensen, L.L. Sørensen; Atmospheric chemistry: reactions (in Danish), *Danish Marine Res. Prog.*, Danish EPA Rep. **24,** 1993, p.48.

Hillamo, R.E., V.-M. Kerminen, W. Maenhaut, J.-L. Jaffrezo, S. Balachandran, C.I. Davidson; Size distributions of atmospheric trace elements at Dye 3, Greenland - I. Distribution characteristics and dry deposition velocities, *Atmos. Environ.* **27**A (1993) 2787–2802.

Injuk, J., H. van Malderen, R. van Grieken, E. Swietlicki, J.M. Knox, R. Schofield; EDXRS study of aerosol composition variations in air masses crossing the North Sea, *X-Ray Spectrom.* **22** (1993) 220–228.

Injuk, J., H. Van Malderen, R. Van Grieken; Study of the heavy metal concentration, deposition and sources of the North Sea aerosols using X-ray emission techniques, in: P.M. Borrell, P. Borrell, T. Cvitaš, W. Seiler (eds), *Proc. EUROTRAC Symp. '92*, SPB Academic Publishing bv, The Hague 1993, pp. 793–795.

Jaffrezo, J.-L., R.E. Hillamo, C.I. Davidson, W. Maenhaut; Size distributions of atmospheric trace elements at Dye 3, Greenland - II. Sources and transport, *Atmos. Environ.* **27**A (1993) 2803–2814.

Jensen, D.R., G. de Leeuw and A.M.J. van Eijk; Work plan for the marine aerosol properties and thermal imager performance trial (MAPTIP), *Technical Document* 2573, Naval Command, Control and Ocean Surveillance Center, San Diego 1993.

Lansens P., Meuleman C. Casais C., Baeyens W.; Comparative study of microwave induced plasma atomic emission spectrometry and atomic fluorescence spectrometry as gaschromatographic detectors for the determination of methylmercury in biological samples., *App. Organometallic Chem.* **7** (1993) 45-51.

Larsen, S.E., J.B. Edson, C.W. Fairall, P.G. Mestayer; Measurements of temperature spectra by a sonic anemometer, *J. Atmos. Ocean. Technol.* **10** (1993) 345–354.

Larsen, S.E., P. Hummelshoej, N.O. Jensen, J.B. Edson, G. de Leeuw, P.G. Mestayer; Deponering af luftbaarne partikler til havoverflader, *Havforskning for Miljoestyrelsen* Nr. 47 (1993), Copenhagen, Denmark, pp. 1–87.

Larsen, S.E.; Air to sea deposition of gases and particles, in: P.M. Borrell, P. Borrell, T. Cvitaš, W. Seiler (eds), *Proc. EUROTRAC Symp. '92*, SPB Academic Publishing bv, The Hague 1993, pp. 689–692.

Leeuw, G. de, A.M.J. van Eijk and G.R. Noordhuis; Modeling aerosols and extinction in the marine atmospheric boundary layer, in: A. Kohnle, W.B. Miller (eds), *Atmospheric Propagation and Remote Sensing II, Proc. SPIE* **1968** (1993) 70–80.

Leeuw, G. de, A.M.J. van Eijk, H. Dekker; Transformation and removal of N-compounds in the marine atmosphere, in: S. E. Larsen (ed), *Progress Report 1992*, *Contract STEP*-CT90-0047, EC, Brussels 1993.

Leeuw, G. de; Aerosol effects on electro-optical propagation over sea, in: M. Oron, I. Shladov and Y. Weissman (eds), *Eighth Meeting on Optical Engineering in Israel: Optical Engineering and Remote Sensing*, *Proc. SPIE* **1971** (1993) 2–15.

Li, C.-L., P.K. Hopke, W. Maenhaut, P. Pacyna; Identification of the potential source locations for elements observed in particles collected at Ny Aalesund, in: *Abstracts 12th Annual Meeting of the American Association for Aerosol Research* (AAAR'93), Oak Brook 1993, p. 224.

Liousse, C., H. Cashier and S.G. Jennings; Optical and thermal analysis of black carbon in different environment: variations of o, the specific attenuation coefficient, *Atmos. Environ.* **27**A (1993) 1203–1211.

Liss, P.S., A.J. Watson, M.T. Liddicoat, G. Malin, P.D. Nightingale, S.M. Turner, R.C. Upstill-Goddard; *Phil. Trans. R. Soc. London* **A 343** (1993) 531–541.

Liss, P.S., G. Malin and S.M. Turner; Production of DMS by marine phytoplankton, in: G. Restelli and G. Angeletti (eds), *Dimethylsulphide: Oceans, Atmosphere, and Climate*, Kluwer, Dordrecht 1993, pp. 1–14.

Liss, P.S.; Marine emissions, in: P.M. Borrell, P. Borrell, T. Cvitaš, W. Seiler (eds), *Proc. EUROTRAC Symp. '92*, SPB Academic Publishing bv, The Hague 1993, p. 254.

Losno, R., J.L. Colin, N. Lebris, G. Bergametti, T.D. Jickells, B. Lim; Solubility of aluminium in rainwater and molten snow, *J. Atmos. Chem.* **17** (1993) 29–43.

Maenhaut, W., G. Ducastel, J.M. Pacyna; Atmospheric aerosol at Ny Aalesund during winter 1989: Multielemental composition, receptor modelling, and comparison with data from previous winters, in: N.Z. Heidam (ed), *Proc. 5th Int. Symp. Arctic Air Chemistry, NERI Technical Report* **70** (1993) 1–11.

Maenhaut, W., G. Ducastel, R.E. Hillamo, T. Pakkanen, J.M. Pacyna; Atmospheric aerosol studies in Southern Norway using size-fractionating sampling devices and nuclear analytical techniques, *J. Radioanal. Nucl. Chem.* **167** (1993) 271–281.

Maenhaut, W., G. Ducastel, R.E. Hillamo, T. Pakkanen; Evaluation of the applicability of the MOUDI impactor for aerosol collections with subsequent multielement analysis by PIXE, *Nucl. Instrum. Methods Phys. Res.* **B75** (1993) 249–256.

Maenhaut, W., K.G. Malmqvist; Particle induced X-ray emission analysis, in: R. Van Grieken, A.A. Markowicz (eds), *Handbook on X-Ray Spectrometry: Methods and Techniques*, Marcel Dekker Inc., New York 1993, pp. 517–581.

Maenhaut, W.; Composition and origin of the regional atmospheric aerosol at great distance from anthropogenic source areas. Assessment of the extent of the anthropogenic perturbation, in: *Belgian Impulse Programme "Global Change" Symp.1993 Proc.* Vol 1, Belgian Science Policy Office, Brussels 1993, pp. 5–30.

Malin, G., S.M. Turner, P. Liss, P. Holligan, D. Harbour; Dimethylsulphide and dimethylsulphoniopropionate in the Northeast Atlantic during the summer coccolithophore bloom, *Deep-Sea Res.* **40** (1993) 1487–1508.

Meuleman C., Casais C., Lansens P., Baeyens W.; A study of the behaviour of methylmercury compounds in aqueous solutions and of gas/liquid distribution coefficients, using head space analysis, *Water Research* **27** (1993) 1431–1446

Mihalopoulos, N., B.C. Nguyen, J.P. Putaud; Seasonal variation of methylsulfonic acid in precipation at Amsterdam Island in the Southern Indian Ocean, *Atmos. Environ.* **27A** (1993) 2069–2073.

Ottley, C.J., R.M. Harrison; Atmospheric dry deposition flux of metallic species to the North Sea, *Atmos. Environ.* **27**A (1993) 685–695.

Pakkanen, T.A., R.E. Hillamo, W. Maenhaut; Simple nitric dissolution method for electrothermal atomic absorption spectrometric analysis of atmospheric aerosol samples collected by a Berner-type low-pressure impactor, *J. Anal. Atom. Spectrom.* **8** (1993) 79–84.

Plass-Duelmer, C., A. Kehim, R. Koppmann, F.J. Johnen and J. Rudolph; Emissions of light nonmethane hydrocarbons from the Atlantic to the atmosphere, *Global Biogeochem. Cycles* **7** (1993) 211–228

Prinn, R., P. Liss, P. Buat-Ménard; Biogeochemical ocean-atmosphere transfers, *Global Biogeochem. Cycles* **7** (1993) 245–246.

Putaud, J.P., S. Belviso, B.C. Nguyen, N. Mihalopoulos; Dimethylsulfide, aerosol and condensation nuclei over the tropical northeastern Altantic Ocean, *J. Geophys. Res.* **98** (1993) 14863–14871.

Quetel, C.R., E. Remoudaki, J.E. Davies, J.-C. Miquel, S. Fowler, C.E. Lambert, G. Bergametti, P. Buat-Ménard; Impact of atmospheric deposition on particulate iron flux and distribution in northeastern Mediterranean waters, *Deep-Sea Res.* **40** (1993) 989–1002.

Ratte, M., C. Plass-Duelmer, R. Koppmann, J. Denga, J. Rudolph; Production mechanism of C_2–C_4 hydrocarbons in sea water: field measurements and experiments, *Global Biogeochem. Cycles* **7** (1993) 369–378.

Rendell, A.R., C.J. Ottley, T.D. Jickells, R.M. Harrison; The atmospheric input of nitrogen species to the North Sea, *Tellus* **45**B (1993) 53–63.

Rojas, C.M. J. Injuk, R.E. Van Grieken, R.W. Laane; Dry and wet deposition fluxes of Cd, Cu, Pb and Zn into the Southern Bight of the North Sea, *Atmos. Environ.* **27**A (1993) 251–259.

Rojas, C.M., R.E. Van Grieken, R.W. Laane; Comparison of three dry deposition models applied to field measurements in the Southern Bight of the North Sea, *Atmos. Environ.* **27**A (1993) 363–370.

Rojas, C.M., R.E. van Grieken, W. Maenhaut; Elemental composition of aircraft-sampled aerosols above the southern bight of the North Sea, *Water, Air Soil Pollut.* **71** (1993) 391–404.

Schulz, M.; ASE North Sea Experiment 1991, in: P.M. Borrell, P. Borrell, T. Cvitaš, W. Seiler (eds), *Proc. EUROTRAC Symp. '92*, SPB Academic Publishing bv, The Hague 1993, p. 693.

Spokes, L., T.D. Jickells, A. Rendell, M. Schulz, A. Rebers, W. Danneker, O. Kruger, M. Leermakers, W. Baeyens; High atmospheric nitrogen deposition events over the North Sea, *Marine Pollut. Bull.* **26** (1993) 698–703.

Spokes, L.J., T.D. Jickells, M. Schulz, A. Rebers, W. Dannecker, W. Baeyens, M. Leermakers; EUROTRAC air-sea exchange experiment: North Sea 4th–27th September 1991 - a wet deposition study, in: P.M. Borrell, P. Borrell, T. Cvitaš, W. Seiler (eds), *Proc. EUROTRAC Symp. '92*, SPB Academic Publishing bv, The Hague 1993, p. 792.

Struyf, H., R. Van Grieken; An overview of wet deposition of micropollutants to the North Sea, *Atmos. Environ.* **27**A (1993) 2669–2687.

Sutton, M.A., D. Fowler, R. I. Smith, M. Eager, C.J. Placee, W.A.H. Asman; Modelling the net exchange of reduced nitrogen; in: J. Slanina, G. Angeletti, G. Beilke,(eds), General assement of biogenic emissions and deposition of notrogen compounds and oxidants in Europe, *Air Pollution Research Report* **47** CEC, Brussels 1993, pp.117–131.

Sutton, M.A., W.A.H. Asman, J.K. Schjørring; Dry deposition of reduced nitrogen; in: G. Lövblad, J.W. Erisman, D. Fowler (eds), M*odels and methods for the quantification of atmospheric input to ecosystems. Nordic Council of Ministers, Rep.* **573** Copenhagen. Denmark 1993, pp. 127–143.

Van Malderen, H., L. De Bock, J. Injuk, Ch. Xhoffer, R. Van Grieken; North Sea aerosol characterization by single particle analysis techniques, in: *Progress in Belgian Oceanographic Research*, Royal Academy of Belgium, Brussels, 1993, pp. 119–135.

1994

Amouroux, D., C. Pecheyran, G. Uher, V. Ulshoefer, O.F.X. Donard, M.O. Andreae; Experimental study of photochemical reactivity in seawater: Impact on selenium speciation, in: P.M. Borrell, P. Borrell, T. Cvitaš, W. Seiler (eds), *Proc. EUROTRAC Symp. '94*, SPB Academic Publishing bv, The Hague 1994, pp. 1060–1063.

Amouroux,D., O. F. X. Donard, S. Rapsomanikis; Sea-to-air selenium fluxes in the eastern Mediterranean Sea, in: P.M. Borrell, P. Borrell, T. Cvitaš, W. Seiler (eds), *Proc. EUROTRAC Symp. '94*, SPB Academic Publishing bv, The Hague 1994, pp. 573–577.

Asman, W.A.H.; Below cloud scavenging coefficients for highly soluble gases, in: P.M. Borrell, P. Borrell, T. Cvitaš, W. Seiler (eds), *Proc. EUROTRAC Symp. '94*, SPB Academic Publishing bv, The Hague 1994, pp. 1111–1113.

Asman, W.A.H.; Emission and deposition of ammonia and ammonium, *Nova Acta Leopoldina NF 70* **288** (1994) 236–297.

Asman, W.A.H., L.L. Sørensen, R. Berkowicz, K. Granby, H. Nielsen, B. Jensen, E.H. Runge, C. Lykkelund, S.E. Gryning,A.M. Sempreviva; Dry deposition processes (in Danish), *Danish Marine Res. Prog. 90, Danish EPA Rep.* **35** 1994, p. 199.

Asman, W.A.H., O. Hertel, R. Berkowicz, J. Christensen, E. H. Runge; Modelling the atmospheric nitrogen input to the Danish coastal waters, in: P.M. Borrell, P. Borrell, T. Cvitaš, W. Seiler (eds), *Proc. EUROTRAC Symp. '94*, SPB Academic Publishing bv, The Hague 1994, pp. 632–634.

Asman, W.A.H., R. Berkowicz, J. Christensen, O. Hertel, E.H. Runge; Atmospreric nitrogen input to the Kattegat (in Danish), *Danish Marine Res. Prog. 90, Danish EPA Rep.* **37** 1994, p. 115.

Asman, W.A.H., R. M. Harrison, C. J. Ottley; Estimation of the net air-sea flux of ammonia over the southern bight of the North Sea, *Atmos. Environ.* **28** (1994) 3647–3654.

Asman, W.A.H., R. M. Harrison, C. J. Ottley; The net air-sea flux of ammonia over the southern bight of the North Sea, in: P.M. Borrell, P. Borrell, T. Cvitaš, W. Seiler (eds), *Proc. EUROTRAC Symp. '94*, SPB Academic Publishing bv, The Hague 1994, pp. 635–638.

Asman, W.A.H., R.M. Harrison, C.J. Ottley; Estimation of the net air-sea flux over the southern bight of the North Sea, *Atmos. Environ.* **28** (1994) 3647–3654.

Asman, W.A.H., S. Matthiesen, F. Vejen; Sequential precipation on Zealand (in Danish), *Danish Marine Res. Prog. 90, Danish EPA Rep.* **54** 1994, p 153.

Berghmans, P., J. Injuk, R. Van Grieken, F. Adams; Microanalysis of atmospheric particles and fibres by electron energy loss spectroscopy, electron spectroscopic imaging and scanning proton microscopy, *Anal. Chim. Acta* **297** (1994) 27.

De Bock, L.A., H. Van Malderen, R.E. Van Grieken; Individual aerosol particle composition variations in air masses crossing the North Sea, *Environ. Sci. Technol.* **28** (1994) 1513–1520.

François, F., W. Maenhaut; Chemical composition and sources of the size-fractionated atmospheric aerosol collected at the research platform Nordsee, in: P.M. Borrell, P. Borrell, T. Cvitaš, W. Seiler (eds), *Proc. EUROTRAC Symp. '94*, SPB Academic Publishing bv, The Hague 1994, pp. 472–476.

Granby, K., S.E. Gryning, O. Hertel; Hydrogen peroxide concentrations in relation to mixing heights at a Danish sea site and a land site: Comparison with a trajectory model, in: P.M. Borrell, P. Borrell, T. Cvitaš, W. Seiler (eds), *Proc. EUROTRAC Symp. '94*, SPB Academic Publishing bv, The Hague 1994, pp. 651–654.

Harrison, R.M.; Chemical transformations in the North Sea atmosphere., in: P.M. Borrell, P. Borrell, T. Cvitaš, W. Seiler (eds), *Proc. EUROTRAC Symp. '94*, SPB Academic Publishing bv, The Hague 1994, pp. 581–587.

Harrison, R.M., Z. Zlatev, C.J. Ottley; A comparison of the predictions of a Eulerian atmospheric transport-chemistry model with experimental measurements over the North sea, *Atmos. Environ.* **28** (1994) 497–516.

Harrison, R.M.,M.I. Msibi, A.-M.N. Kitto, S. Yamulki; Atmospheric chemical transformations of nitrogen compounds measured in the North Sea Experiment, September, 1991, *Atmos. Environ.* **28** (1994) 1593–1599.

Hillamo, R.E., T. Mäkelä, M. Mäkinen, W. Maenhaut; Small deposit area low pressure impactor: Design, calibration and field tests, in: R.C. Flagan (ed) *Absts. of 4th Int. Aerosol Conf. Vol.2,* Los Angeles 1994, pp. 579–580.

Injuk, J., L. Breitenbach, R. Van Grieken, U. Wätjen; Performance of a nuclear microprobe to study giant marine aerosol particles. *Microchim. Acta*, **114/115** (1994) 313–321.

Injuk, J., R. Van Grieken, L. Breitenbach, U. Wätjen; Study of individual particle types and heavy metal deposition for North Sea aerosols, in: P.M. Borrell, P. Borrell, T. Cvitaš, W. Seiler (eds), *Proc. EUROTRAC Symp. '94*, SPB Academic Publishing bv, The Hague 1994, p. 711.

Jickells, T., T. Church, A. Veron, R. Arimoto; Atmospheric inputs of manganese and aluminium to the Sargasso Sea and their relation to surface water concentrations, *Marine Chem.* **46** (1994) 283–292.

Kane, M.M., A.R. Rendell, T.D. Jickells; Atmospheric scavenging processes over the North Sea, *Atmos. Environ.* **28** (1994), 2523–2530.

Katsaros, K.B., J. DeCosmo. R.J. Lind, R.J. Anderson, S.D. Smith, R. Kraan, W. Oost, K. Uhlig, P.G. Mestayer, S.E. Larsen, M.H. Smith, G. de Leeuw; Measurements of humidity and temperature during the HEXOS main experiment. J. *Atmos. Ocean. Tech.* 11 (1994) 964–981.

Larsen, S.E., G. de Leeuw, A. Flossmann, E. Ljungström, P.G. Mestayer; Final report on the STEP project CT90-0047; transformation and removal of N-compounds in the marine atmosphere. Risø National Laboratory, Roskilde, Denmark 1994, 75 pp.

Larsen, S.E., J.B. Edson, P. Hummelshøj, N.O. Jensen, G. de Leeuw, P.G. Mestayer; Deponering af luftb årne partikler til havoverflader. Havforskning for Miljstyrelsen No. 47. Danish Environmental Protection Agency, Copenhagen 1994, 79 pp.

Larsen, S.E., J.B. Edson, P. Hummelshøj, N.O. Jensen, G. de Leeuw, P.G. Mestayer; Laboratory study of the particle dry deposition velocity over water, in: P.M. Borrell, P. Borrell, T. Cvitaš, W. Seiler (eds), *Proc. EUROTRAC Symp. '94*, SPB Academic Publishing bv, The Hague 1994, pp. 698–670.

Leeuw, G. de, A.M.J. van Eijk, P.G. Mestayer, S.E. Larsen; Aerosol dynamics near the air - sea interface, in: P.M. Borrell, P. Borrell, T. Cvitaš, W. Seiler (eds), *Proc. EUROTRAC Symp. '94*, SPB Academic Publishing bv, The Hague 1994, pp. 707–710.

Leeuw, G. de, G.J. Kunz, S.E. Larsen, F. Aa Hansen; Measurements of CO_2 fluxes and bubbles from a tower during ASGASEX. *Proc. 2nd Int. Conf. on Air-Sea Interaction and on Meteorology and Oceanography of the Coastal Zone*, Lisbon, Portugal, *Amer. Meteor. Soc.* 1994, pp. 267-268.

Lim, B., T.D. Jickells, J.L. Colin, R. Losno; Solubilities of Al, Pb, Cu and Zn in rain sampled in the marine environment over the North Atlantic and Mediterranean Sea, *Global Biogeochem. Cycl.* **8** (1994) 349–362.

Liss, P.S., G. Malin, S.M. Turner, P.M. Holligan; Dimethyl sulphide and Phaeocystis: A review, *J. Marine Syst.* **5** (1994) 41–53.

Maenhaut, W., F. François, I. Salma, J. Cafmeyer, C. Gilot; Regional and global atmospheric aerosol studies using the "Gent" stacked filter unit and other aerosol collectors, with multi-elemental analysis of the samples by nuclear related analytical techniques, in: *Applied Research on Air Pollution using Nuclear-Related Analytical Techniques*, NAHRES-19, IAEA, Vienna 1994, pp. 59–77.

Maenhaut, W., F. François, J. Cafmeyer; The "Gent" stacked filter unit (SFU) sampler for the collection of aerosols in two size fractions: Description and instructions for installatiion and use, in: *Applied Research on Air Pollution using Nuclear-Related Analytical Techniques*, NAHRES-19, IAEA, Vienna 1994, pp. 249–263.

Maenhaut, W., G. Ducastel, K. Beyaert, J. E. Hanssen; Chemical composition of the summer aerosol at Ny Ålesund, Spitsbergen, and relative contribution of natural and anthropogenic sources to the non-sea-salt sulfate, in: P.M. Borrell, P. Borrell, T. Cvitaš, W. Seiler (eds), *Proc. EUROTRAC Symp. '94*, SPB Academic Publishing bv, The Hague 1994, pp. 467–471.

Maenhaut, W., V. Havranek, G. Ducastel, J.E. Hanssen; PIXE analysis of cascade impactor samples collected at the Zeppelin background station in Ny Ålesund, Spitsbergen, in: J.L. Duggan, I:L. Morgan (eds), *Absts. of 13th Int. Conf. On the Application of Accelerators in Research and Industry*, Denton, Texas 1994, p. 152.

Malin, G., P.S. Liss, S.M. Turner; Dimethyl sulphide: production and atmospheric consequences, in: J.C. Green, B.S.C.. Leadbeater (eds), *The Haptophyte Algae*, Systematics Association Special Volume **51**, Clarendon Press, Oxford 1994, pp. 303–320

Mestayer, 1994, Dry deposition of particles to ocean surfaces, *Proc. 2nd Int. Conf. on Air-Sea Interaction, and Meteorology and Oceanography of the Coastal Zone*, Lisbon, Portugal, *Amer. Meteor. Soc.* 1994, pp. 243–244.

Mestayer, P.G., B. Tranchant; Modélisation du Flux d'Eau de l'Océan vers l'Atmosphere Sous Formes Vapeur et Liquide (Embruns), Atelier de Modélisation de l'Atmosphère 1994, *Centre International de Conférences de Météo France*, Toulouse, Volume d'Actes 1994, pp. 179–186.

Mestayer, P.G., A.M. J. Van Eijk, G. De Leeuw; Simulation of spray droplet dynamics over waves, *Proc. 2nd Int. Conf. on Air-Sea Interaction, and Meteorology and Oceanography of the Coastal Zone*, Lisbon, Portugal, *Amer. Meteor. Soc.* 1994, pp. 228–229.

O'Connor, T.C., F.M. McGovern, S.G. Jennings, C. Philipp; Aitken nuclei measurements and evidence of gas-to-particle conversion processes at Mace Head, Ireland, in: P.M. Borrell, P. Borrell, T. Cvitaš, W. Seiler (eds), *Proc. EUROTRAC Symp. '94*, SPB Academic Publishing bv, The Hague 1994, pp. 1206–1209.

Otten, P., J. Injuk, R. Van Grieken; Elemental concentrations in atmospheric particulate matter sampled on the North Sea and the English Channel. *Sci. Tot. Environ.* **155** (1994) 131–149.

Rebers, A., M. Schulz, L. Spokes, T. Jickells, M. Leermakers; A regoinal scale case study of wet deposition in the North Sea area, in: P.M. Borrell, P. Borrell, T. Cvitaš, W. Seiler (eds), *Proc. EUROTRAC Symp. '94*, SPB Academic Publishing bv, The Hague 1994, pp. 621–625.

Schulz, M.; An environmental issue: heavy metal deposition - trace metals in the aerosol. in: P.M. Borrell, P. Borrell, T. Cvitaš, W. Seiler (eds), *Proc. EUROTRAC Symp. '94*, SPB Academic Publishing bv, The Hague 1994, pp. 32–33.

Schulz M., C. Richter, A. Rebers, G. Peters, W. Dannecker; A case study on the scavenging of atmospheric trace constituents based on vertical profiles of rain droplet spectra and sequential rain analysis in: *Proc. 3rd Int. Symp. on Tropospheric Profiling*, Hamburg, Vol. 1 1994, pp. 95–97.

Schulz, M., T. Stahlschmidt, F. François, W. Maenhaut, S.E. Larsen; The change of aerosol size distributions measured in a Lagrangian-type experiment to study deposition and transport processes in the marine atmosphere, in: P.M. Borrell, P. Borrell, T. Cvitaš, W. Seiler (eds), *Proc. EUROTRAC Symp. '94*, SPB Academic Publishing bv, The Hague 1994, pp. 702–706.

Smith, S.D., K.B. Katsaros, W.A. Oost, P.G. Mestayer; The impact of the HEXOS programme, *Proc. 2nd Int. Conf. on Air-Sea Interaction, and Meteorology and Oceanography of the Coastal Zone*, Lisbon, Portugal, *Amer. Meteor. Soc.* 1994, pp. 226–227.

Sørensen, L.L., H. Nielsen, K. Granby, W.A.H. Asman; Diffusion scrubber technique for measuring ammonia; *Nat. Environ. Res. Inst. Tech. Rep.* **99** Roskilde, Denmark 1994, p. 54.

Sørensen, L.L., K. Granby, H. Nielsen, W.A.H. Asman; Diffusion scrubber technique used for measurements of atmospheric ammonia, *Atmos.Environ.* **28** (1994) 3619–3628.

Sørensen, L.L., O. Hertel, M. Schulz, R. M. Harrison; Fluxes of nitrogen gases to the North Sea, in: P.M. Borrell, P. Borrell, T. Cvitaš, W. Seiler (eds), *Proc. EUROTRAC Symp. '94*, SPB Academic Publishing bv, The Hague 1994, pp. 626–631.

Souza Sierra, M.M. de, O.F.X. Donard, M. Lamotte, C. Belin, M. Ewald. Fluorescence spectroscopy of coastal and marine waters. *Marine Chem.* **47** (1994) 127–144.

Spokes, L. J., T.D. Jickells; Metal speciation in the atmosphere, in: A. Ure, C. Davidson (eds), *Chemical Speciation in Natural Systems*, Blackie, 1985, pp. 137–168.

Spokes, L.J., T.D. Jickells; Solubilisation of aerosol trace metals by cloud processing: A laboratory study. *Geochim. Cosmochim. Acta* **58** (1994) 3281–3287.

Sutton, M.A., W.A.H. Asman, J.K. Schørring; Dry deposition of reduced nitrogen, *Tellus* **46B** (1994) 255–273.

Treiger, B., J. Injuk, I. Bondarenko, P. Van Espen, L. Breitenbach, U. Wätjen, R. Van Grieken; Non-linear mapping of microbeam proton induced X-ray emission data for source identification of North Sea aerosols, *Spectrochim. Acta* **49**B (1994), 345–353.

Uher, G., M.O. Andreae; The diurnal cycle of carbonyl sulfide in seawater: A simple empirical model, in: P.M. Borrell, P. Borrell, T. Cvitaš, W. Seiler (eds), *Proc. EUROTRAC Symp. '94*, SPB Academic Publishing bv, The Hague 1994, pp. 563–567.

Ulshöfer, V.S., O. R. Flöck, G. Uher, S. Rapsomanikis, M.O. Andreae; On COS: Photoprodution and supersaturation in the Aegean Sea in summer, photoproduction, undersaturation, and depth profiles in the North Atlantic in winter, diurnal cycles of COS, in: P.M. Borrell, P. Borrell, T. Cvitaš, W. Seiler (eds), *Proc. EUROTRAC Symp. '94*, SPB Academic Publishing bv, The Hague 1994, pp. 568–572.

Van Grieken, R., L. de Bock. J. Injuk, H. Van Malderen; Atmospheric deposition of heavy metals in the North Sea as studied by micro- and trace analysis, in. S.P. Varnavas (ed), *Environmental Contamination*, CEP Consultants Ltd., Edinburgh 1994, pp. 284–286.

1995

Amouroux, D., O.F.X. Donard; Determination of hydrogen peroxide in estuarine and marine waters by flow injection with fluorometric detection. *Oceanologica Acta* **18** (1995) 353–361.

Asman, W.A.H., R. Berkowicz; Atmospheric nitrogen deposition to the North Sea; *Marine Pollut. Bull.* (1995).

Asman, W.A.H.; A gridded inventory of European ammonia emissions and its possible use in abatement strategies. in: *Proc. Workshop on the potential for abatement of ammonia emissions from agriculture and assicated costs,* Culham U.K. 1995.

Asman, W.A.H.; Parameterization of below-cloud scavenging of highly soluble gases under convective conditions, *Atmos. Environ.* **29** (1995) 1359–1368.

Baeyens W. and M. Leermakers, Particulate, dissolved and methylmercury budgets for the Scheldt estuary (Belgium and the Netherlands), in: W. Baeyens, R. Ebinghaus, O. Vasiliev (eds), *Global and Regional Mercury Cycles: Sources, fluxes and mass balnances*, NATO ARW Series, Kluwer Publishers 1995.

Bergin, M.H., J.E. Dibb, C.I. Davidson, J.L. Jaffrezo, S.N. Pandis, R. Hillamo, W. Maenhaut, H.D. Kuhns, T. Makela; The contributions of snow, fog and dry deposition to the summer flux of anions and cations at Summit, Greenland, *J. Geophys. Res.* **100D** (1995) 16275–16288.

Flöck, O. R., M. O. Andreae; Vertical profiles of dissolved carbonyl sulfide (COS) and photochemical and non-photochemical formation and destruction of COS and methylmercaptane (MeSH), *Annales Geophysicae* **13** Supplement II (1995) C391.

François, F., W. Maenhaut, J.-L. Colin, R. Losno, M. Schulz, T Stahlschmidt, L. Spokes, T. Jickells; Intercomparison of elemental concentrations in total and size-fractionated aerosol samples collected during the Mace Head experiment, April 1991, *Atmos. Environ.* **29** (1995) 837–849.

Hertel, O., J. Christensen, E. Runge, W.A.H. asman, R. Berkowicz, M.F. Hovmand, Ø. Hov; Development and testing of a new variable scale air pollution model ACDEP, *Atmos. Environ.* **29** (1995) 1267–1290.

Injuk, J., R. Van Grieken; Atmospheric concentrations and deposition of heavy metals over the North Sea. a literature review, *J. Atmos.Chem.* **20** (1995) 179–212.

Jickells, T.; Atmopsheric input of metals and nutrients to the oceans: Their magnitude and effects, *Marine Chem.* (1995).

Kunz, G.J., G. de Leeuw, S.E. Larsen, F. Aa. Hansen (1995a). Eddy correlation fluxes of momentum, heat, water vapor and CO2 during ASGASEX. in: W.A.Oost (ed), *Proc. ASGASEX workshop,* Report TR-174, KNMI, deBilt 1994, pp. 12–16.

Kunz, G.J., G. de Leeuw, S.E. Larsen, F. Aa. Hansen; Over-water eddy correlation measurements of fluxes of momentum, heat, water vapor and CO_2. In Air-Water Gas Transfer: in: B.Jähne, E.C.Monahan (eds), *3rd Int. Symp. on Air-Water Gas Transfer*, Heidelberg , Aeon Verlag and Studio, Hanau 1995, pp. 685–701.

Larsen, S E, J.B.Edson, P.Hummelshøj, N O Jensen, G.deLeeuw and P G Mestayer, 1995: Dry deposition of particles to ocean surfaces, OPHELIA,42,193–204.

Larsen,S.E., F.Aa.Hansen; Comparison between heat fluxes from the drag formulation and the eddy correllation during the NOSE: The North Sea Experiment, Sept.1991. Risø-I-report 1995.

Leermakers M., Meuleman C., Baeyens W.; Speciation of mercury in the Scheldt estuary Water, *Air and Soil Pollution* **80** (1995) 641–652.

Leermakers M., Ebinghaus R., Kubala J., H. Kock; Determination of atmospheric mercury above the North Sea during the North Sea Experiment, *Water, Air and Soil Pollution*.

Leermakers M., Baeyens W.; Evasion fluxes and formation rates of Hg in the Scheldt estuary, *Marine Pollution Bulletin*, submitted.

Leermakers M., Baeyens W.; Formation rates of MeHg in estuarine waters, *Marine Pollution Bulletin*, submitted.

Liousse C., C. Deveaux, F. Dulac, H. Cachier; Aging of savanna biomass burning aerosols: consequences on their optical properties, *J. Atmos. Chem.* **22** (1995) 1–17.

McArdle, N.C., P.S. Liss; Isotopes and atmospheric sulphur, *Atmos. Environ.* **29** (1995) 2553–2556.

Mestayer, P.G., A.M.J. VanEijk, G. De Leeuw, B. Tranchant, Numerical simulation of the dynamics of sea spray over the waves, *J. Geophys. Res.* **101C** (1995) 20771–20797.

Otten, P., J. Injuk, R. Van Grieken; Vertical sulphur dioxide, ozone and heavy metal concentration profiles above the Southern Bight of the North Sea, *Israel Chem.* (1995).

Plass-Dülmer, C., R. Koppmann, M. Ratte, J. Rudolph; Light nonmethane hydrocarbons in seawater: An overview, *Global Biogeochem. Cycles* **9** (1995) 79–100.

Ratte, M., C. Plass-Dülmer, C., R. Koppmann, J. Rudolph; Biological, chemical and physical parameters affecting the concentration of light hydrocarbons in sea water. Implications for the NMHC production mechanism, *Tellus* B (1995).

Uher, G., G. Schebeske, S. Rapsomanikis, M. O. Andreae; Measurements of dimethyl sulfide in surface waters of the northeast Atlantic Ocean, *Annales Geophysicae* **13** Supplement II (1995) C395.

Ulshöfer, V. S., G. Uher, M. O. Andreae. 1995. The seasonal cycle of carbonyl sulfide in surface waters of the Northeast Atlantic Ocean: measurements and model calculations, *Annales Geophysicae* **13** Supplement II (1995) C395.

Ulshöfer, V. S., G. Uher, M. O. Andreae; Evidence for a winter sink of atmospheric carbonyl sulfide in the northeast Atlantic Ocean, *Geophys. Res. Lett.* **22** (1995) 2601–2604.

Xhoffer, C., R. Van Grieken, W. Jacob, P. Buseck; Quantitative microanalysis of individual salt aerosol particles using electron energy loss spectroscopy *Spectrochim. Acta* B (1995).

Theses

M. Sc. / Diploma

Corn, M. Relations entre la dynamique de l'ecosystème pélagique et l'évolution spatio-temporelle des composés soufrés (DMPS et DMS): Modélisation et observations, University of Paris VI, 1992.

Ducastel, G.; Studie van het atmosferisch aerosol in zuidelijk Noorwegen door middel van gefractioneerde monsterneming en protonengeïnduceerde X- straal emissie analyse, University of Gent, 1989.

Haster, T. Größenklassierte Kurzzeitprobenahme von Aerosolen und Untersuchung ihrer Spurelementgehalte mittels TRFA University of Hamburg, 1991.

Liousse, C.; Propriétés optiques du carbone-suie, University of Paris VII, 1990.

Polle, C.; Contribution a l'etude de la source marine d'hydrocarbures legers non methaniques, University of Paris VI, 1990.

Rebers A.; Einflüsse der Na-depositionssammlung auf die Erfassung partikulär gebundener und gelöster Regeninhalts-stoffe, Fachbereich Chemie, University of Hamburg, 1991.

Saye, G.; Bepaling van enkele belangrijke componenten in atmosferische aerosolen met ionen chromatografie, University of Gent, 1992.

Tisserant, M.; Analyse de especes Fe(II) et Fe(III) dans les precepitations, D.E.A. de chimie de la pollution atmospherique et physique de l'environnement, University of Paris VII, 1993.

Vassalli, I. Mesure du pouvoir complexant d'une eau de pluie. D.E.A. de chimie de la pollution atmospherique et physique de l'environnement, University of Paris VII, 1993.

Ph. D.

Amouroux, D.; Etude du cycle biogéochimique du sélénium à l'interface océan-atmosphère, Université de Bordeaux I, 1995.

Bremond, M.P.; Le carbone suie atmospherique: aspects geochimiques et methodologiques, University of Paris VII, 1989.

Cornille, P.; Chemische karakerizering en brononderzoek van het atmosferisch aerosol, University of Gent, 1991.

Ducastel, G.; Het atmosferisch aerosol in zuidelijk Noorwegen en het Arctisch gebied: chemische samenstelling, deeltjesgrootte, bronnen en brongebieden, University of Gent, 1994.

Ducret, J.; Incorporation of carbonaceous particles into wet precipitation (in French), University of Paris VII, 1993.

Hummelshøj, P.; Dry Deposition of Particles and Gases, Risø National Laboratory, Roskilde 1992.

Injuk, J.; Assessment of atmospheric pollutant fluxes to the North Sea by X-ray emission analysis Universtiy of Antwerp, 1995.

Kane, M.M.; Processes contributing to the atmospheric deposition of trace metals to the North Sea, University of East Anglia, Norwich 1992.

Le Briss, N.; Contribution to the study of trace metals dissolved/particulate partition in wet precipitations (in French), University of Paris VII, 1993.

Lim, B.; Trace metals in North Atlantic precipitation, University of East Anglia, Norwich 1991.

Liousse, C.; Optical properties and remote sensing of tropical combustion aerosol (in French), University of Paris VII, 1993.

Losno, R.; Chimie d'elements mineraux en trace dans les pluies mediterraneennes, University of Paris VII, 1989.

McArdle, N.C.; The use of stable sulphur isotopes to distinguish between natural and aanthropogenic sulphur in the atmosphere, University of East Anglia, Norwich 1993.

Msibi, M.I.; An investigation of inorganic nitrogen compounds in the atmosphere, University of Birmingham, 1992.

Nightingale, P.D.; Low molecular weight halocarbon in sea water. University of East Anglia, Norwich 1991.

Otten, P.; Transformatie, concentraties en depositie van Noordzee-aerosolen University of Antwerp, 1991.

Ottley, C.J.; Estimation of atmospheric pollutant dry deposition to the North Sea, University of Essex, 1991.

Plass-Dülmer, C.; Die Bedeutung der Ozeane als Quelle leichter Kohlenwasserstoffe in der Atmosphare (The role of oceans as source for light hydrocarbons in the atmosphere), University of Cologne, 1992.

Ratte, M.; Untersuchungen zum Produktionsmechanismus leichter Nichtmethan-Kohlenwassterstoffe im Meerwasser, RWTH Aachen, 1993.

Rendell, A.R.; The wet deposition of trace metals and nutrients to the southern North Sea, University of East Anglia, Norwich 1992.

Rojas Palma, C.; Study of concentration and deposition of tropospheric aerosols using X-ray emission techniques University of Antwerp, 1992.

Rouault, M.P.; Modelisation numerique d'une couche limite unidimensionelle stationnaire d'embruns University of Aix-Marseilles II, 1991.

Schulz, M.; Raeumliche und zeitliche Verteilung atmosphärischer Einträge von Spurenelementen in die Nordsee, Fachbereich Chemie, University of Hamburg, 1993.

Souza Sierra, M.M.; Caracterisation et reactivite de la matiere organique fluorescente dissoute dans les milieux cotiers et marins, University of Bordeaux, 1992.

Steiger, M. ; Anthropogene und natürliche Quellen urbaner und mariner Aerosole charakterisiert und quantifiziert mit Hilfe der Multielementanalyse und chemischen Receptormodellen, Fachbereich Chemie, University of Hamburg, 1991.

Wouters L.; Laser microprobe mass analysis of individual environmental particles Universtiy of Antwerp, 1991.

Xhoffer, C.; Chemische karakterisatie van individuele deeltjes door electronen probe X-stralen microanalyse en electronen energieverlies spectrometrie University of Antwerp, 1993.

Transport and Chemical Transformation of Pollutants in the Troposphere

Volume 9, Part II

The TRACT Experiment

Franz Fiedler (Karlsruhe) Coordinator

and

Peter Borrell (Garmisch-Partenkirchen)

TRACT Steering Group

Domenico Anfossi	Turin
Rainer Friedrich	Stuttgart
Fritz Gassmann	Villigen
Franco Girardi	Ispra
Niels Otto Jensen	Roskilde
Anne Jochum	Oberpfaffenhofen
Paul Lightman	Ascot
Volker Mohnen	Garmisch-Partenkirchen
Bruno Neininger	Zürich
Robert Rosset	Toulouse
Eberhard Schaller	Cottbus
Alwig Stingele	Ispra
Heinz Wanner	Bern

Chapter 5

TRACT: Transport of Air Pollutants over Complex Terrain

Franz Fiedler[1] and Peter Borrell[2]

[1]Universität Karlsruhe, Forschungszentrum Karlsruhe,
Inst. für Meteorologie und Klimaforschung, D-76128 Karlsruhe, Germany
[2]P&PMB Consultants, Ehrwalder Straße 9, D-82467 Garmisch-Partenkirchen, Germany

Introduction

Understanding the mass balance of the many chemical constituents either emitted to the atmosphere from anthropogenic or biogenic sources or formed within the atmosphere belongs to the most challenging problems in atmospheric sciences. Depending of the size of the volume for which a system with open boundaries is defined, different processes come into play determining the local distribution and temporal variation of the concentrations within the system.

Most of the European areas with large cities, dense population and heavy industrialisation have hilly and mountainous terrain shape. Therefore the fate of pollutants emitted to the atmosphere cannot be described and followed by simplified methods without taking into account the many effects of the mountainous terrain.

Aims of the subproject TRACT

It was the principal aim of the EUROTRAC Subproject TRACT (Transport of Air Pollutants Over Complex Terrain) to address the spectrum of processes which are relevant to describe the transport, the turbulent diffusion and in part also some chemical aspects within the lower atmosphere over mountainous terrain.

In the past detailed studies of transport and turbulent diffusion of air pollutants over complex terrain have mainly been made over smaller distances in the order of tens of kilometres. In almost all studies, simple configurations of the land surfaces were chosen to meet the conditions of horizontal homogeneity which underlie the theories of the model concepts used. In other studies, transport and diffusion processes could only be taken into account by using mean-flow trajectories. When the transport and the mixing of pollutants is considered over longer distances the influence of the horizontal inhomogeneities of the earth surface plays an important role since at each barrier the air is deflected, both vertically and horizontally. At the same time the level of turbulence and convective activity in the lower atmosphere changes dramatically. However the parameterisation assumptions presently used in numerical models incorporate these effects very poorly. They must be improved by detailed and accurate observations of the processes governing atmospheric transport over complex terrain. But without detailed analyses of the phenomena involved little progress in the parameterisation can be expected.

It was clear from the beginning of the project that an observational data base of a much broader scope would be needed to perform a detailed analysis of the processes involved.

First it would be necessary to take into account all the features which are connected to the daily cycle of the lower atmosphere, especially within the atmospheric boundary layer. This makes it necessary to observe the atmosphere in an area through which the air moves within a single day. Over this time scale neither the individual sources nor the source types of the chemical species can be considered to be constant. The chemical species also undergo complex transformations within such scales of time and space. Therefore, a realistic field study over typical distances of 300 km × 300 km cannot be undertaken without a reasonable balance between observing the meteorological conditions over the complex terrain and taking into account the chemical transformation of the species during the transport.

Particular attention was paid to the following points.

* The effects of orography on atmospheric transport and exchange processes within the planetary boundary layer over complex terrain. In particular, turbulent dispersal, channelling and mountain induced wind systems were investigated in different scales ranging from industrialised regions to the transalpine scale (TRANSALP sub-group).

* Handover of air pollutants from the atmospheric boundary layer to the free troposphere. This process is especially important for the long-range transport of air constituents and may contribute significantly to the transalpine exchange. Secondary flow systems along mountain sides are a particularly important mechanism for the handover processes.

* To establish mass balances of pollutants (especially SO_2, NO and NO_2) in mesoscale regions including the respective emissions and deposition.

In the project plan the work was focussed from the beginning on preparing a large field campaign to obtain the observational database necessary for the detailed understanding of the processes involved.

Within TRACT, special emphasis was given to the interaction of field observations and mesoscale modelling. Therefore the observations were prepared in such a way that consistent data sets for the needs of model initialisation would be provided. The data would also provide a basis for model evaluation.

The subproject organised four field campaigns; three, which formed the project, TRANSALP, were tracer experiments to examine the channelling of flows through the high Alps. The other, the TRACT campaign itself was focussed on the meteorological and chemical problems associated with complex terrain, and held in the region comprising south-western Germany, south-eastern France and north-western Switzerland. The next section deals with the TRANSALP campaigns. The remainder, which constitutes the majority of this report, is devoted to the TRACT campaign.

TRANSALP

The Alps constitute an appreciable topographical and meteorological break between northern and central Europe and the Mediterranean. The peaks, glaciers and upper valleys are sensitive ecosystems while many of the lower valleys are sources of appreciable pollution. An understanding of the air flow into the region and through and between the valleys is necessary to any measures intended to reduce the impact of pollutants in the Alps.

The project TRANSALP (Transport of tracer gases across the Alps) was a sub-group of TRACT set up to study transport of air masses through very complex terrain by means of tracer release experiments. Three possible processes for transalpine transport of trace substances were considered:

* injection from the planetary boundary layer into the free troposphere;
* channelling of air through the north to south directed valleys; for example the Reuss, Gotthard and Leventina valleys;
* coupling of the air masses between adjacent valleys when the boundary layer height exceeds that of the intervening mountains.

The project consisted of field campaigns backed by modelling activities [1, 2].

a. TRANSALP Experimental campaigns

There were three field campaigns.

1989 To investigate the split in a northerly flow between the Leventina and Blenio valleys in Ticino in southern Switzerland

1990 To investigate the transport in a northerly flow from the Leventina valley across the Gotthard pass between southern Switzerland and Italy and the Swiss plateau

1991 To investigate the transport in a southerly flow between the Swiss plateau and southern Switzerland and Italy

In each case a chemically inert perfluorocarbon tracer, perfluoro-methylcyclohexane (C_7H_{14}) or perfluorodimethylcyclohexane (C_8H_{16}), was released, and its dispersion and transport tracked with at sampler stations situated in the valleys downwind from the release site. In the 1990 experiment there were some 40 sampling sites; in 1991 more than 60. Fig. 5.1 shows the distribution in 1990.

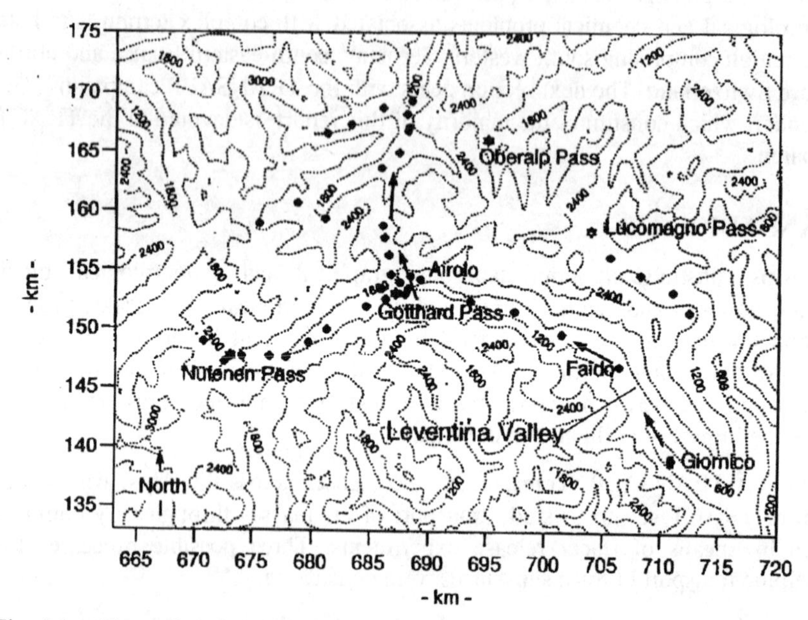

Fig. 5.1: The area of the 1990 TRANSALP campaign, showing the tracer release points (squares) and sampling locations.

The samplers for the tracer were supplemented in 1990 and 1991 by a motor glider (METAIR) which flew pre-defined tracks along the valleys of interest to investigate the vertical distribution of the tracer. A variety of meteorological measurements were made, many with the help of the Swiss weather service [2].

b. TRANSALP experimental results

The 1989 TRANSALP campaign

The main tracer release in 1990 campaign was made during a general anti-cyclonic situation so that the flow is principally determined by daily south to north valley breezes, the "Inverna", that blow up from the Po valley to the central Alps. The release took place in the Leventina valley about 5 km south of the bifurcation between the Leventina and Blenio valleys. The tracer was found in both valleys, and comparison between equivalent samplers indicated that about 20 % of the tracer found its way into the Blenio. More detailed analysis showed that the centre line of the plume to be along the south west side of the Levantino.

The results are consistent with a clear channelling of the flow, the plume being shifted towards the south western side due to differential heating of the valley sides. The plume enlarges more quickly than would be expected from experiments in flat terrain, presumably due to the effects of thermal eddies in the confined valley [2].

The 1990 TRANSALP campaign

The 1990 release was again made during an anti-cyclonic situation under "Ivernone" wind conditions in which the normal southerly "Iverna" valley breezes are supplemented by strong southerly flows aloft. The release was made in the Leventina valley about 25 km south of the Gotthard pass. As Fig. 5.1 shows, the sampling stations were placed so as to detect flows across the Gotthard, Nufenen, Oberalp and Lucomagno passes. The arrival of the tracer was seen at most stations but the concentrations were smaller than hoped for because of wind speeds were lower than forecast. The motor glider detected the tracer in the surrounding valleys and very high tracer concentrations above the top of the Gotthard pass.

It is clear from the results that channelling of the tracer occurred across the Gotthard, Nufenen and Lucomagno passes. As tracer was also found high above the valley bottom as well as much to the north of the Gotthard pass, some transfer to the free troposphere also occurred. Some evidence for night time flow reversal was also obtained [2].

The 1991 TRANSALP campaign

The weather conditions in 1991 were chosen to ensure that there would be a flow from north to south. The tracer was released on the eastern shore of Lake Lucerne about 45 km north of the Gotthard pass. Samplers were placed in valleys to the south of the release site and also in the valleys that run east and west. Three motor gliders were also employed to take samples aloft. The results obtained at some of the sampling sites across the Gotthard are shown in Fig. 5.2 [3].

From the results it was possible to identify two transport routes under north-south flow conditions. The speed of transport was nearly identical with the average valley wind speed. There is also transport in the west to east direction which indicates the importance of local thermal wind fields in determining the dispersion. The results at Airolo in Fig. 5.2 show an interesting effect: the tracer arrives in the valley about an hour before it does at the two more elevated sites suggesting that the flow is channelled to the bottom and that the valley "fills up" with tracer. The valley drains to the East. Flights in the South of the area on the following day show a general distribution of tracer up to 2500 m a.s.l. within the area confirming the trans-alpine transfer of pollutants [2, 3].

c. *Modelling activities in TRANSALP*

AS the experiments were carried out in a limited area with non-reactive tracers, the models were used to produce the wind fields and to simulate the dispersion of the tracer. The diagnostic wind field model, CONDOR, was used to drive the Lagrangian dispersion model, ARCO, and the wind field model, MINERVE, was used with the dispersion model, SPRAY. The combinations produced satisfactory agreement both with each other and, considering the complexity of the terrain, with the experiments [4, 5]. Some wind field simulations were also done with the RAMS model [6].

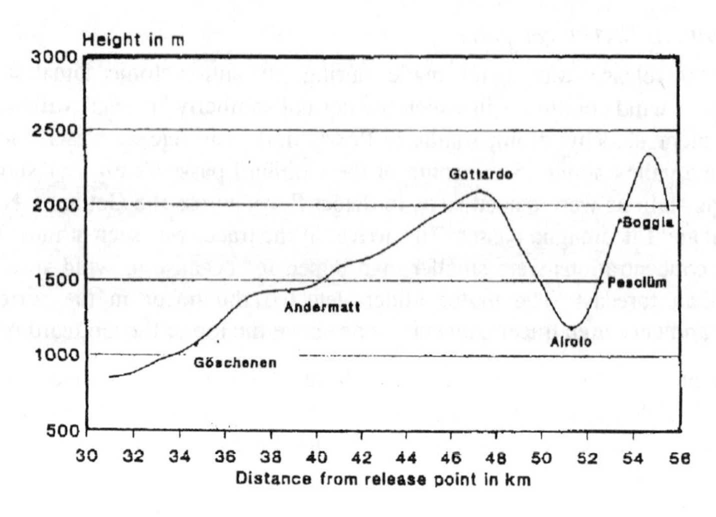

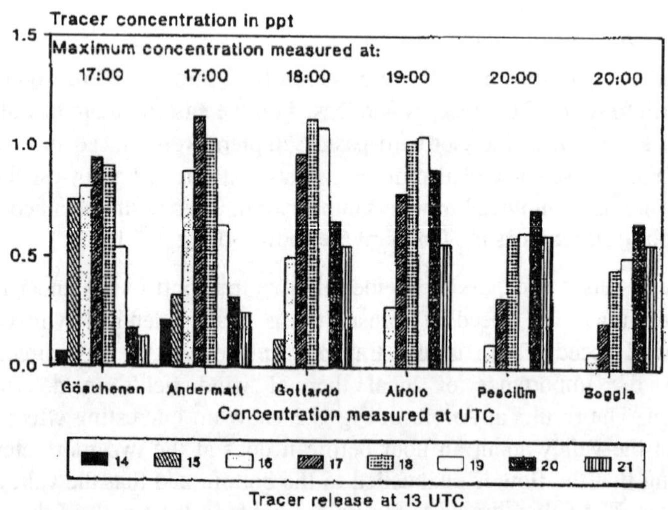

Fig. 5.2: Tracer plume evolution across the St. Gotthard pass, as seen at 6 ground-level sites.

d. Conclusions from TRANSALP

The campaigns provide direct evidence for transport of air pollutants through alpine valleys and also over the ridges. The initial dispersion is by channelling along the valley bottoms but there is evidence for transfer to the free troposphere as well. However as the processes are totally dependent on the overall and local weather situations for the time of day the results cannot be regarded as typical. Much more work will be required to estimate for example budgets for the transfer between alpine valleys or for transport into the Alps from the surrounding regions.

The TRACT field experiment

As a typical region for European conditions, the area of south-west Germany with some parts of eastern France and northern parts of Switzerland was chosen. At the western side the Vosges mountains form a natural boundary. The foothills of the Alps formed the southern boundary of the experimental area. Within this region, the mountains range up to heights of about 1500 m. It is intersected by the broad valley of the Rhine running from north to south and by the Danube valley running from south-west to north-east. There are also many smaller valleys as well. The width of the valleys allows different meteorological phenomena like channelling and flow partitioning to develop together with some other smaller scale processes which are important to the dispersion of air pollutants in mountainous terrain. These processes include mountain and valley winds and the initiation of convection at mountain tops which contribute to vertical mixing, a process often known as "mountain venting".

The area chosen had been used for previous studies such as the TULLA experiment [7] and so a reasonably good data base for the meteorological conditions and anthropogenic emissions from industries and traffic was already available. Thus the TRACT measurements started with a better understanding of the area than would have been the case for a completely new area.

The selected area is shown in Fig. 5.3. There are always two contradictory influences when choosing an experimental site. First the area should be large enough to give the different air pollutants, stemming from either single intense sources or from industrial regions or from road traffic, time to mix completely and fill the whole boundary layer and to start chemical reaction. One would like to choose as large an area as possible, to provide enough time for the mixing and chemical reactions.

However with the inhomogeneity of the atmospheric conditions, created by the terrain shape, land use, the scattered locations of the emission sources by industry and the roads, it is necessary to make as many accurate observations of the atmospheric system as possible. Therefore, if the area is chosen is too large, the available instruments and manpower will not be sufficient to provide the required information base.

It is noticeable that in many field campaigns the paucity of measurements is then replaced by a broad spectrum of assumptions or speculation.

a. *The observation systems*

Three types of observation system were used in the TRACT experiment:

1. Continuous surface observations of meteorological and chemical variables, partly with stations in an operative meteorological network and partly with additional temporary stations;

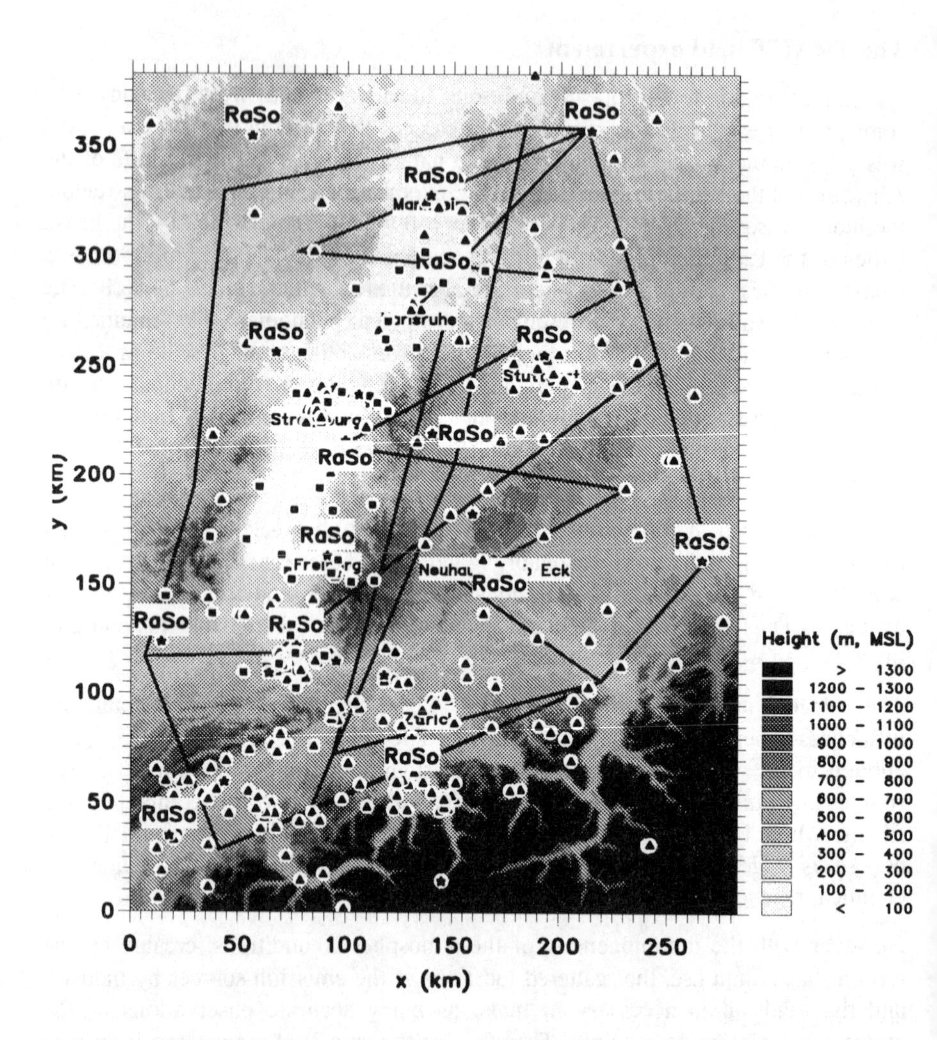

Fig. 5.3: The TRACT area comprises Baden-Württemberg in Germany, the northern part of Switzerland and the eastern part of France. The points show the measuring stations in the area, the majority of which are operated by the national networks. Flight tracks and radio sonde stations are also shown.

2. A reasonably dense network of vertical soundings, indicated in Fig. 5.3 by RaSo, for meteorological variables such as wind speed, wind direction, temperature, moisture and pressure. Measurements were made at frequent intervals in the intensive measurement periods;

3. Ten aircraft mainly equipped with chemical instruments measuring concentrations of different species. Five of these flew along prescribed tracks, indicated by the black lines in Fig. 5.3; the remainder were employed in the nested area.

Before the direct field measurements were made a period (7th–9th September) was devoted to quality control and quality assessment. It was the first time that a large effort had been devoted to instrument calibration and inter-comparison.

Ground based measurements from the operational and temporary stations were made for the whole measurement period. They provided the background information for both meteorological and chemical variables.

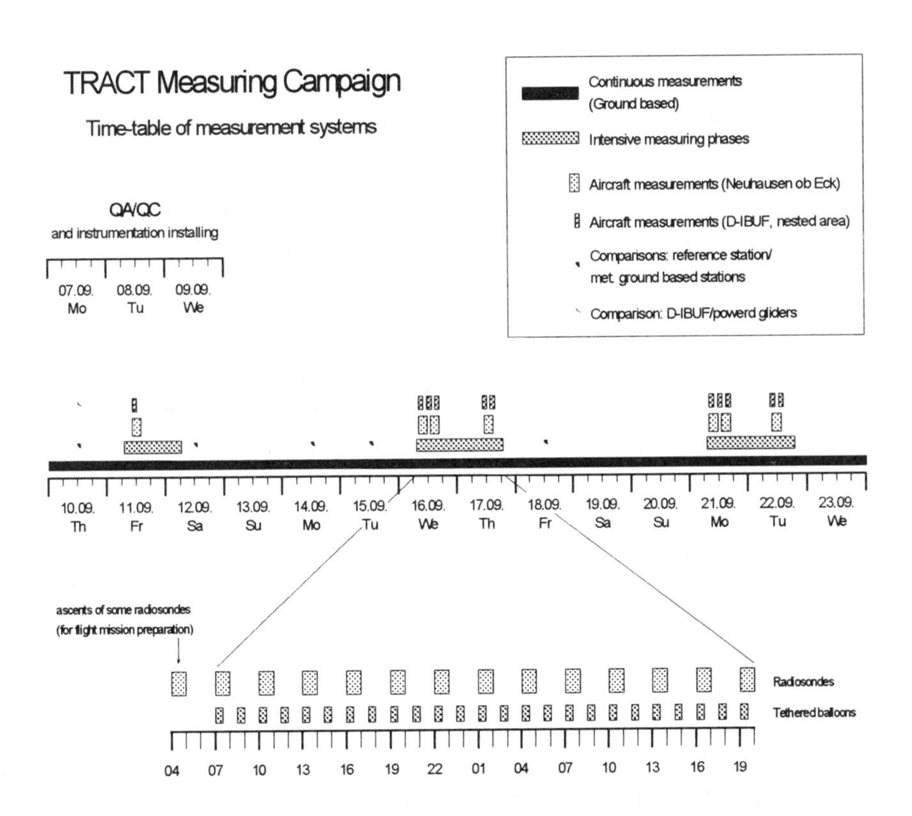

Fig. 5.4: Time table of the TRACT experiment (7th to 23rd September 1992) indicating continuous measurements and embedded intensive measuring phases

Three intensive measuring phases were included within the field phase starting on 10th (24 h), 16th (36 h) and 21st (36 h) September. During these intensive periods a network of 11 radiosonde stations and other vertical soundings were activated to measure the atmospheric conditions at three hourly intervals (see Fig. 5.5). The details are summarised in Table 5.1.

Table 5.1: The vertical sounding stations for TRACT (see Fig. 5.5 for their locations.

No.	Station	Height in m	Category	Measurement system
1	Idar-Oberstein	376	Valley	Radiosonde
2	Wertheim	342	Valley	Radiosonde
3	Mannheim	96	Valley	Radiosonde
4	Bruchsal	110	Valley	Radiosonde
5	Oberbronn	274	Slope	Radiosonde
6	Hegeney	180	Valley	Tethered balloon
7	Stuttgart	315	Slope	Radiosonde
8	Erlenhof	127	Valley	Pilot balloon Sodar
9	Sasbach	137	Valley	Tethered balloon
10	Hernenberg	370	Slopoe	Tethered balloon
11	Brandmatt	650	Slope	Tethered balloon
12	Hornisgrinde	1135	Mountain	Sodar
13	Musbach	695	Mountain	Radiosonde, Rass, Lidar
14	Offenburg-Ortenberg	150	Valley	Radiosonde, Tethered balloon, Sodar
15	Freiburg	300	Valley	Radiosonde
16	Neuhausen ob Eck	802	Plateau	Radiosonde
17	Memmingen	580	Plateau	Radiosonde, Sodar
18	Schauinsland	1284	Mountain	Radiosonde
19	Belfort	369	Valley	Radiosonde
20	Lörrach	290	Valley	Radiosonde
21	Wallbach	350	Valley	Sodar
22	Härkingen	430	Valley	Tethered balloon
23	Schwarzhäusern-Battenweid	430	Valley	Tethered balloon
24	Merenschwand	385	Valley	Radiosonde
25	Walperswil	441	Valley	Sodar
26	Siselen	442	Valley	Radiosonde
27	Payerne	491	Valley	Radiosonde

In order to study the interactions of the atmosphere and the orography in more detail, an additional nested area was installed at a cross section of the Rhine valley (indicated by the square around Strasbourg in Fig. 5.5). Along this cross section line additional measurements with tethered balloons were made at half hourly intervals. The nested area is shown in greater details in Fig. 5.6 (see Table 5.2)

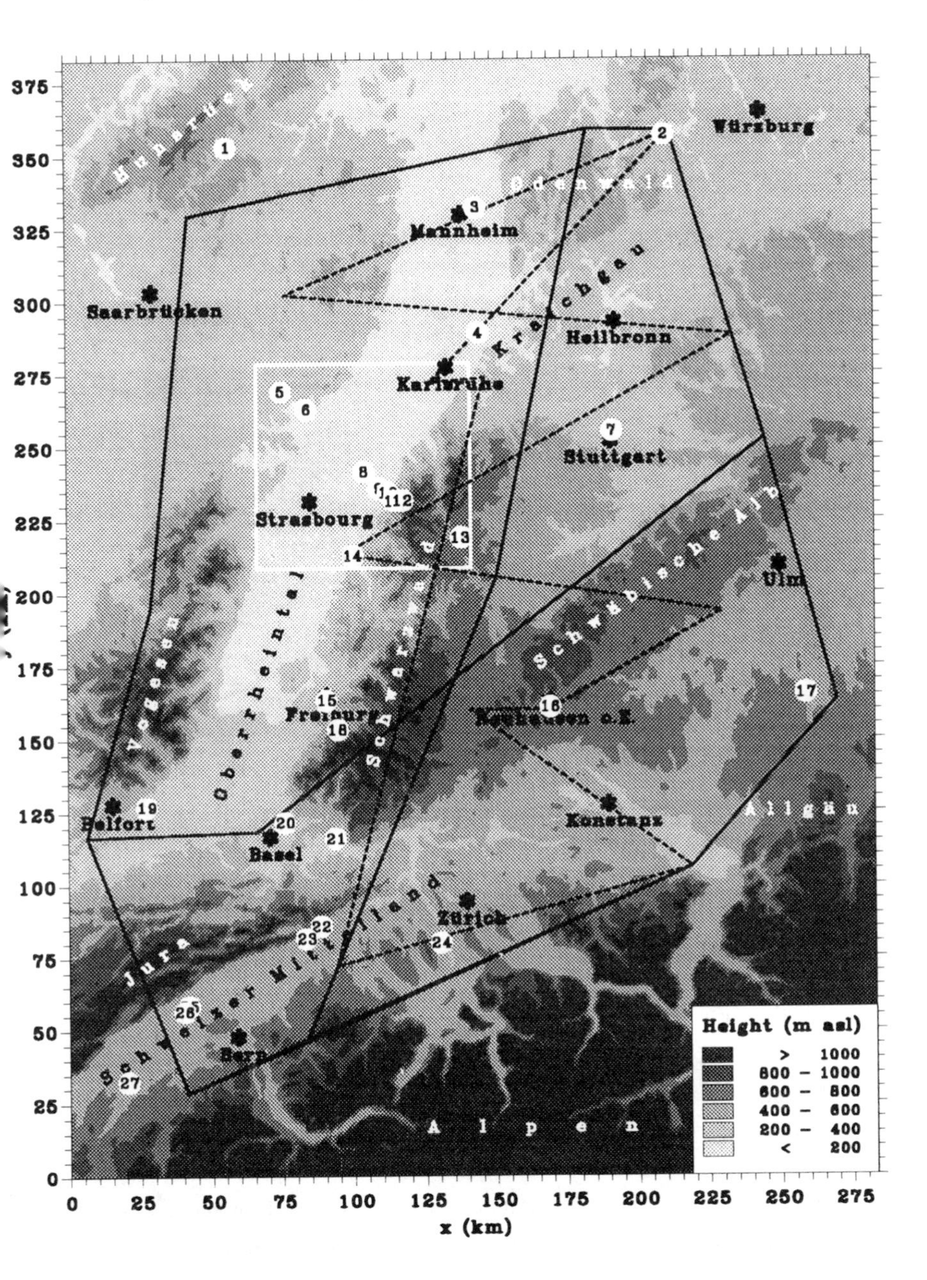

Fig. 5.5: The locations of the vertical sounding stations during the intensive measuring phases of TRACT. For the names of the stations and the type of sounding, please refer to Table 5.1.

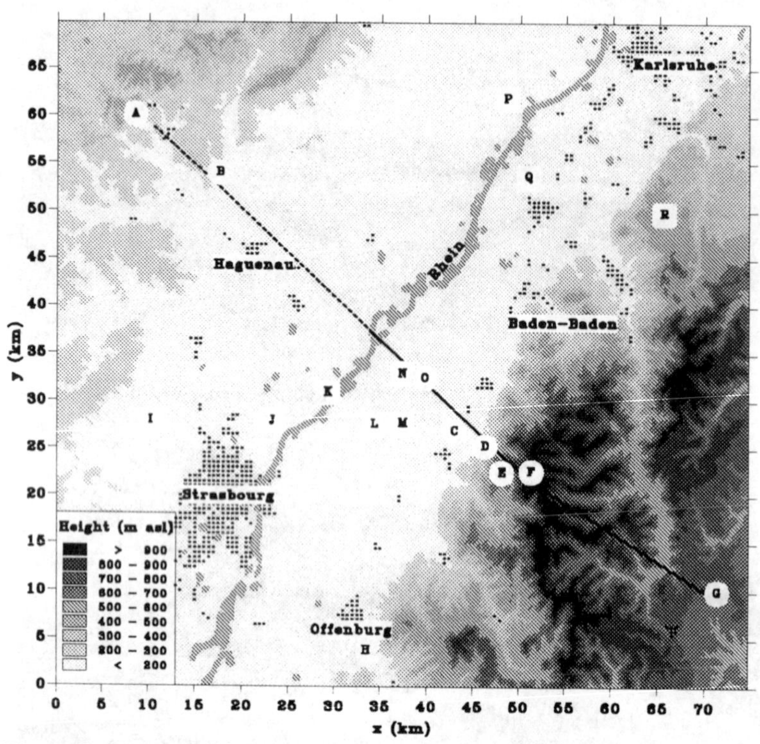

Fig 5.6: The nested area across the Rhine valley north of Strasbourg with the northern area of the Black Forest.

Table 5.2: The measuring stations in the TRACT nested area (the locations are shown in Fig. 5.6).

No.	Location	Altitude in m a.s.l.	Measuring systems
A	Oberbronn	274	radiosonde, met. surface station
B	Hegeney*	180	tethersonde, met. surface station
C	Sasbach*	137	tethersonde, met. + chem. Surface
D	Hornenberg*	370 337	station, tethersonde met. surface station
E	Brandmatt	650	tethersonde, chem. surface station
F	Hornisgrinde	1135	Sodar, met. + chem. surface station
G	Musbach	695	radiosonde, Rass, Lidar, met. + chem. surface station
H	Offenburg-Ortenberg	150	radiosonde, tethersonde, Sodar
I	Pfettisheim*	155	met. surface station
J	La Wantzenau*	128	met. surface station

Table 5.2 (continued)

K	Gambsheim	128	met. + chem. surface station
L	Freistett*	131	met. surface station
M	Gamshurst	131	met. + chem. surface station
N	Scherzheim	128	met. + chem. surface station
O	Unzhurst	127	pilot balloon, Sodar, met. + chem. surface station
P	Lauterbourg*	110	met. surface station
Q	Plittersdorf*	113	met. surface station
R	Freiolsheim*	487	met. surface station

Ten aircraft, which constituted the most expensive and most sophisticated measurement systems available to the campaign, were active during parts of the intensive periods. Five aircraft flew along the flight tracks within the main experimental area (see black lines in Figs. 5.3 and 5.5). One aircraft and four motor gliders operated within the nested area.

In order to minimise the sampling time and so to obtain as near a simultaneous measurement as possible, the main area was sub-divided into two sub-areas. From the aircraft measurements, the inflow of the air into and outflow of air from the area, together with the conditions inside the experimental area was determined. However it was also necessary to observe the top of the boundary layer and the interaction between the free atmosphere and the underlying boundary layer. Therefore, depending on the weather conditions, the aircraft flew along the tracks of the western and the eastern box, starting at Neuhausen ob Eck (Fig. 5.5 station 16). During synoptic flow conditions from the north-west or the south-east, the area was sub-divided into northern and southern boxes.

The vertical structure of the concentration distribution was obtained by one aircraft flying a curtain pattern between the lowest level (~ 150 m above ground) to a height a little above the boundary layer. A second aircraft measured the conditions in approximately the centre of the boundary layer. A further aircraft sampled the concentrations just above the boundary layer flying cruising along the track indicated in Fig. 5.5 by the dashed line.

A comprehensive description of the co-ordinated measurements during the TRACT experiment is given in the Field Phase Report of TRACT [8]. Full descriptions of the preparatory phase is given in the general TRACT Operational Plan [9], the TRACT Aircraft Operational Plan [10] and a Quality Assurance Project Plan for Aircraft Measurement Systems [11].

The TRACT operation centre was located nearly in the middle of the experimental area at the Neuhausen ob Eck air field (station 16 in Fig. 5.5), from where all aircraft for the main investigation area operated. For the nested area along the cross-section through the Rhine valley a local operation centre was established at the airport at Offenburg. (Fig. 5.6, station H)

b. *Data bases*

In such an extensive field activity as TRACT a variety of observational data are collected. In addition to having to compile and archive the data, it is necessary to inter-compare the data sets from different groups for completeness and accuracy. So five different data sub-centres were created to treat the various groups of data:

- background data (from operational and temporal ground based networks)
- vertical sounding data from radiosondes and tethered balloon systems
- aircraft data (mainly chemical species)
- emission data
- turbulent flux and energy balance data

After compilation and processing all the data were collected together into the main TRACT data base. The complete data base is located at the Institute of Meteorology and Climate Research of the University of Karlsruhe. It is accessible via internet and may also be obtained on a CD-ROM [12].

The data formed the starting point of the individual analyses of the studies from groups participating in TRACT. The data has also been used in various other subsequent air pollution studies, because it forms a useful comprehensive data source for evaluations of combined atmospheric and chemical model studies.

c. *Quality assurance and control*

The results and conclusions from most field experiments are based on the comparison of measurements from a large number of ground stations and airborne platforms. It is essential, therefore, that the measurements meet the highest expectations with respect to their accuracy, precision, completeness, comparability and representativeness; experience shows that the necessary data quality can only be obtained through the implementation of a rigorous quality assurance programme. The project description [13] stated the quality assurance principles of the programme, somewhat before these became so fashionable, and the programme was fully implemented in the campaign.

Quality control measures were implemented specifically for [8]:

- aircraft measurements (main area aircraft);
- aircraft measurements (nested area aircraft);
- radiosondes;
- ground-based meteorological measurements; and
- ground-based chemical measurements.

Aircraft measurements (main area aircraft)

A detailed quality assurance plan, setting data quality objectives for the principal quantities to be measured, was drawn up for the five aircraft operating in the main area [14] Table 5.3.

Table 5.3: Data quality objectives for the main area aircraft campaign.

Quantity	Data quality objective
temperature	± 0.5 K
dew point	± 1.0 K
O_3	± 20 % or 2 ppb (whichever is greater)
SO_2	± 30 % or 2 ppb (whichever is greater)
NO	± 30 % or 1 ppb (whichever is greater)
NO_2	± 20 % or 1 ppb (whichever is greater)
NO_y	± 30 % or 2 ppb (whichever is greater)
H_2O_2	± 35 % or 0.2 ppb (whichever is greater)

A ground-based mobile calibration laboratory was established[14] to provide known flows of the calibrated gases to the aircraft. At the beginning of the campaign the analytical equipment on each aircraft was calibrated with samples from the mobile laboratory, an onerous task which for the five machines took nearly two days of round the clock working [11].

An intercomparison was then carried out with formation flights of all five aircraft at three different altitudes to check the measurement system under actual flight conditions. At the end of each flight mission the instruments on pairs of aircraft were again inter-compared with "fly-bys".

Two aircraft were equipped to take grab samples of non-methane hydrocarbons. These were subsequently analysed by gas chromatography using for calibration samples obtained from the National Centre for Atmospheric Research (NCAR) in Colorado.

The ozone and NO_y measurements met the data quality objectives without restriction. The values found for sulfur dioxide and nitric oxide were generally close to or below the detection limits, and there were some difficulties with hydrogen peroxide. The results showed some surprising deviations in the "standard" measurements of altitude, temperature and relative humidity [14].

Aircraft measurements (nested area aircraft)

The four motor gliders and the single normal aircraft operating in the nested area measured ozone together with a number of other standard parameters. An intercomparison was made with a two formation flight by the motor gliders. The normal aircraft could not fly sufficiently slowly and so flew the same track as the motor gliders several times during the flight period. The ozone measurements met the data quality objectives set for the aircraft in the main area. Once again some deviations were found in the "standard" parameters [8, 15].

Radiosondes

Radiosondes were used at five sites to obtain vertical profiles of the temperature, pressure, relative humidity and the horizontal wind speed and direction. Two types of sonde were used and a single intercomparison between the two types showed that the parameters were within the manufacturers' specifications [8].

Ground-based meteorological measurements

An appreciation of the meteorology was essential to TRACT and so an effort was made to compare the instruments used at the six principal stations. A mobile station was set up to measure:

total radiation,	relative humidity,	solar radiation,
wind speed,	surface radiation temperature,	wind direction,
air temperature,	air pressure, and	wet bulb temperature.

It was then driven to the various sites, at each of which a three-hour intercomparison was made. The results, although pinpointing one or two problems, were satisfactory [8, 16].

Ground-based chemical measurements within the nested area

For the temporary stations within the nested area, a calibration laboratory was set up at Ehrlenhof by the Association pour la surveillence et l'étude de la pollution atmosphérique en Alsace. The various instruments were brought initially to the laboratory for comparison with the accepted standards and then were brought again after the campaign was finished. The results obtained for ozone and sulphur dioxide were excellent (with a few percent) but problems were experienced with the nitrogen oxides which, however, still appear to be within the data quality objectives for the aircraft in the main area [8, 17].

Conclusions on quality assurance

Quality assurance is always a carefully defined compromise between the optimism of the investigators concerning their own measurements and the reality of making measurements under difficult field conditions. It allows one to set realistic goals and, when the results are analysed, to see which differences are truly significant and which lie within the noise. For TRACT, effective measures were taken which ensured that the data obtained was "adequate for the purpose for which it was intended" [18].

Emission inventory

In regional studies of air quality, the emissions of pollutants from anthropogenic sources are of vital importance, and a field study of the size of TRACT would be incomplete without a detailed analysis of the emissions within the area of investigation as a function of time and space [19, 20]. The TRACT area is a densely populated area and includes industrial centres such as Stuttgart, Mannheim-Ludwigshafen, Karlsruhe, Strasbourg, Freiburg and Basel. These areas are connected by roads forming important traffic routes running north to south along the Rhine valley or from west to east across the northern mountains of the Black Forest.

The emission inventory was constructed so as to provide a data set consistent with the chemical modules incorporated into atmospheric models both in temporal and spatial resolution and also from the variety of the chemical species included.

Therefore the emission inventory concentrated mainly on those species which are precursors for ozone and other photo-oxidants in the boundary layer, since the experiment was designed to be conducted under fair weather conditions. Nitrogen oxides, carbon monoxide, sulfur dioxide and non-methane hydrocarbons were determined at hourly intervals for the whole region on a 1 km by 1 km grid. Statistics on road traffic counts, housing density, numbers of inhabitants and industrial activity were used to calculate the emissions which were then interpolated to obtain the gridded data. Large point sources were identified with a special survey and analysed individually.

In all a detailed and reasonably reliable emission data set was obtained for the whole period from 9th to 22nd September, 1992 in hourly time steps with a spatial resolution of 1 km by 1 km.

In this way, the concept of an integrated study was developed which included, as a function of time and space, the calculation of the emissions inside the investigation area, the analysis of meteorological transport and mixing conditions and the development of the concentration fields during the day.

Modelling activities

Understanding observed phenomenological distributions of air pollutants on a regional scale is, in the current view, not possible without the use of well-developed models. It is recognised that the behaviour of the concentrations of different species is the result of the combined activity of atmospheric transport on the mean wind from the source to the receptor, of turbulent diffusion and of chemical transformation during transport. Additionally, solar radiation gives rise to photochemical reactions, and the source distributions and deposition play important roles. The conditions are still more complex if the area is bounded by complex terrain.

The measurements in TRACT were organised in such a way that all data bases required for the use of atmospheric models including the chemical transformation were available [9]. As well as the atmospheric variables are needed for initialisation of the model, terrain and land use data were available for the TRACT area at high resolution.

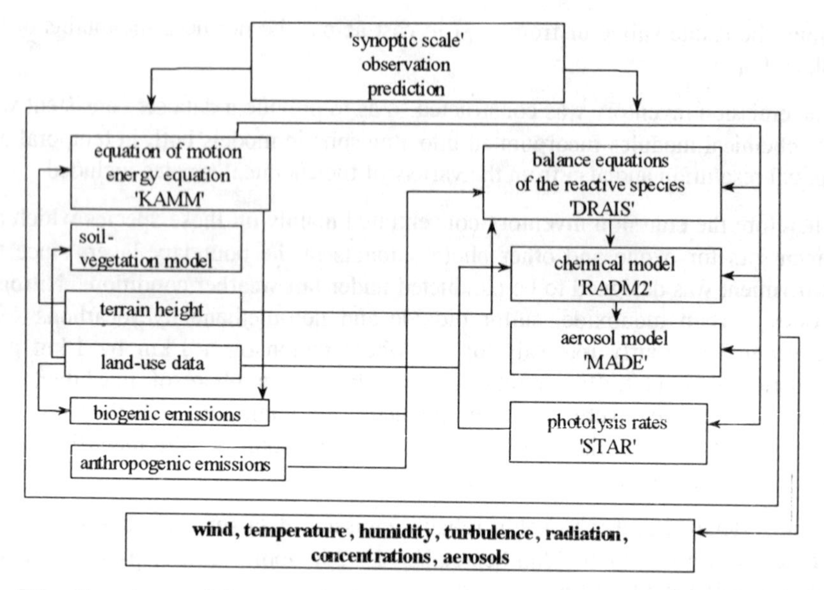

Fig 5.7: The scheme of the atmospheric and chemical model system KAMM used for the TRACT project.

As the horizontal resolution must be as high as possible in order to avoid smearing effects of the emissions in grid boxes that are too large close to intensive point and line sources, a non-hydrostatic model system KAMM (Karlsruhe Atmospheric Mesoscale Model [21, 22]) was combined with the chemical code RADM2, modified for European conditions [23]. The complete model system with its different sub-modules is shown in Fig. 5.7. Particular emphasis is given to the harmonised parameterisation of turbulence with respect to turbulent transport, momentum, heat and moisture and the turbulent transport of chemical species. The same inter-dependence is present for the interaction at the surface between the atmospheric variables and the evaporation and transpiration of vegetation and the deposition and biogenic emissions of various chemical species. Model simulations are virtually the only means of apportioning the relative contribution of the different processes to the observed concentration fields.

Sensitivity studies of various sorts have been made in order to clarify the way in which complex terrain influences the transport, diffusion and even the chemical transformation processes. In Fig. 5.8 an interesting example is given for the general flow structure within the experimental area of TRACT and particularly in the Rhine valley. In the figure the flow at 15 m above ground (independent of the terrain height) is shown for an ideal westerly synoptic flow. It is seen that the topography deforms the flow field at lower heights dramatically. In the south-west there is an inflow through the Burgundy Gate. There the flow splits with the main branch flowing to the north about 150 km along the Rhine valley. A second branch passes along the southern end of the Black Forest and then flows to the north-east seeking a way between the Black Forest and the Swabian mountains.

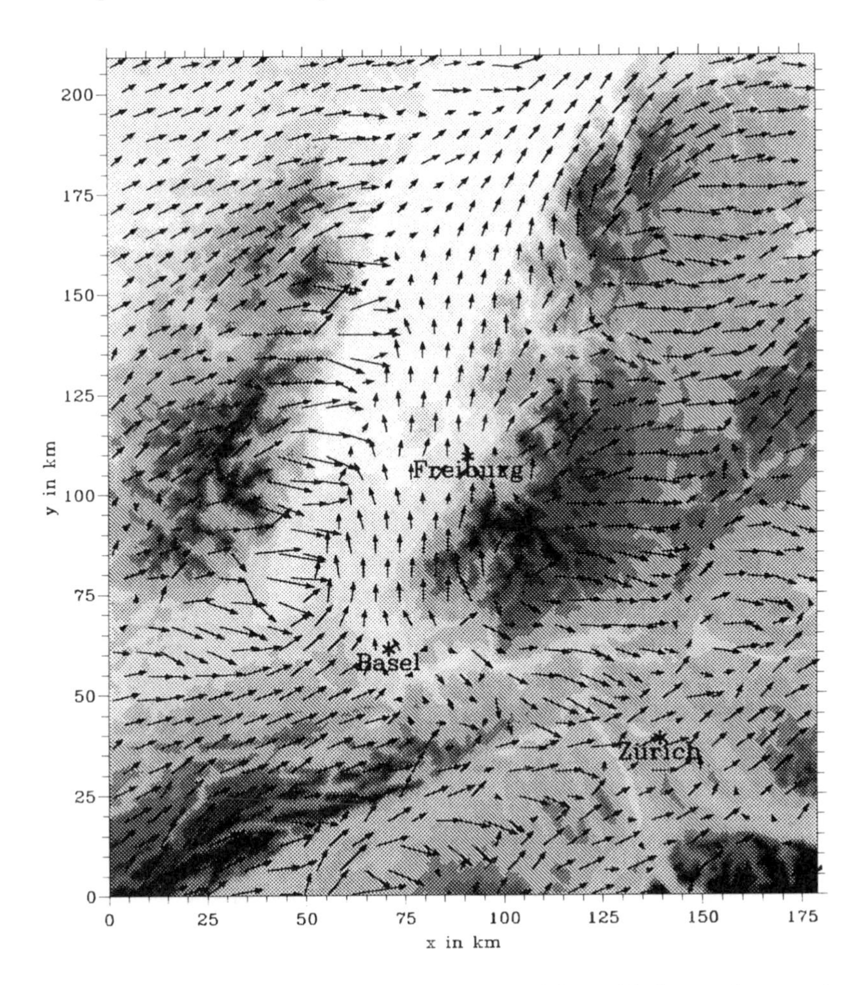

Fig. 5.8: Simulated wind field at 15 m above ground level for an ideal westerly synoptic flow.

There is a similar picture for the case of the TRACT intensive measuring period on 16th and 17th September, 1992. In Fig. 5.9 the modelled wind field for the 16th September at 12:00 CET is shown. In general the same structure is visible in comparison to the ideal case presented in Fig. 5.8, but details are much more complicated. This model result has been obtained by using the large scale synoptic conditions to drive the mesoscale model by splitting up the meteorological conditions into a basic state (synoptic large scale field) and deviations from the basic state which is the result of the mesoscale model.

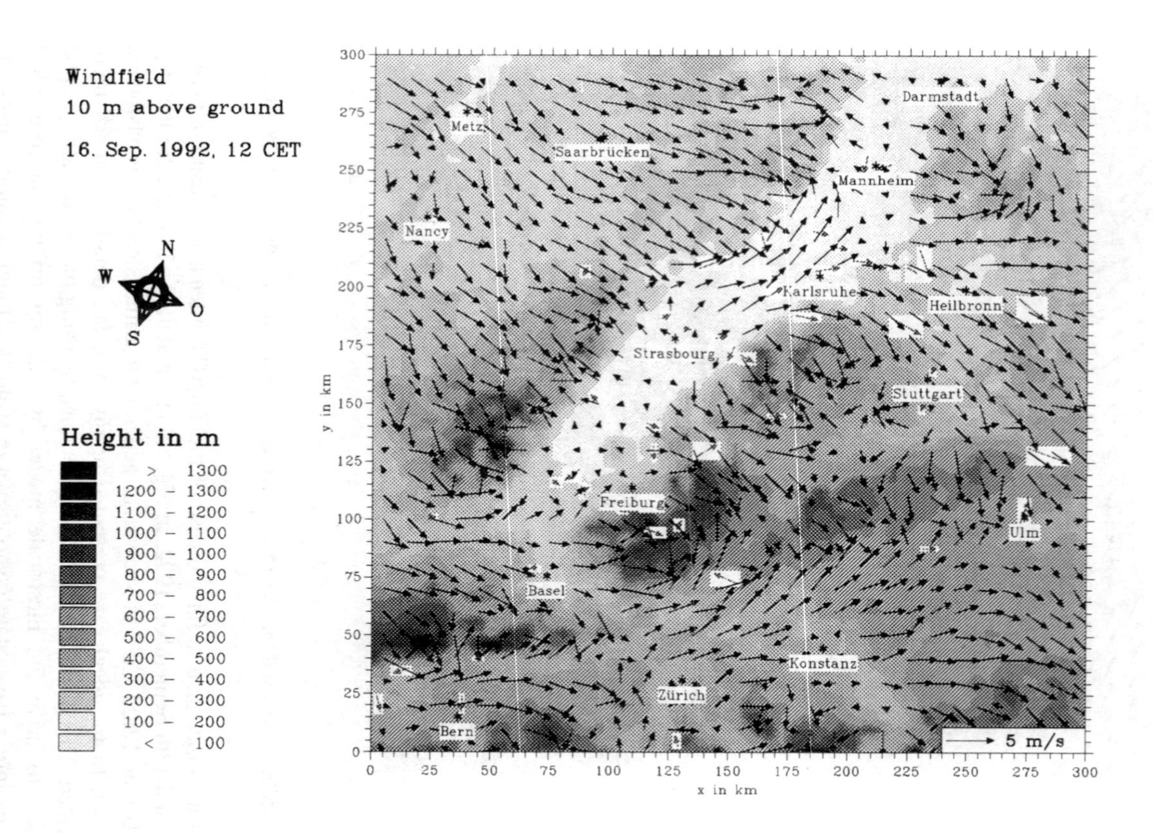

Fig. 5.9: Simulated wind field at 10 m a.g.l. at 12:00 CET for the intensive measuring phase of 16th–17th September.

The north-westerly flow around Metz, Nancy and Saarbrücken is consistent with the synoptic situation. At the Burgundy Gate, west of Basel, the inflow south of the Vosges appears again. At the southern edge of the Black Forest a strong flow develops following partly the Danube valley towards Ulm and partly crossing the Swabian mountains towards Stuttgart. How complex the flow is, which develops in mountainous terrain, is visible from the stagnating air between the Vosges and Black Forest, west of Freiburg and also in the northern Rhine valley around Mannheim. The arrows in the white insets show the observed wind measurements at the same time. Although many of the ground observations are influenced by buildings, trees, hedges or forest and so are not representative for a comparison of the calculated wind field within a grid box, it is astonishing how well the model shows up the details of the wind flow in the TRACT area.

Another interesting feature is the converging flow region on the lee side of the Black Forest, west of Stuttgart, where it is expected from the wind flow alone that the emissions from the Karlsruhe area and the Stuttgart area must overlap. It can also be concluded from the meteorological conditions that in areas with identical emission rates but differing in wind speed and direction, the observed concentrations and the ongoing chemical transformations will develop quite differently. Therefore the concentration pattern shows a large variability.

Due to pressure disturbances connected to mountain ranges smaller scale secondary flow systems often develop both in front of and, in a more pronounced way, on the lee side of mountains. These larger flow features are much more likely to produce vertical transport than the general smaller scale turbulence.

Lakes, such as Lake Constance, act as disturbances in the surface, and combined with mountains around, are a strong influence on secondary flows. A model simulation for this area is shown in Fig. 5.10 where the potential temperature and the wind field is presented. These secondary flows are intensified, when the model resolution is improved.

Fig. 5.11 shows an example from a model run with a horizontal resolution of 250 m. At the bottom of smaller valleys, a counter flow develops as part of a standing eddy with a horizontal rotation axes. In complex terrain it is through these eddies that the emitted material is removed from lower levels of the atmosphere and handed over to higher levels. Mountain effects are not included in the usual parameterisations employed in numerical models which are theoretically only then valid for horizontally homogeneous terrain. It is clear that models which have a coarser grid size cannot resolve these features, and must create errors in the transfer due to vertical mixing. The hand over due to these secondary eddies, connected to mountains, must be considered as an important mechanism for effective and rapid exchange, inside the boundary layer, of air close to ground with air at higher levels.

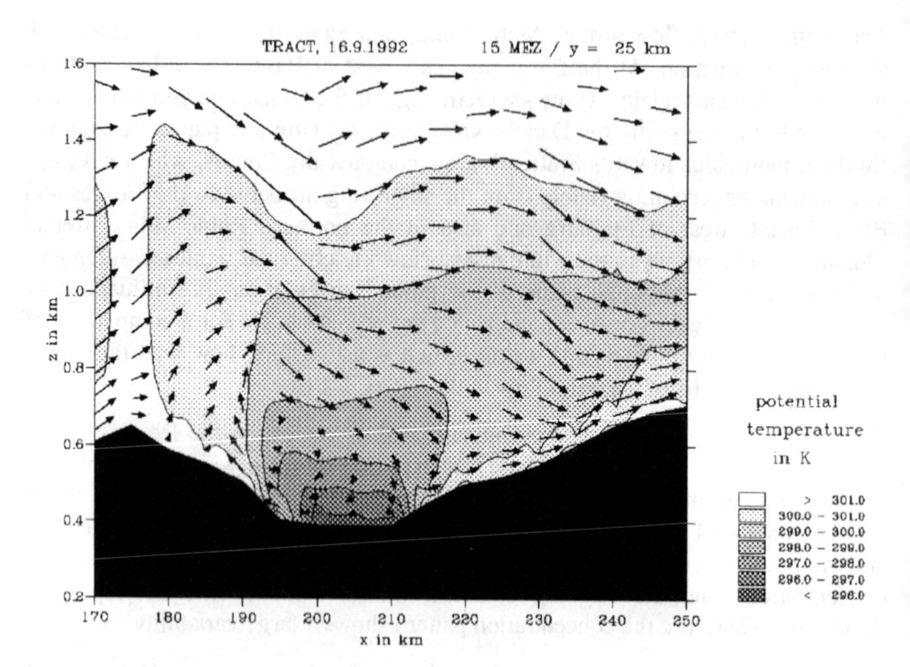

Fig. 5.10: Simulated potential temperature and wind field for a vertical cross section at Lake Constance with an intensive secondary flow system.

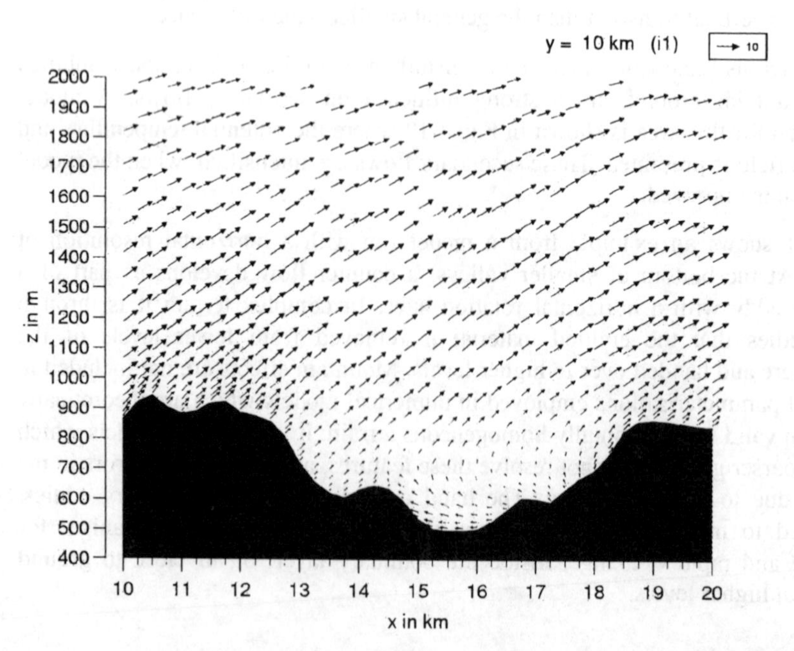

Fig. 5.11: Simulated vertical cross section of the wind over the Murgtal with a resolution of 250 m.

Sensitivity studies with numerical models have also been carried out in order to determine the effect of terrain on the deposition of material at the surface [24]. The results indicated that, during night hours, the deposition values calculated with coarser model resolution, without the secondary flow systems involved, may be a factor of two too low.

Similar studies have been undertaken to study the influence on biogenic emissions of temperature variations connected to mountainous terrain [25]. For summer conditions with high temperatures, comparison of the emission rates were made for a particular terrain type and the corresponding mountainous terrain, keeping other conditions constant.

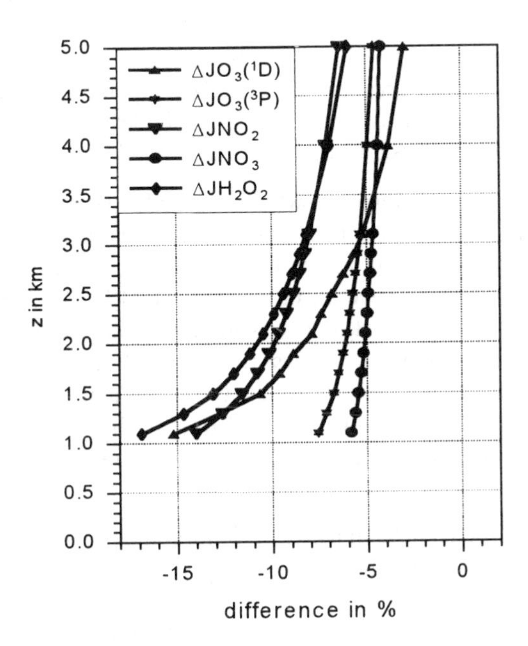

Fig. 5.12: Simulated vertical profiles of the relative difference of the photolysis rate $\Delta J = (J_{Schauinsland} - J_{Offenburg})/J_{Offenburg}$ (17th September, 1992, 12:00 CET).

Since the surface temperatures at different heights in the mountains vary dramatically, compared with the temperatures over flat terrain, biogenic emission rates are larger when real topography is taken into account. As a consequence the contribution of biogenic emissions to ozone production is larger with real topography than with a flat terrain. Similar studies indicate that deposition rates are also a function of land use [25].

An unexpected effect of topography is the change of the photolysis rate in mountainous terrain [27, 28]. Fig. 5.12 shows that the photolysis rates at the mountain top are reduced by up to 15 % compared to the photolysis rates at the same height over flat terrain.

Observations and results in TRACT

In complex terrain the atmospheric conditions are variable so that observations should be made with a high spatial and temporal resolution. It was fortunate that the TRACT field campaign overlapped with REKLIP, the Regional Climate Project of the Rhine valley [29]. REKLIP operated a surface network of 36 well-equipped micrometeorological stations for about five years. These were used together with the operational networks of the weather services, the environmental agencies and private operators, and also the TRACT temporary stations to provide a reasonably dense measurement system. Thus a much clearer insight in the wind flow conditions was obtained than would have been possible in earlier times.

Fig. 5.13 summarises, in the form of wind roses, the day-time surface wind observations for the whole TRACT area for the whole period. The general structure seen in the observations is similar to that found from the model results. It can be seen that the mountains have an influence up to 150 km downstream. In particular, the flow around the Black Forest is more pronounced than that around the northern and southern side of the Vosges. It is known from climatological data that the area around Basel is often not foggy when the neighbouring areas are completely covered with fog. The divergence of flow in the horizontal directions leads to a downward compensating flow trough in which low level clouds and fog are dispersed.

It can also be seen that the Kraichgau between Heilbronn in the east and Karlsruhe and Mannheim in the west is an important pathway for the transport of pollutants. Another path leads from Basel towards Ulm in the southern area along the valley of the Danube. During the day, when strong convection strongly links the flow close to the ground with that of the higher altitudes. The influence of the terrain is seen even at heights above the summits of the mountains. The flow in the lower levels follows the valley axes and convection then forces the flow at higher altitudes into the same direction.

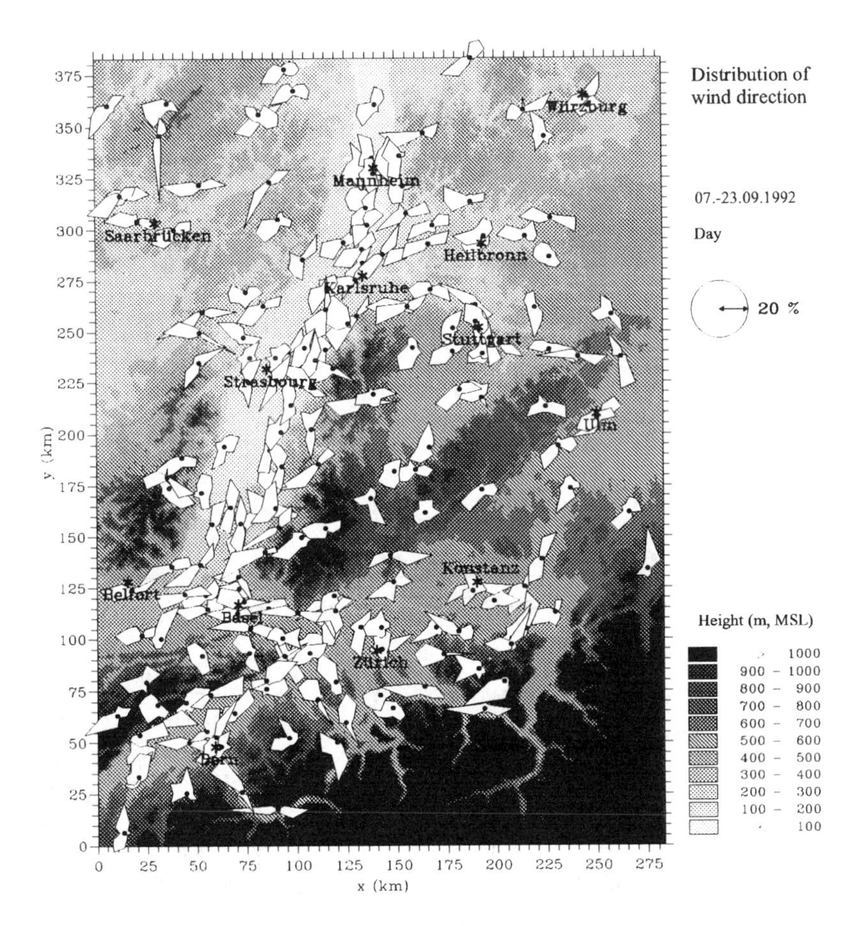

Fig. 5.13: Frequency distribution of the wind direction from ground station for the time period of the TRACT experiment during the day.

The vertical structure of the atmospheric boundary layer is of vital importance to the concentrations of the air pollutants. Under neutral conditions the height of the boundary layer is determined by the strength of the geostrophic wind speed (and of the Coriolis force). Under non-neutral conditions the temperature structure plays a dominant role.

One important problem that is not yet fully understood is the height development of the boundary layer over complex terrain during the course of the day. Under stable conditions the height of the boundary layer usually follows the terrain height provided the mountain slopes are not too steep and the summits do not break through the boundary layer.

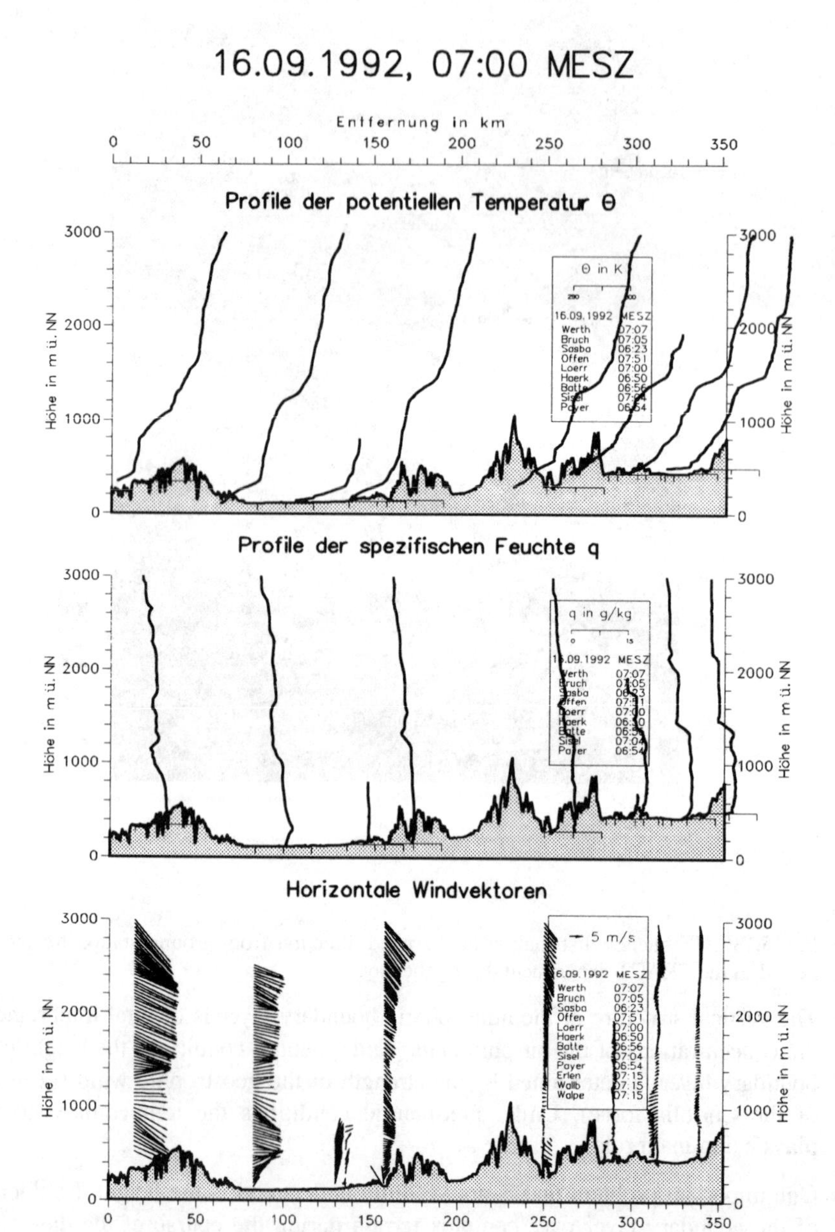

Fig. 5.14: Measured vertical profiles of potential temperature, specific humidity and horizontal wind vector along the north-south cross section from Wertheim to Payerne on 16th September, 1992 at 7:00 CEST.

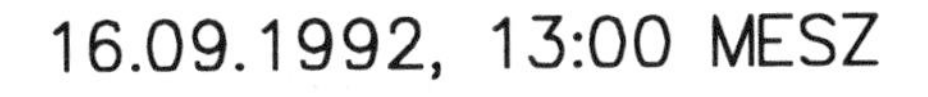

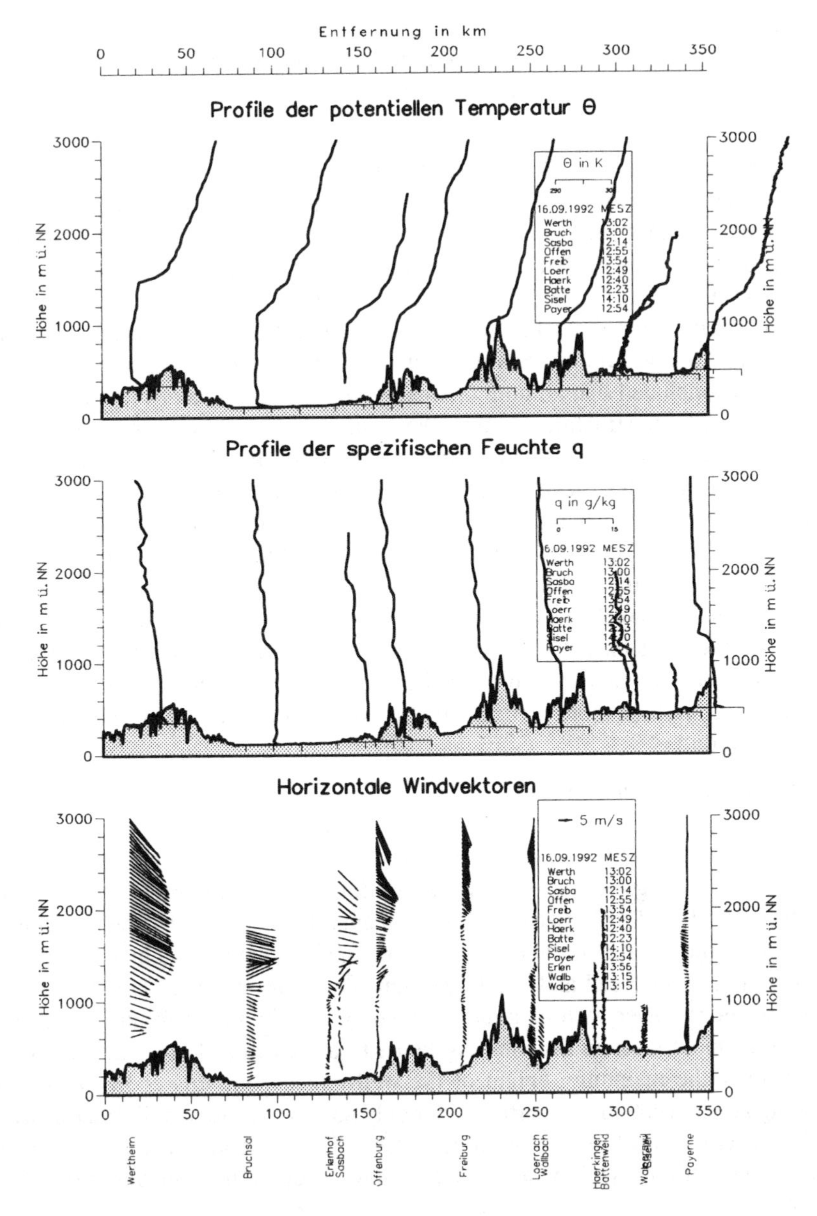

Fig. 5.15: Measured vertical profiles of potential temperature, specific humidity and horizontal wind vector along the north-south cross section from Wertheim to Payerne on 16th September at 13:00 CEST. The location of the measurement station is always at the left end of the temperature scale.

Much emphasis was given to resolving the boundary layer structure during the intensive measuring phases by launching radiosondes at 11 stations simultaneously, by using tethered balloons and also by flying some of the aircraft in a curtain-like pattern. An example of the vertical structure of the boundary layer is given in Fig. 5.14 for the morning hours and Fig. 5.15 for the conditions at noon. Vertical profiles of potential temperature, specific humidity and of the horizontal wind velocity are plotted along cross-sections through the named stations.

In the morning a shallow stable layer, of approximately 200 m in height, follows the terrain; the bulk of the boundary layer with less stable conditions is influenced by the residual layer from the convection of the previous day. The pronounced boundary between the friction layer and the free troposphere is marked by a strong inversion at around 1200 m a.s.l. The temperature structure is strongly correlated with specific humidity, the increase in potential temperature within the capping inversion being strongly correlated with a decrease in the specific humidity. It may be concluded that all non- or weakly-reactive chemical species would show a similar vertical structure. Within the boundary layer there are always marked changes of wind speed and direction. During the day, when a well-mixed boundary layer has developed potential temperature and specific humidity are nearly constant with height. The transition to the free troposphere is indicated by a strong inversion layer with the more stable conditions in the free troposphere above it. From the results it can be clearly seen that the boundary layer height decreases from north to south. At Wertheim it reaches 1200 m whereas at Freiburg and Lörrach it is only 900 m. The complexity of conditions in the real atmosphere is illustrated by this example: in the mesoscale there are interactions between synoptic conditions, the vertical structure of the boundary layer and the influence of the terrain.

The 16th and 17th September during the intensive measuring phase turned out to be the "golden period" for the analysis, with the synoptic conditions remaining quite stable and almost no horizontal advection. This may be seen from the three-hourly soundings at the radiosonde stations at Oberbronn (Fig. 5.16) and at Musbach (Fig. 5.17). Between these ascents, between 7:00 in the morning and 16:00 in the afternoon, there is almost no shift in potential temperature, specific humidity, wind speed and wind direction. The differences are largely confined to the boundary layer with its strong diurnal cycle. The boundary layer is converted from a stable layer in the morning hours to a well-mixed layer in the afternoon. However the maximum height of the well-mixed zone differs by about 150 m, being about 1000 m at Oberbronn but only 850 m at Musbach. Further analyses of the meteorological conditions during TRACT and of accompanying spatial and temporal variations of chemical species may be found in Koßmann [30].

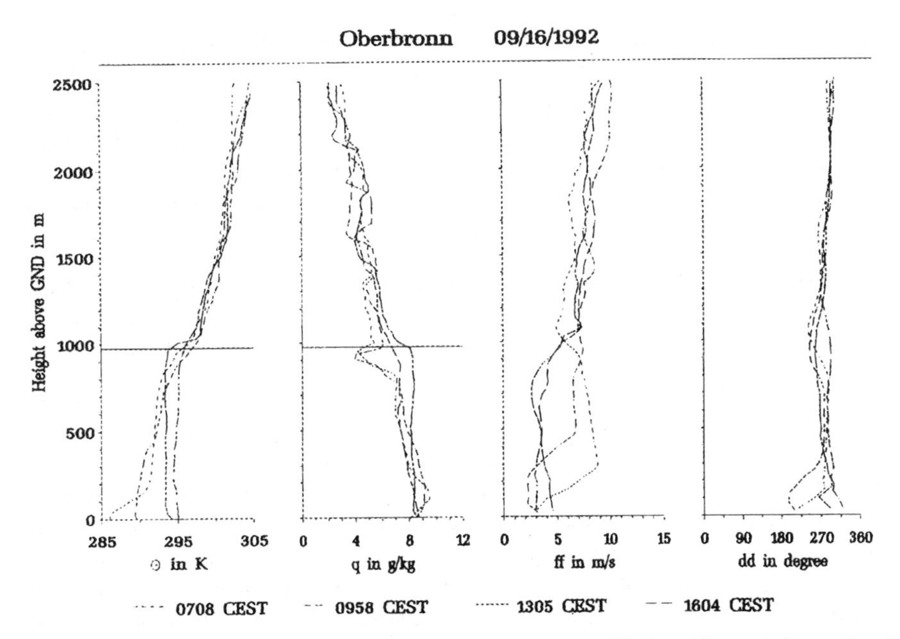

Fig. 5.16: Temporal evolution of potential temperature, specific humidity, wind speed and wind direction on 16th September at Oberbronn, 274 m a.s.l. The solid line indicates the boundary layer height at 13:00 CEST.

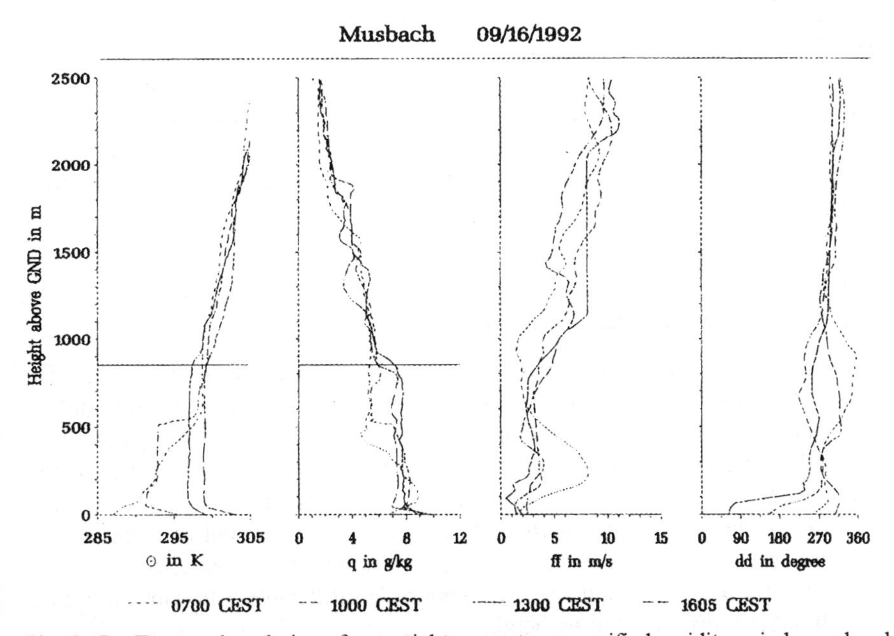

Fig. 5.17: Temporal evolution of potential temperature, specific humidity, wind speed and wind direction on 16th September at Musbach, 694 m a.s.l. The solid line indicates the boundary layer height at 13:00 CEST.

The influence of the changing vertical mixing of ozone, and its consequences for the daily variation, during TRACT campaign, has been analysed by Löffler-Mang *et al.* [31], particularly for frontal passages or during the presence of valley wind systems. It was found that many dynamic and thermodynamic atmospheric processes are involved in the observed temporal changes of the concentrations. So the effects of these processes must first be separated before chemical production or removal may be identified in the observations.

It is this strong interaction between the dynamics and the chemistry which makes an integrative observation programme with an equally emphasis on the meteorological and chemical measurements essential to a proper understanding.

How important the spatial resolution is for understanding the transport and diffusion processes over complex terrain, is illustrated by the flight simulation in Fig. 5.18. The diagrams show profiles of the potential temperature (upper) and the specific humidity (lower). The boundary layer height is indicated by the dashed line. It can be clearly seen that, over the mountain plateau there is a well-mixed layer. With easterly winds the layer is transferred over the Rhine valley where it shifts outside the atmospheric boundary layer. This is a frequently occurring event and an important process for the efficient handover of material from the boundary layer to the free troposphere.

Fig. 5.19 shows the turbulence field for the same situation, and indicates another important aspect. Over the mountains the turbulence level is high and there is strong mixing associated with it in all directions. But as soon as the air moves over the Rhine valley the turbulence level is much reduced. Thus material transferred outside the boundary layer may then be transported much longer distances without strong vertical mixing and with little deposition to the surface. In this case, the boundary layer structure clearly follows the topographic structure, a phenomenon which is much more easily detected from the intensity of the turbulence than from the structure of mean quantities.

The handover process of polluted air from the boundary layer to the free atmosphere is also strongly influenced by secondary flow systems originating from pressure and temperature disturbances connected to the terrain structure.

In Fig. 5.20 the analysis of ozone is shown for the Lake Constance area. The simulation shows clearly that, during the conditions of a westerly flow aloft, a weak counter flow develops at the lower levels. The counter flow brings down air with only low concentrations of ozone and carries air upward on the western side. The boundary layer determined from the model is indicated by the black line. However, as in other cases, the boundary layer inversion is not an impermeable lid: there is a constant flow through this layer. During the day, advective processes and vertical mixing by turbulence are responsible for the different daily courses of air pollutants at different terrain heights.

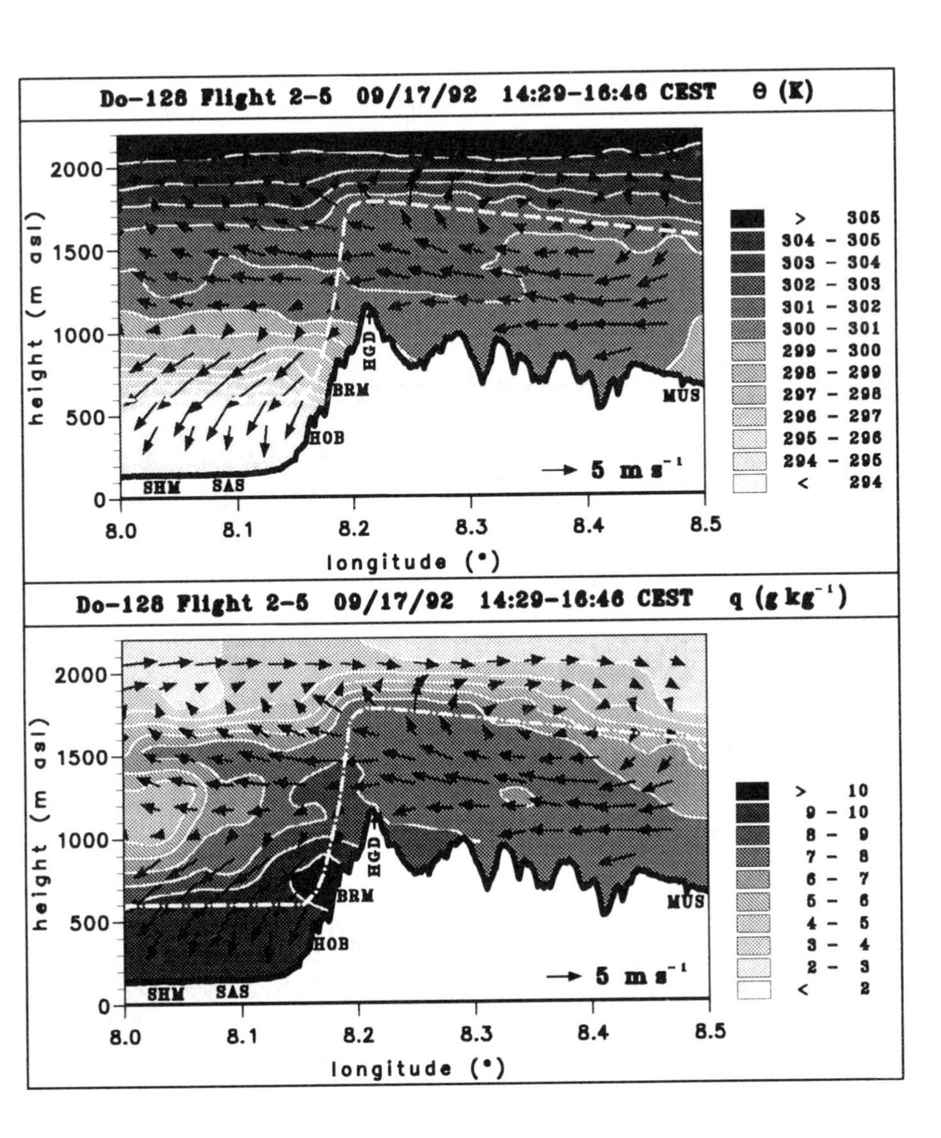

Fig. 5.18: Interpolated cross sections of potential temperature, specific humidity and the horizontal wind vector from the Rhine river to Musbach in the Black Forest during the afternoon of 17th September 1992. The white dashed line indicates the top of the convective boundary layer as derived from aircraft measurements and vertical sounding systems. The horizontal range of the cross sections is about 36 km.

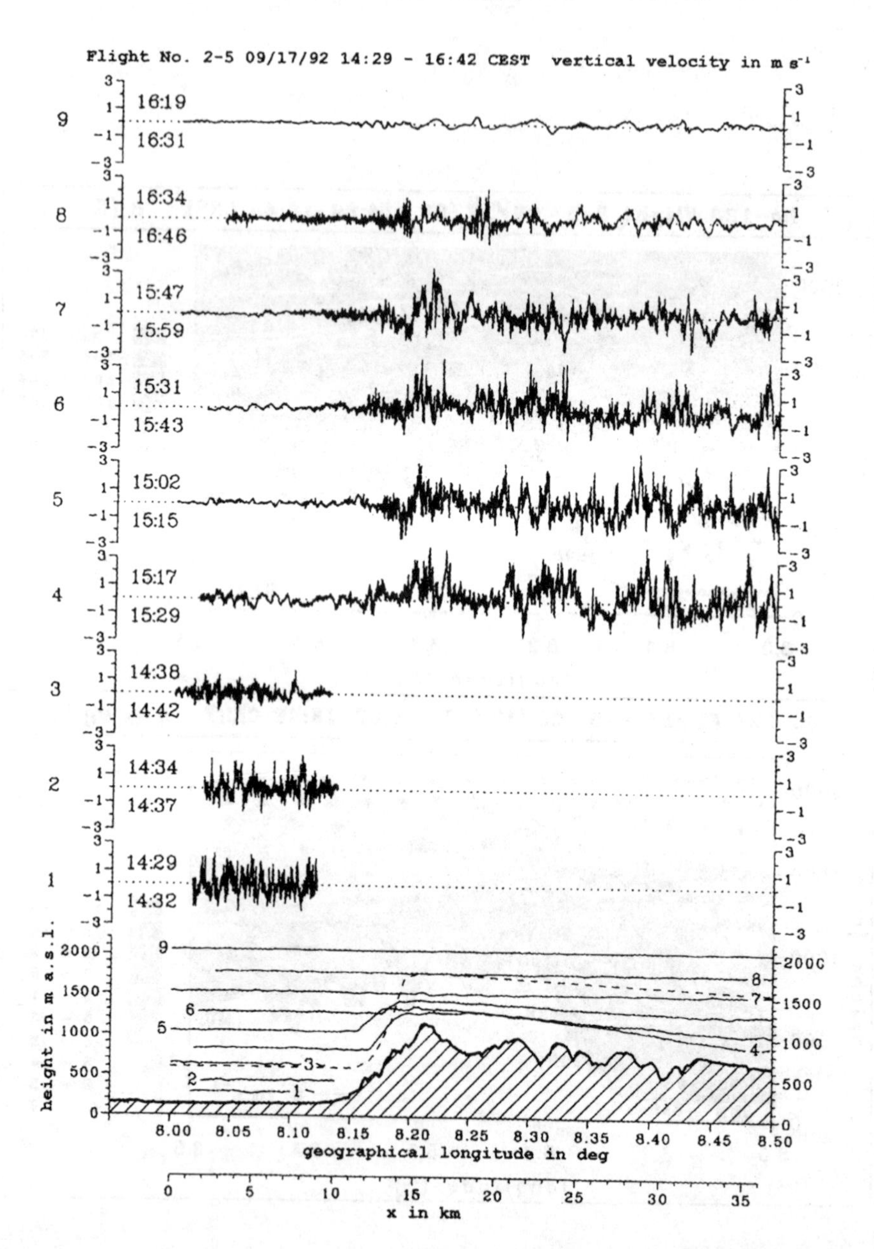

Fig. 5.19: Flight pattern and vertical wind component along the cross section from the Rhine river to Musbach in the Black Forest during the afternoon of 17th September. The frequency of recording was 25 Hz. Solid lines above the orography (shaded) indicate aircraft flight legs. The starting point of each flight leg is marked by the leg number in the graph of the flight pattern. The dashed line above the orography represents the top of the convective boundary layer as derived from aircraft measurements and vertical sounding systems. Additionally, the leg numbers and the times of beginning and end of the legs are shown on the left side of the w-component graphs.

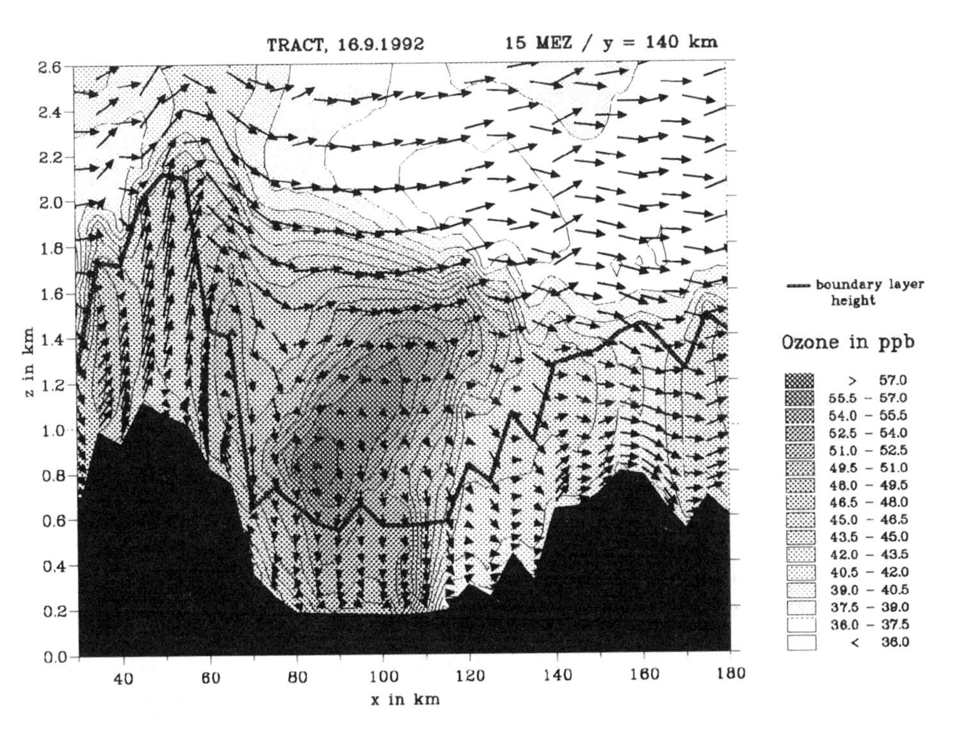

Fig. 5.20: Simulated ozone concentration for a vertical cross section in the Rhine valley. The black line indicates the boundary layer height.

In Figs. 5.21 and 5.22 the time variation of ground level concentrations of O_3, NO_2 and of NO and CO are plotted for the intensive measuring period, 16th–17th September, at four different locations. Schwarzwald-Süd and Freudenstadt are located within hills of the Black Forest whereas Rastatt and Kehl-Hafen are within the polluted region of the Rhine valley. The station with the highest elevation, Schwarzwald-Süd, shows high ozone concentrations for the whole day with only a weak variation during the whole period. In contrast, at the other stations which are located closer to the sources of NO and CO, there are sharp reductions in ozone during the night. The principal reason for the constant ozone concentration is the absence of deposition or the replacement of ozone by downward transport from higher altitudes of that lost by a weak deposition flux. The weaker thermal stability at higher levels is also an important influence: it permits a stronger coupling of the layers at different elevations and produces a more or less constant level of the concentrations during the whole day. It is only in the late afternoon that the vertical growth of the boundary layer within the Rhine valley reaches the Schwarzwald-Süd. Then an increase in the ozone level is observed for a few hours.

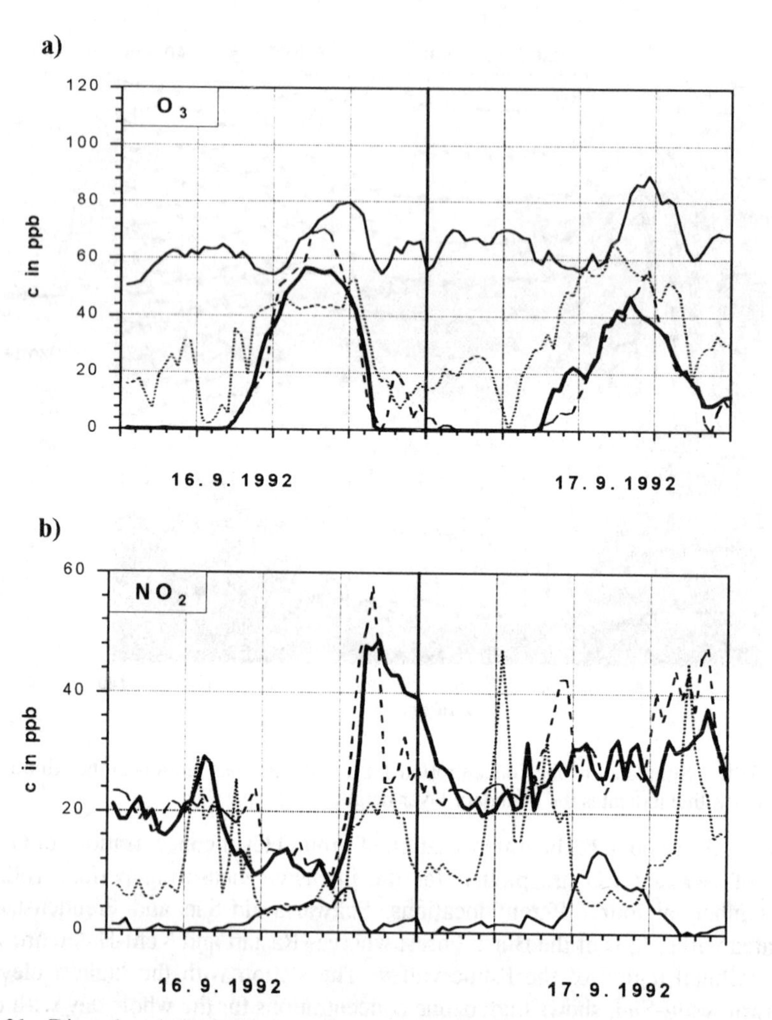

Fig. 5.21: Diurnal variation of measured ground level concentrations of a) O_3 and b) NO_2 at the stations Rastatt —, Kehl-Hafen ---, Freudenstadt ·····, and Schwarzwald-Süd — during the intensive measuring phase (16th–17th September, 1992).

An important question concerning the processes which determine the decrease of ozone in the lower layers of the boundary layer was addressed by Güsten *et al.* [32] with measurements in the nested area. From frequent captive balloon ascents during the late afternoon and night, the decrease of the ozone level was observed and it was possible to quantify the deposition of ozone at the ground using a profiling method. It appears that there is a discrepancy between the values of the ozone deposition determined directly from the deposition velocity and those determined from the isochron method. The deposition velocity gives values which are only between 20 to 50 % of those obtained by profiling. As well as deposition, chemical reactions of ozone with NO contribute to the depletion of ozone, which is confined to the shallow layer of at most 200 m above the ground.

a)

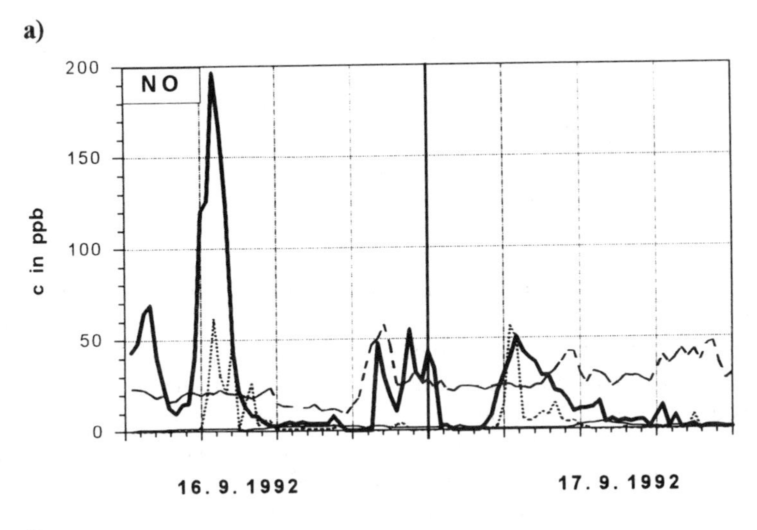

b)

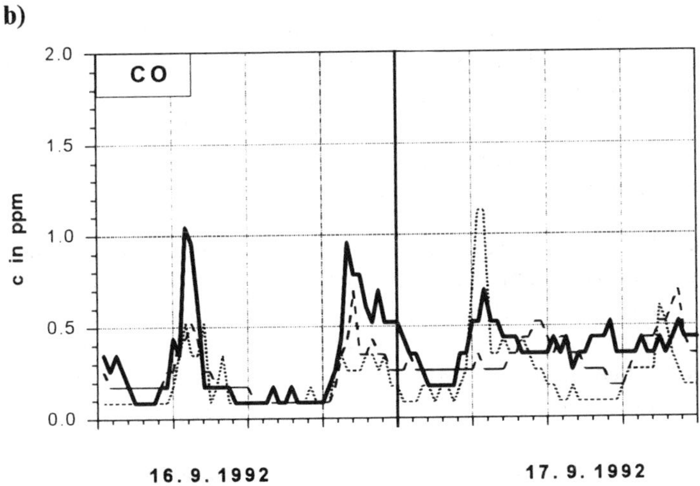

Fig. 5.22: Diurnal cycles of measured ground level concentrations of a) NO and b) CO at the stations Rastatt ——, Kehl-Hafen ---, Freudenstadt ····· and Schwarzwald-Süd — during the intensive measuring phase (16th–17th September, 1992).

This depletion processes for ozone during the daily cycle, obtained in a modelling study, are depicted in Figs. 5.23 and 5.24 for the nested-area cross section through the Rhine valley with the Hornisgrinde mountain of the northern Black Forest to the east. In the morning hours for the shallow layer up to 200 m, all ozone has been removed by deposition or by chemical reaction. In the upper levels higher ozone values still remain in the well mixed region left over from the previous day.

The mountain summits are surrounded by air with higher concentrations, so that the ozone values stay high throughout the whole day as shown in Fig. 5.21. In the early morning hours when vertical mixing is initiated again, the lower levels are filled up quickly by the downward transport of air with higher ozone concentrations. In addition, photochemical production of ozone in the whole domain of the boundary layer leads to a further increase of the ozone concentrations.

The increase of ozone near the ground, from almost zero to the maximum values seen in the afternoon (Fig. 5.22) is about 60 to 70 % due to the vertical mixing and about 30 to 40 % due to photochemical production within the boundary layers. However, in the proximity of strong sources of the precursors, especially NO, there is a net removal of ozone close to the ground throughout the day due to chemical reaction. Thus the increase of ozone in the morning hours observed within cities and close to roads with heavy traffic is simply due to the vertical mixing which more than compensates the photochemical loss.

Within the nested area there was sufficient information concerning the patchy ground structure to enable studies to be undertaken of the effective deposition and effective fluxes of other physical quantities such as momentum, heat and moisture. Hasager and Jensen [33] developed a physical model for the calculation of representative fluxes of momentum within areas which form the grid size of present day mesoscale or weather forecast models.

As well as taking the vertical soundings of meteorological variables, the aircraft were able to provide some insight into the vertical structure of some of the chemical components. Standard species, such as ozone, could be measured continuously particularly in a curtain flight patterns. Volatile organic compounds (VOC) were sampled at prescribed locations by collecting air samples in small cans. A list of the VOC species analysed is shown in Fig. 5.25, the measurements in this case being made at a height of 500 m above the terrain. The figure also shows the rate constants for reaction with OH. As expected there is a rough anti-correlation between the measured VOC concentrations and the OH rate constants. From the measurements it was also found, that species with high OH rate constants showed a stronger decrease with height than those with low OH rate constants. For some of these species empirical formulae for the profiles could be derived. An overview of the findings is presented by Koßmann *et al.* [34].

Figs. 5.26, 5.27 and 5.28 show the vertical structure of meteorological variables and chemical species obtained from three successive ascending and descending curtain flights in the Swiss part of the TRACT experimental area between the Zürich and Zuger lakes. The left hand box shows the potential temperature, specific and relative humidity together with wind speed and direction. In the middle box, the vertical profiles of NO, NO_2 and NO_x (left three curves)and of O_3 and O_x (right two curves) are shown. Finally, the right hand box shows vertical profiles for SO_2 and H_2O_2.

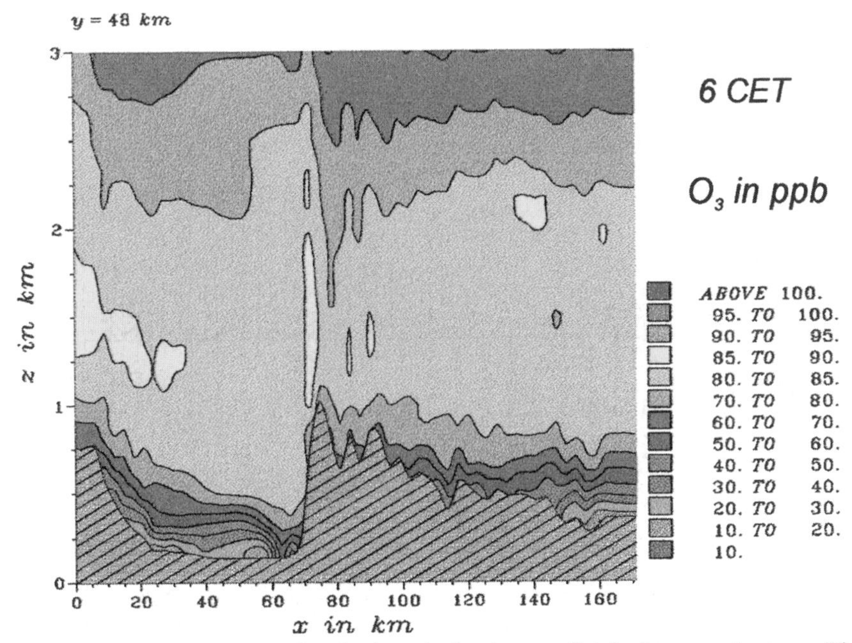

Fig. 5.23: Vertical cross section for the calculated ozone distribution over the upper Rhine valley (Vosges Mountains in the west, Black Forest in the east) at 6:00 CET with the depleted shallow layer close to the ground.

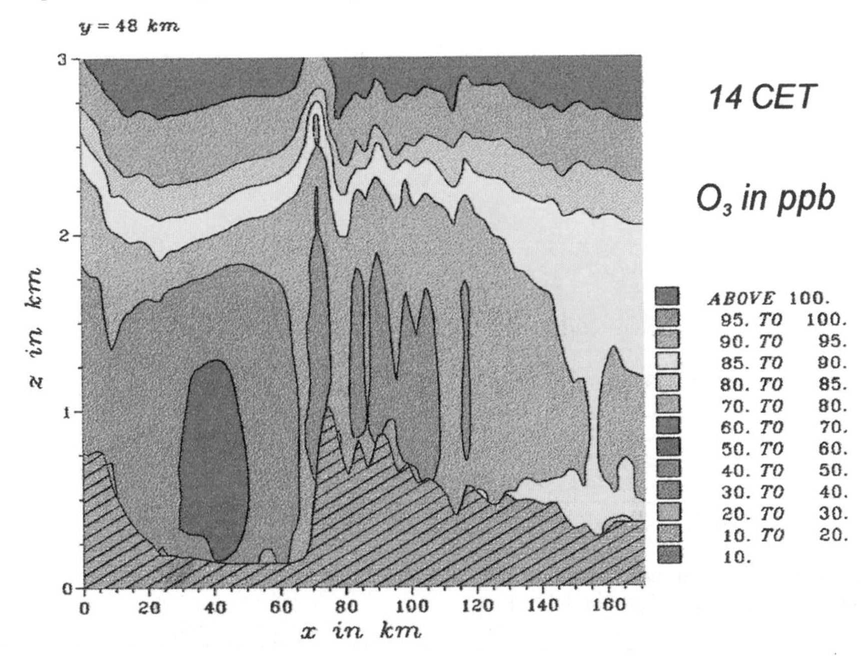

Fig. 5.24: Vertical cross section for the calculated ozone distribution over the upper Rhine valley (Vosges Mountains in the west, Black Forest in the east) at 14:00 CET at the time of maximum ozone concentration.

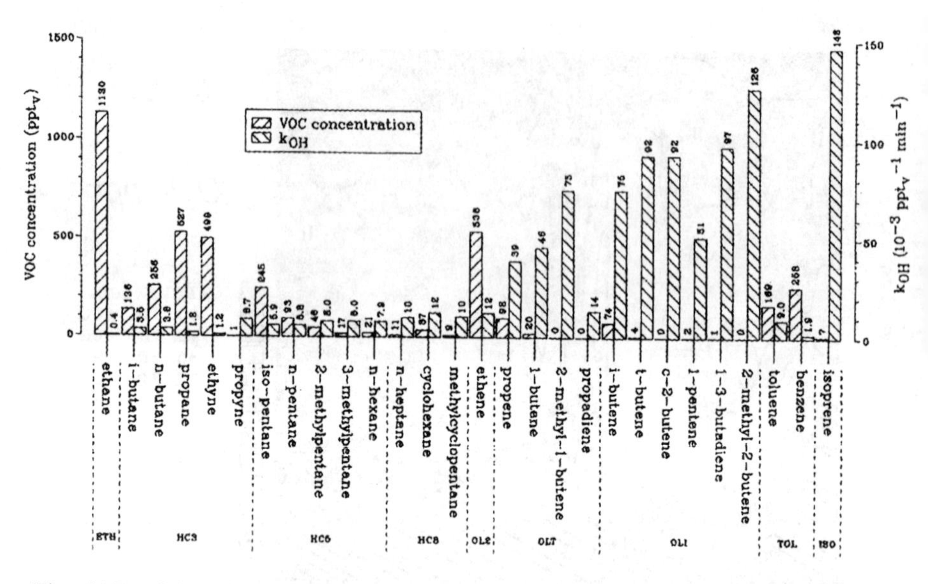

Fig. 5.25: Mean VOC concentrations of the 14 air samples with NO_x load ≤ 2 ppb collected below 500 m a.g.l. Additionally, the OH rate constants k_{OH} and the corresponding VOC classes of the individual species in the chemical gas-phase mechanism RADM2 are shown.

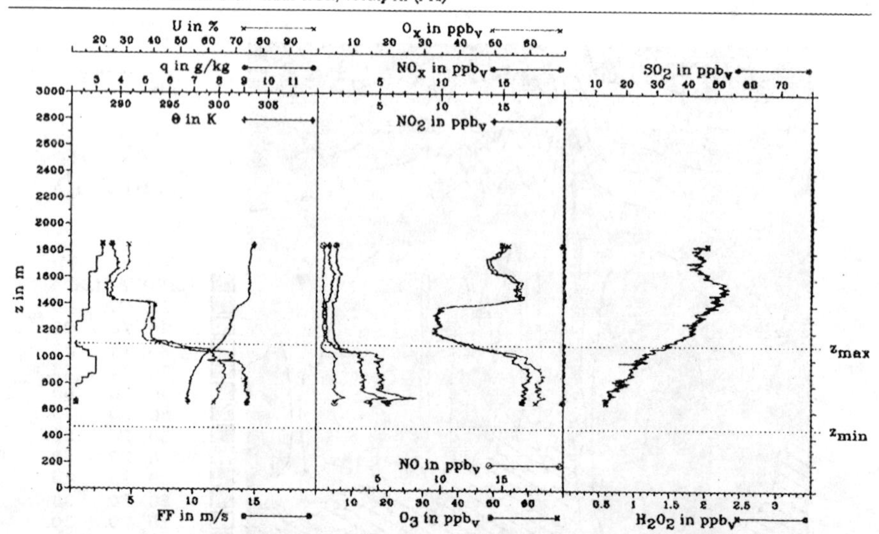

Fig. 5.26: Vertical profiles along the flight tracks of meteorological and chemical variables taken close to Rüthi in Switzerland.

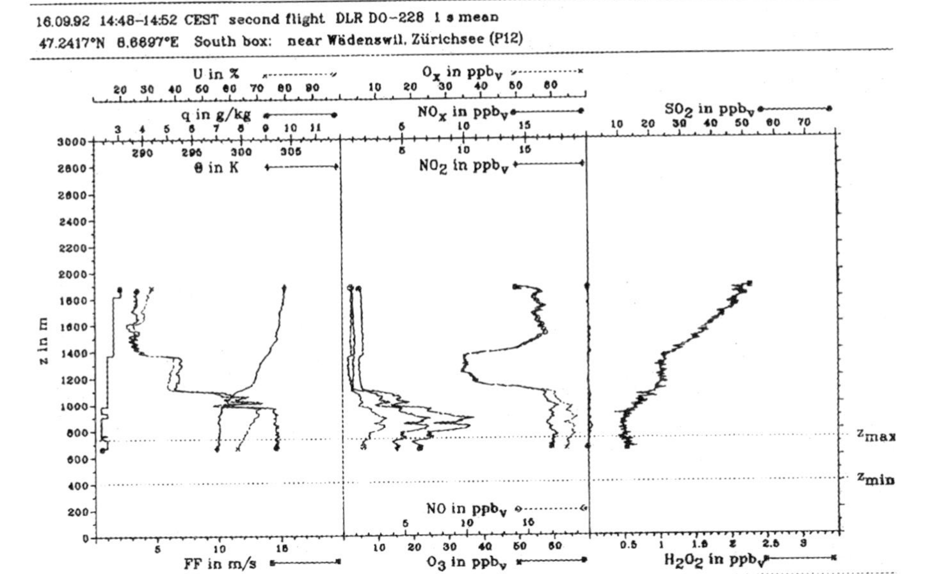

Fig. 5.27: Vertical profiles along the flight tracks of meteorological and chemical variables taken close to Lake Zürich.

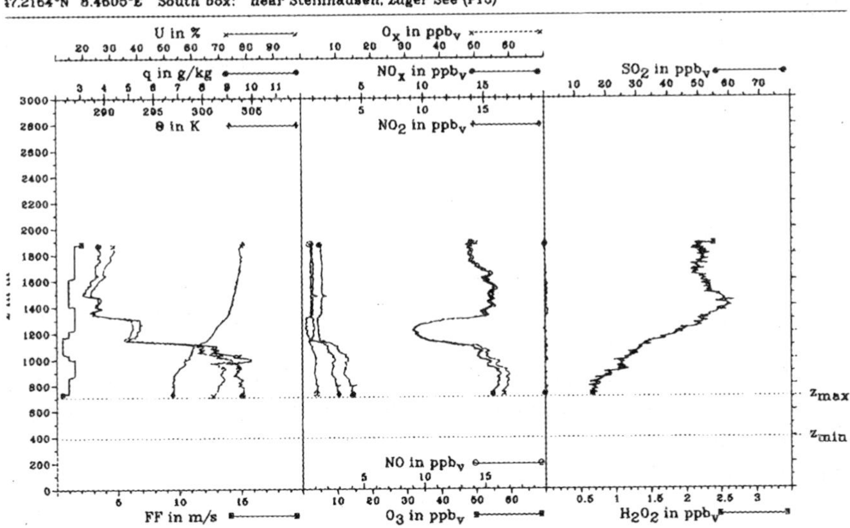

Fig. 5.28: Vertical profiles along the flight tracks of meteorological and chemical variables taken close to the Zuger lake.

In all three figures the potential temperature increase corresponds to a decrease in specific humidity, and a rich vertical structure is visible. The nitrogen components have a similar structure to the specific humidity, except in Fig. 5.27, where high concentrations were found in the middle of the boundary layer.

Ozone shows a marked vertical structure in all three figures. underneath the strong capping inversion in the boundary layer, the ozone concentrations are much smaller, due presumably to chemical reaction since the specific humidity stays almost constant within this layer. Although the concentrations of NO and of NO_2 are rather low in this height range, all three vertical profiles are similar; NO is somewhat higher than NO_2 indicating that there is a plume from a strong source within this area.

In contrast to the other species, the concentrations of H_2O_2 increase with height. However, as profiles from the descending flights, P11 and P13, and the ascending flight, P12, show, it is clear that the measuring instrument had too slow a response time for the speeds chosen for aircraft ascent and descent. Finally, it should be mentioned that sulfur dioxide concentrations were very low during the whole measuring campaign.

These field measurements provided data not only for intensive analysis of the observed phenomena but also for model evaluation, and the observed vertical structure of the different chemical species is a sensitive and clear indicator of the quality of the models being used.

Furthermore the correlations between the concentrations of the various chemical components is noteworthy, as is shown for the concentrations of O_3 and of NO_x in Fig. 5.29, which shows results from flights made during the intensive measuring period, 11th September, 1992. At the beginning very low NO_x values are correlated with O_3 mixing ratios in the range from 50 to 70 ppb. As time progresses the NO_x concentrations increase and are correlated with a small decrease of the observed ozone values.

Conclusions and reservations

a. TRACT: a successful campaign

The aim of the extensive field campaign of the EUROTRAC subproject TRACT was to study the processes of transport, turbulent mixing and chemical transformation of air pollutants during episodic conditions, particularly those that are involved in the daily cycles. The season chosen guaranteed a reasonable probability of fair weather conditions which meant that not only transport and diffusion processes could be studied but also the interaction during the day with gas phase chemical reactions.

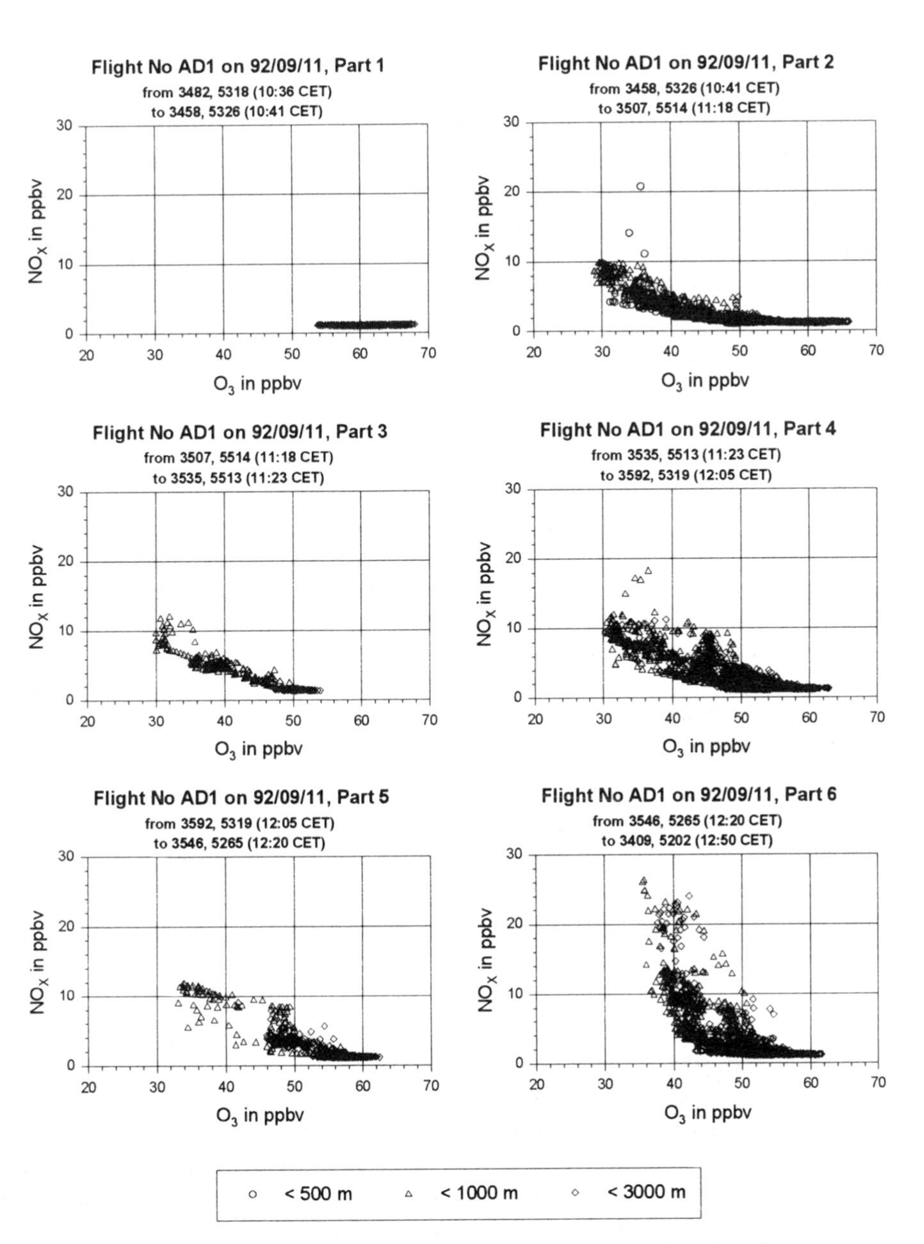

Fig. 5.29: Relation of measured NO_x ($NO + NO_2$) and O_3 during aircraft ascents and descents at various times of the flight on 11th September, 1992.

Although, as is natural for field studies of this size, the synoptic conditions were not completely ideal for the whole observation period, a well-balanced study was achieved. The meteorological conditions during days on which strong thermal stability within the atmospheric boundary layer at night moved over to well mixed conditions connected to a rapidly growing boundary layer during the day were fully documented in a three-dimensional way. The TRACT study may be considered as one of the largest field studies of the boundary layer structure over complex terrain so far undertaken.

The relatively large number of vertical soundings from radiosondes and also from tethered balloon sondes with a three-hourly time step during the intensive measuring phases has provided an interesting insight into the vertical and horizontal structure of the boundary layer over mountainous terrain. A particular advantage was that a large number of ground based stations both from operational networks and from temporary stations were available in the area and these allowed very detailed studies to be made of the conditions close to the ground. Although aircraft stay in the air only for a few hours, which is a short time compared to the total length of an intensive measuring period, they provide valuable information about the three-dimensional distributions of the chemical species and meteorological variables. In particular, the availability of a large number of different measurements over the whole area gives us confidence about the specific details found in the various smaller areas, because those details appear in a consistent manner in many of the measured variables.

A reasonably clear picture has now emerged from TRACT about how the topography influences the horizontal flow system in the mesoscale domain of 300 km × 300 km. It was possible to demonstrate that the flow is influenced by the mountains more than 150 km downstream of the sources which has considerable influence for the mixing of emissions of different chemical species produced in a specific area.

Since the topography produces a variety of secondary flow conditions, quantitative results were derived from the observations for the exchange of air both in the vertical and horizontal directions. There are some examples where the exchange of polluted air in the vertical, due to secondary flow conditions, is more effective than that due to the usual small scale turbulence. Furthermore, it was shown how the turbulence level is distributed over some of the mountainous regions with a strong increase of turbulent intensity over the mountain barriers. Connected with the channelling and splitting of the horizontal flow through the mountains and the turbulence level connected to it, the different height development of the boundary layer could be quantitatively studied.

As well as an understanding of the physics of the boundary layer structure, valuable insight was gained into the three-dimensional distribution of a number of chemical species, particularly the nitrogen oxides and ozone and some hydrocarbons. The distribution is a function of the source locations and of the three-dimensional flow system.

Special studies were carried out which demonstrated the influence of a mountainous terrain on the three dimensional spatial distribution of the chemical species as a function of time. These studies include the influence on the general flow structure, on the deposition at the ground and on the photolysis rate, as well as the influence of the temperature level at different elevations on the chemical transformations These special studies show, that large errors may occur account if the dynamics of the relief on the mesoscale are not taken into account.

In summary, TRACT has provided a comprehensive and consistent data set. All the data needed to initialise complex mesoscale chemical-transport models, have been derived from the field study and are as complete as possible. The data range from larger scale synoptic data to the land use and emission data of the most important species. The TRACT data have already be used in subsequent model evaluation studies.

b. Field campaigns: some reservations

While the successful results obtained from this complex field study have been emphasised, it is worthwhile mentioning some of the difficulties associated with large field campaigns in general and TRACT in particular.

The complexity of field measurements is often underestimated, particularly when chemical reactions and competing physical processes determine the final distribution of the different chemical species. Frequently the meteorological conditions are poorly determined and the effects of advection and turbulent phenomena on the measurements are no sufficiently appreciated. Too often, a poor compromise is made between the making enough measurements to determine the participating processes and making those which from the available resources are available. This often restricts the quantification of the different processes involved.

Also it is necessary to devise the appropriate measuring strategy for the particular aim of the field study in question, be it to determine a mass balance, to do a process study, to estimate quantities for a parameterisation or just to study the phenomena under particular conditions. In most of the field campaigns a mixture of aims is addressed by the different groups taking part but the overall strategy must include the measurements appropriate to the aims addressed.

Although much valuable data were collected and analysed by the large number of groups participating in TRACT, the accuracy of the results, say the mass balances could have perhaps been improved by examining the sampling requirements in a more rigorous way. For example the fluxes across the outer boundaries of the TRACT area were provided by the aircraft and the radio soundings during the hours of aircraft operation. But these times when aircraft are available were short compared to the time of a whole daily cycle. Even better perhaps would have been to make more measurements of the chemical species within the three-dimensional space of the research area.

Finally, atmospheric scientists should be always aware of the real scientific goal: that is to do only those measurements which are responsive to the effects being studied and to take into account the complexity of the system without making unjustified compromises. It is perhaps the willingness to make such compromises in field measurements that has condemned atmospheric science to progress more slowly than some other disciplines.

Acknowledgements

We would like to thank the many colleagues who took part in the TRACT and TRANSALP field campaigns, who submitted their data to the data bases in good time and who have been evaluating and publishing their results since. We are particularly indebted to the many colleagues, from the national networks and from the REKLIP campaign, for making their data available and so give us a more comprehensive data set with which to work. We must acknowledge the generous help from the many national funding agencies which supported the principal investigators involved. Among them are the following:

Bundesministerium für Bildung und Forschung (BMBF, Germany)

European Commission DG-XII, Brussels

Amt für Wehrgeophysik, Traben-Trabach

ASPA, Strasbourg

Bundesministerium für Verteidigung

Bundeswehr, Standort Neuhausen ob Eck

Deutscher Wetterdienst, Offenbach/Stuttgart/Hohenpeißenberg/Mannheim

Fischer Flug Inc., Offenburg

Meteo France, Illkirch

Schweizerische Meteorologische Anstalt, Zürich/Payerne

References

1. Anfossi, D., P. Gaglione, G. Graziani, K. Nodop, A. Stingele A. Marzorati, G. Graziani; 1996, The TRANSALP tracer campaigns: lessons learned and possible project evolution, in P.M. Borrell, P. Borrell, T. Cvitaš and W. Seiler (eds), *Proc. EUROTRAC Symp. '96*, Computational Mechanics Publications, Southampton, Vol. 1, pp. 751–756.
2. Ambrosetti, P., D. Anfossi, S. Cieslik, G. Graziani, R. Lamprecht, A. Marzorati, K. Nodop, S. Sandroni, A. Stingele, H. Zimmermann; 1998, Mesoscale transport of atmospheric trace constituents across the central Alps: TRANSALP tracer experiments, *Atmos. Environ.* **32**, 1257–1272.
3. Nodop, K., P. Ambrosetti, D. Anfossi, S. Cieslik, P. Gaglione, R. Lamprecht, A. Marzorati, A. Stingele, H. Zimmermann; 1992, Tracer experiment to study air flow across the Alps, in: P.M. Borrell, P. Borrell, T. Cvitaš and W. Seiler (eds), *Proc. EUROTRAC Symp. '92*, SPB Academic Publishing bv, The Hague, pp. 203–207.
4. Desiato, F., S. Finardi, G. Brusasca, M.G. Morselli; 1998, TRANSALP 1989 experimental campaign - I. Simulation of 3-D flow with diagnostic wind field models, *Atmos. Environ.* **32**, 1141–1156.

5. Anfossi, D., F. Desiato, G. Tinarelli, G. Brusasca, E. Ferrero and D. Sacchetti, 1998, TRANSALP 1989 experimental campaign - II. Simulation of a tracer experiment with Lagrangian particle models, *Atmos. Environ.* **32**, 1157–1166

6. Martilli, A. and G. Graziani; 1998, Mesoscale circulation across the Alps: preliminary simulation of the TRANSALP 1990 simulations, *Atmos. Environ.* **32**, 1241–1256.

7. Fiedler, F., G. Adrian, M. Bär, J. Franck, K. Höschele, W. Hübschmann, K. Nester, T. Pfeiffer, P. Thomas, B. Vogel, S. Vogt, O. Walk; 1991, Transport und Umwandlung von Luftschadstoffen im Lande Baden-Württemberg und aus Anrainerstaaten (TULLA), KfK-PEF 88.

8. Zimmermann, H.;, 1995, *Field phase report of the TRACT field measurement campaign*, EUROTRAC ISS, Garmisch-Partenkirchen, pp 196.

9. Fiedler, F.; 1992, TRACT Operational Plan. Inst. Meteorol. Klimaforsch., Universität Karlsruhe/Forschungszentrum Karlsruhe, 66pp.

10. Wachs, P., U. Corsmeier, F. Fiedler; 1992, Subproject TRACT: Aircraft Operational Plan, Inst. Meteorol. Klimaforsch., Universität Karlsruhe/Forschungszentrum Karlsruhe.

11. Mohnen, V., U. Corsmeier, F. Fiedler; 1992, Quality Assurance Project Plan: Aircraft Measurement Systems, Inst. Meteorol. Klimaforsch., Universität Karlsruhe/Forschungszentrum Karlsruhe 184pp.

12. Zimmermann, H., 1999, contact *zdb@imk.fzk.de* to access the TRACT data.

13. Fiedler, F.; 1989, *TRACT Project Description,* EUROTRAC ISS, Garmisch-Partenkirchen.

14. Kanter, H-J., F. Slemr, and V. Mohnen, 1996, Airborne chemical and meteorological measurements made during the 1992 TRACT experiment: quality control and assessment, *J. Air and Waste Management Association* **46**, 710–724

15. Koßmann, M., U. Corsmeier, R. Vögtlin, A.M. Jochum, C. Strodl, H. Willeke, B. Neininger, W. Fuchs, W. Graber; 1994, Aircraft intercomparison in the nested area during the TRACT campaign, *EUROTRAC Annual Report 1993, part 10, TRACT*, EUROTRAC ISS, Garmisch-Partenkirchen, pp. 94–102

16. Kalthoff, N, O. Kolle; 1993, Intercomparison of meteorological surface measurements during the TRACT campaign in summer 1992, *EUROTRAC Annual Report 1992, part 2, TRACT,* EUROTRAC ISS, Garmisch-Partenkirchen, pp. 104–119.

17. Heinrich, G., J. Weppner, H. Güsten, G. Clauss, H. Antz, A. Target; 1994, Quality assurance for trace gas sensors used in the nested area of TRACT, *EUROTRAC Annual Report 1993, part 10, TRACT,* EUROTRAC ISS, Garmisch-Partenkirchen, pp. 103–119.

18. Mohnen, V., 1996 Quality Assurance in Atmospheric Measurements and Assessments, in: P.M. Borrell, P. Borrell, T. Cvitaš, K. Kelly, W. Seiler (eds), *Proc. EUROTRAC Symp. '96,* Computer Mechanics Publications, Southampton, vol. 2, pp. 417–423.

19. Obermeier, A., J. Seier, C. John, P. Berner, R. Friedrich; 1996, TRACT: Erstellung einer Emissionsdatenbasis für TRACT, Forschungsbericht des Instituts für Energiewirtschaft und Rationelle Energieanwendung, Band 27, Stuttgart, ISSN 0938-1228.

20. Berner, P., R. Friedrich, C. John, A.Obermeier; 1996, Generation of an emission data base for TRACT, in: P.M. Borrell, P. Borrell, T. Cvitaš, K. Kelly, W. Seiler (eds), *Proc. EUROTRAC Symp 96.* Computational Mechanics Publ., Southhampton, Vol. 1, pp. 791–799.

21. Adrian, G., F. Fiedler; 1991, Simulation of unstationary wind and temperature fields over complex terrain and comparison with observations. *Beitr. Phys. Atmos.* **64**, 27–48.

22. Fiedler, F., I. Bischoff-Gauß, N. Kalthoff, G. Adrian; 1999, Modeling of the Transport and Diffusion of a Tracer in the Freiburg-Schauinsland Area, *J. Geophys. Res.*, in press.

23. Vogel, B., Riemer, N., Vogel, H., Fiedler, F.; 1999, Findings on NOy as an Indicator for Ozone Sensitivity based on Different Numerical Simulations, J. Geophys. Res. , 3605–3620.

24. Uhrner, U.; 1997, Physikalische Ursachen für regionale Unterschiede in der nächtlichen Ozonverteilung, Diplomarbeit, Inst. Meteorol. Klimaforsch., Universität Karlsruhe/Forschungszentrum Karlsruhe.

25. Vogel, B., F. Fiedler, H. Vogel; 1995,.Influence of topography and biogenic volatile organic compounds emission in the state of Baden-Wuerttemberg on ozone concentrations during episodes of high air temperatures, *J. Geophys. Res.* **100**, 22907–22928.

26. Vogel, H., Bär, M., Fiedler, F.; 1991: Influence of different land-use types on dry deposition, concentration and mass balances: a case study, in: P.M. Borrell, P. Borrell, T. Cvitaš, W. Seiler (eds), *Proc. of EUROTRAC Symp. 92.*, SPB Academic Publishing bv, The Hague, pp. 559–562.

27. Ruggaber, A., R. Dlugi, R. Forkel, W. Seidl, H. Hass, T. Nakajima, B. Vogel, M. Hammer, 1995, Modelling of radiation quantities and photolysis frequencies in the troposphere, in: A. Ebel, N. Moussiopoulos (eds), *Air Pollution III, Volume 4: Observation and Simulation of Air Pollution,* Computational Mechanics Publications, Southampton,
pp. 111–118.

28. Vogel, B., M. Hammer, N. Riemer, H. Vogel, A. Ruggaber;, 1997, A method to determine the three-dimensional distribution of photolysis frequencies in the regional scale, in: P.M. Borrell, P. Borrell, K. Kelly, T. Cvitaš, W. Seiler (eds), *Proc. EUROTRAC Symp. '96*, Computational Mechanics Publications, Southampton, pp. 745–749.

29. Fiedler, F.; 1995, Klimaatlas Oberrhein Mitte-Süd 1995, Herausgeber: Oberrheinische Universitäten (Basel, Freiburg, Strasbourg, Karlsruhe),

30. Koßmann, M., R. Vögtlin, U. Corsmeier, B. Vogel, F. Fiedler, H-J Binder, N. Kalthoff, F. Beyrich; 1998, Aspects of the convective boundary layer structure over complex terrain, *Atmos. Environ.* **32**, 1323–1348

31. Löffler-Mang, M., M. Koßmann, R. Vögtlin, F. Fiedler; 1997: Valley Wind Systems and their Influence on Nocturnal Ozone Concentrations, *Beitr. Phys. Atmos.* **70**, 1–14.

32. Güsten, H., G. Heinrich, D. Sprung; 1998: Nocturnal Depletion of Ozone in the Upper Rhine Valley, *Atmos. Environ.* **32**, 1195–1202.

33. Hasager, C. B., N. O. Jensen; 1999, Surface-Flux aggregation in heterogeneous terrain, *Quart. J. Roy. Meteorol. Soc.* **125**, 2075–2102.

34. Koßmann, M., H. Vogel, B. Vogel, R. Vögtlin, U. Corsmeier, F. Fiedler, O. Klemm, H. Schlager; 1997, The composition and Vertical Distribution of Volatile Organic Compounds in Southwest Germany, Eastern France and Northern Switzerland during the TRACT Campaign in September 1992, *Physics and Chemistry of the Earth* **21**, 429–433.

Chapter 6

Data Generation and Quality Assessment

6.1 Generation of an Emission Data Base for TRACT

J. Seier, P. Berner, R. Friedrich, C. John and A. Obermeier.

Institute for Energy Economics and the Rational Use of Energy,
University of Stuttgart, Heßbrühlstraße 49a, D-70565 Stuttgart, Germany

Within the framework of this subproject of TRACT, hourly data of anthropogenic emissions of NMVOC species, NO_x, CO and SO_2 are generated in grids of 1 km × 1 km for the period from 9th to 22nd September 1992. The investigation area includes the German federal state of Baden-Württemberg, southern parts of the federal states of Rheinland-Pfalz and Hessen, as well as northern parts of Switzerland and the departments Haut-Rhine and Bas-Rhine in France. The emission inventory is needed as an input for modelling the atmospheric transport and transformation of air pollutants in this region.

Emissions of major point sources are calculated using plant-specific data. These data are obtained by a special survey carried out among several hundred plant operators as well as by an analysis of emission data bases of public authorities. Road traffic outside of settlements is modelled as line sources, using, for example, detailed road network data and traffic counts. Other sources of anthropogenic emissions are treated as area sources, *e.g.* temperature data and statistics of production, employees and inhabitants.

Results show that within the TRACT measuring campaign, emissions of NO_x adds up to about 17,100 tons (transportation sector 79 %, firing installations 21 %). Anthropogenic emissions of NMVOC adds up to about 20,700 tons (transportation sector 35 %, firing installations 4 %, solvent use and other stationary sources 61 %). Daily NMVOC emissions on Saturdays and Sundays are 60–65 % lower than on working days, whereas NO_x emissions are only 30–40 % lower than on working days. Furthermore, hourly emissions during daytime are 10 to 20 times higher than during night-time. Significant differences can be identified also for the spatial distribution of NO_x and NMVOC emissions. Areas of high NO_x emissions can be found along highway intersections and at locations of power plants and large industrial firing installations. On the other hand, NMVOC emissions mainly occur within densely settled areas.

In summarising these results, it can be stated that the ratio of NO_x and NMVOC emissions shows very large variations in space and time. This ratio is thought to be of major importance concerning the formation of tropospheric ozone. Therefore, the results underline the necessity for modelling emissions and air quality in a high temporal and spatial resolution.

Aims of the research

The aim of this contribution to TRACT is to generate hourly emission data for the air pollutants NO_x, SO_2, CO and NMVOC (speciated in 28 groups) for the period from 9th to 22nd September 1992 in grids of 1 km $\times$ 1 km. The investigation area includes the German federal state of Baden-Württemberg, southern parts of the federal states of Rheinland-Pfalz and Hessen, northern parts of Switzerland and the departments Haut-Rhine and Bas-Rhine in France.

The emission inventory serves as an input data base for regional air quality models. One of the main objectives of TRACT is to evaluate existing air quality models by comparing calculated ambient air concentrations with appropriate measurements. Therefore, the emission data base has to be as accurate as possible.

Principal methods

The inventory preparation procedures focus on sources of anthropogenic emissions. Among the most important sources are road traffic, firing installations, solvent use and the distribution of mineral oil products. Biogenic emissions are not regarded within this project, as they are calculated in the framework of atmospheric modelling. Emission calculations are carried out by the authors of this paper for the German and French parts of the investigation area. Emission data for Switzerland are provided by two research groups in Basel and Bern [1] in close cooperation with the authors.

In order to achieve a high accuracy of the emission data, stationary sources with high emissions are treated preferably as point sources. Therefore, a survey among several hundred operators of large firing installations and plants with a high annual input of solvents and solvent containing products has been carried out in the German part of the TRACT area, asking for actual hourly emission data during the TRACT episode. In addition, data concerning annual emissions and other information of some 1500 plants located in the German part of the inventory are included in the emission data base. Those plant data are derived from emission declarations, which are collected and made available by public authorities. Concerning point sources in France, annual emission data are derived from the REKLIP data base [2].

To obtain hourly values for point sources, for which only annual emission data are available, simulation models are used which take into account the dependencies of emissions on ambient temperature, production data and working hour regulations for different economic sectors and sizes of enterprises.

Road traffic on classified roads outside of settlements is regarded as a line source. The location of roads as well as average traffic densities are known from data bases of public authorities. 33 vehicle categories are distinguished. Using measured speed distributions for all road classes and vehicle categories, annual emissions have been calculated. Hourly values are obtained using a time series for each vehicle category. These time series are derived from hourly values of traffic densities at more than 100 sites in the investigation area.

Road traffic inside settlements and on non-classified roads outside of settlements, smaller firing installations, plants with a lower annual input of solvents as well as numerous types of other stationary sources are regarded as area sources. Annual emission data per community are calculated using statistical data like the number of employees, vehicles or inhabitants on the one hand, and data like solvent use per employee/inhabitant and emission factors on the other hand.

As an example, the provision of detailed data on the solvent use per employee or inhabitant (distinguishing more than 100 types of solvent containing products and some 50 industrial and commercial branches) is based on the analysis and combination of a large variety of statistical data for Germany and its federal states (*e.g.* national production and foreign trade statistics, industrial waste statistics, material balances). These detailed values are also used for the French part of the investigation area, which means that they are combined with French statistics of employees, inhabitants *etc.* Statistical data for the French part are provided by the REKLIP data base.

Hourly values of area source emissions are obtained using typical time series (*e.g.* working shift regulations for different branches and sizes of enterprises, ambient temperature data and branch specific production rates. To obtain a spatial resolution of 1 km × 1 km, emissions calculated for a community are allocated to its settled areas. Data on the location of settled areas are derived from satellite based land-use maps.

Calculations of emission data for Switzerland are performed by research groups in Basel and Bern [1]. For most source categories similar methods to the ones described above are used. However, for most industrial branches Swiss values for the solvent use per employee are higher than the values used for Germany and France.

Principal results

Emissions of ozone precursors NMVOC and NO_x during the period from 9th to 22nd September 1992 are shown in Table 6.1.1. Emissions of NO_x add up to 17,100 t. With a share of 79 % they are mainly caused by the transport sector. Firing installations including process furnaces cause about 21 % of the total emissions. Anthropogenic NMVOC emissions add up to about 20,700 t. The transport sector accounts for 35 % of these emissions, a share of 62 % is caused by solvent use and other processes, whereas the contribution of firing installations is quite small (4 %). Significant differences concerning the breakdown by sectors are found for the different parts of the TRACT area. Whereas in the German and French part, solvent use and other processes account for 49 % to 59 % of the anthropogenic NMVOC emissions, the contribution of this source category reaches 81 % in the Swiss part. This is partly due to the considerable emissions of the chemical industry in the area of Basel. However, also solvent use emissions are considered to be significantly higher than in Germany and France.

Table 6.1.1: NO_x and NMVOC emissions in the different parts of the investigation area during the period from 9th to 22nd September 1992

pollutant	source category	French part [t]	German part [t]	Swiss part [t]	Total area [t]
NMVOC	traffic	1,000	5,530	660	7,190
	firing installations	140	580	20	740
	other sources	1,120	8,730	2,930	12,780
NO_x	traffic	1,840	10,360	1,330	13,530
	firing installations	640	2,540	410	3,590

A speciation of daily NMVOC emissions on Wednesday 16th September 1992, which sum up to about 1,800 tons/day, are shown in Table 6.1.2. The split into VOC categories and the allocation of substances to these categories follows recommendations of Middleton *et al.* [3]. Emissions of alkanes, alkenes, aromatics and aldehydes are mainly caused by vehicles. On the other hand, emissions of alcohols, ketones, esters and halogenated hydrocarbons mainly are caused by solvent use operations and other types of processes.

Hourly values of total NO_x emissions in the investigation area are shown in Fig. 6.1.1. On working days the highest emissions occur during the rush hour. The ratio between highest and lowest hourly emissions reaches one order of magnitude. At weekends, daily emissions are 30–40 % lower than on working days. This is a consequence of reduced industrial activities and the absence of heavy duty vehicles during the weekend.

Table 6.1.2: Profile of daily NMVOC emissions on Wednesday, 16th September 1992

emission category	emissions [kg]	emission category	emissions [kg]
ethane	11,220	phenols and cresols	700
propane	7,130	styrenes	990
alkanes (r = 2.5-5.0)	72,130	formaldeyde	13,640
alkanes (r = 5.0-10.0)	221,110	other aldehydes	9,260
alkanes (r = 10.0-20)	181,130	acetone	32,070
alkanes (r > 20.0)	99,850	other ketones	49,220
alkanes/aromatics mixture	3,680	organic acids	4,360
ethene	40,450	acetylene	21,020
propene	14,400	halogenated hydrocarbons	41,350
primary alkenes	9,820	unreactive NMVOC	102,720
internal alkenes	6,480	other NMVOC (r < 2.5)	88,000
mixture of primary/ internal alkenes	39,180	other NMVOC (r = 2.5–5.0)	64,360
benzene	18,810	other NMVOC (r= 5.0–10.0)	202,180
aromatics (r < 20.0)	72,920	other NMVOC (r > 10.0)	13,690
aromatics (r > 20.0)	167,650	unidentified/ unassigned NMVOC	178,170

r = OH rate constant ranges in 1,000/(ppm min)

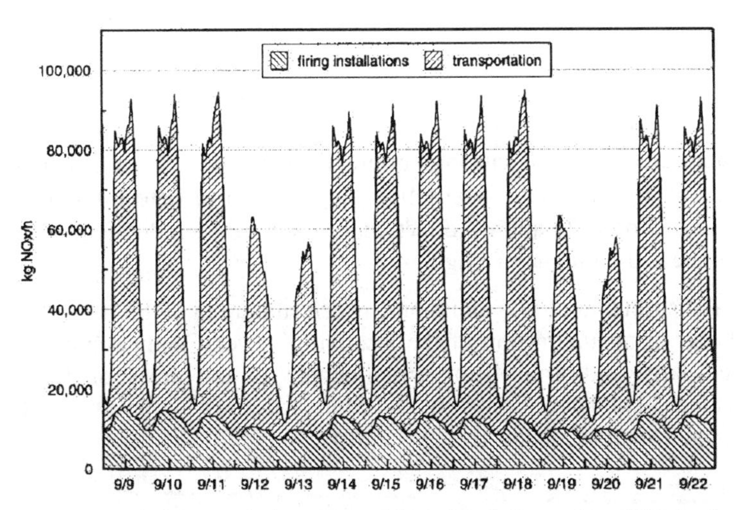

Fig. 6.1.1: Hourly NO_x emissions from 9th to 22nd September 1992 in the TRACT investigation area

Hourly anthropogenic emissions of NMVOC are shown in Fig. 6.1.2. Maximum values on working days do not necessarily occur during the rush hour. This is due to the important role of solvent use and gasoline evaporation, both types of processes mainly occurring during the daytime. The ratio between highest hourly emissions on working days and lowest nocturnal emissions reaches a factor of 20. As the temporal variation of industrial NMVOC emissions primarily depends on working hour regulations, daily emissions during weekend days are about 60 % lower than on working days. As a consequence, NMVOC emissions exceed NO_x emissions especially during daytime on working days, whereas NO_x emissions exceed NMVOC emissions during night-time and at the weekends.

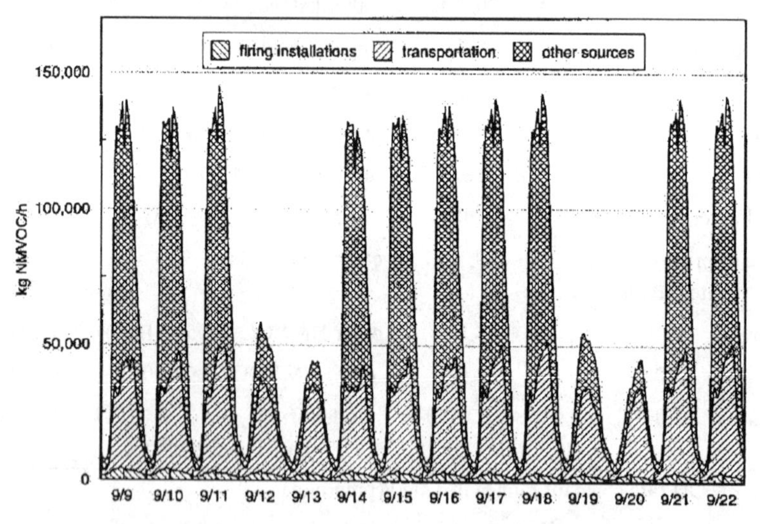

Fig. 6.1.2: Hourly NMVOC emissions from 9th to 22nd September 1992 in the TRACT investigation area

The spatial distribution of the emissions of SO_2, NO_x, CO and NMVOC is shown in Figs. 6.1.3 and 6.1.4 for the hour between 13:00 and 14:00 on Wednesday, 16th September. Emission data are originally calculated for a resolution of 1 km × 1 km. For a better visualisation they are shown here for gridcells of 5 km × 5 km. In order to allow a direct comparison between different pollutants the same scale is used in the two figures.

The spatial distribution of SO_2 emissions, which sum up to 19 t/h between 13:00 and 14:00, gives an impression of the location of several point sources (power plants and large industrial firing installations) in the areas of Karlsruhe, Ludwigshafen, Mannheim, Straßburg and Stuttgart. Emission levels in cells not including point sources are low (below 5 kg/h).

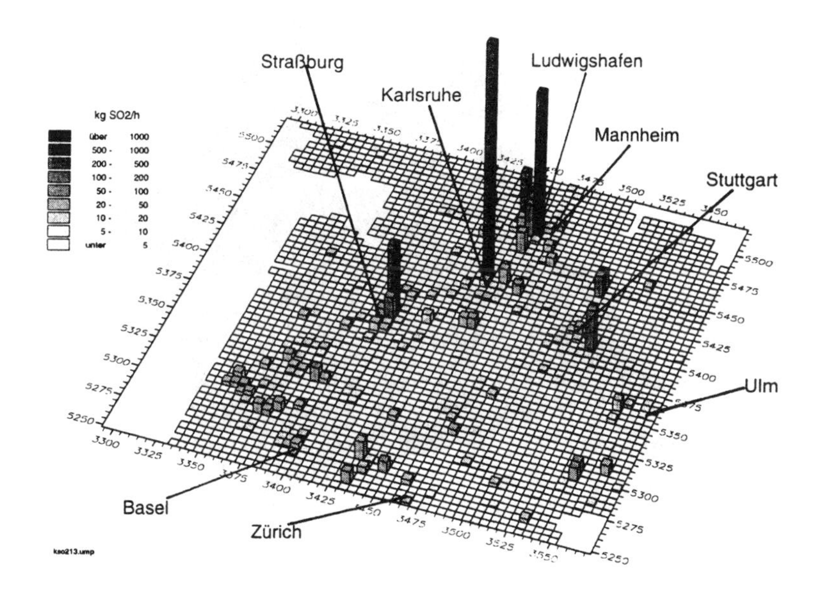

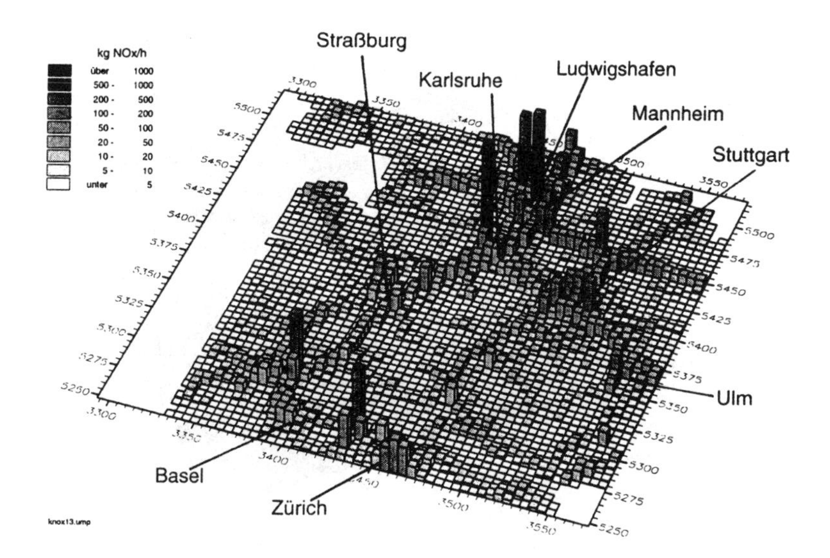

Fig. 6.1.3: Spatial distribution of hourly emissions of SO_2 (top) and NO_x (bottom) on Wednesday, 16th September between 13:00 and 14:00.

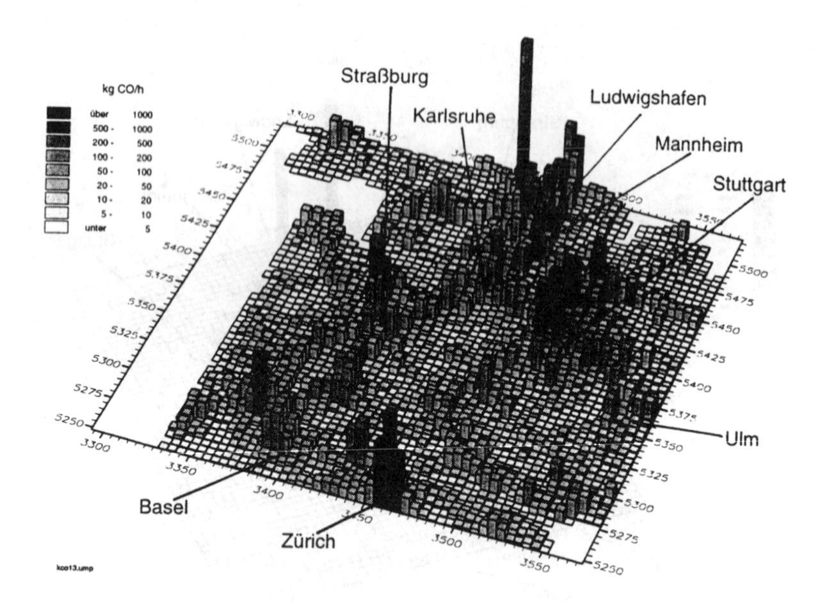

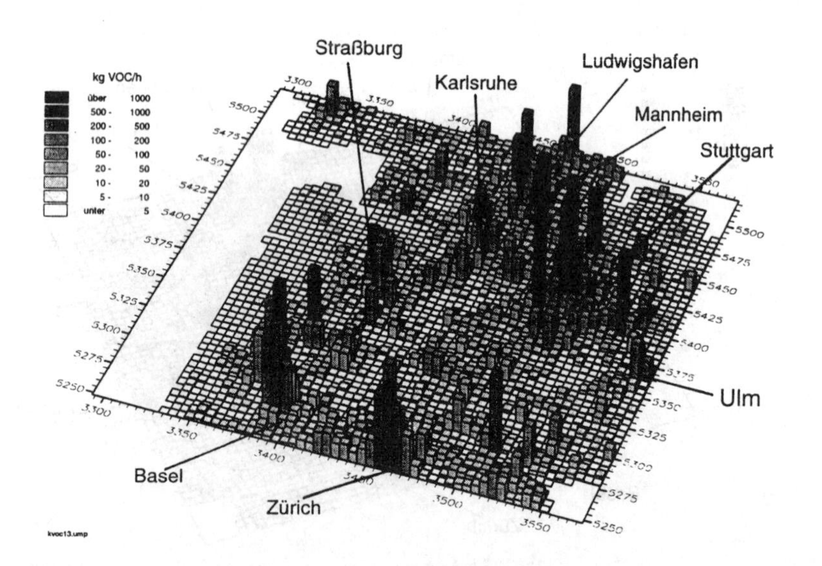

Fig. 6.1.4: Spatial distribution of hourly emissions of CO (top) and NMVOC (bottom) on Wednesday, 16th September between 13:00 and 14:00.

The spatial distribution of hourly NO_x emissions, which add up to 79 t/h, indicates clearly the location of large firing installations as well as the location of highly frequented highway intersections. For example, the highway leading from Mannheim via Karlsruhe and Strasbourg to Basel or the highway from Ulm via

Stuttgart to Mannheim can be depicted. High NO_x emissions near Zürich are mainly caused by highways as well as by the airport.

The spatial distribution of CO emissions, which add up to 181 t/h and which are mainly caused by vehicles, indicates the location of highways as well as the location of densely populated urban areas.

Anthropogenic NMVOC emissions, which add up to 138 t/h, primarily occur within urban areas. The emissions inside of the cities mainly result from road traffic (exhaust gases and gasoline evaporation) and from industrial, commercial and residential solvent use. However, the location of large point sources (mineral oil refineries, paint application plants, chemical plants *etc.*) can also be identified.

To give an impression of the spatial distribution of emissions at different times, Fig. 6.1.5 shows the spatially resolved NO_x emissions for the night between 23.00 and 24:00 on Wednesday, 16th September. The emissions at this hour add up to 23 t/h. The total emissions from large point sources is almost as high as between 13:00 and 14:00 (see lower part of Fig. 6.1.3). However, emissions on highways and other road categories are much lower during night-time, even if the location of the highways still can be detected.

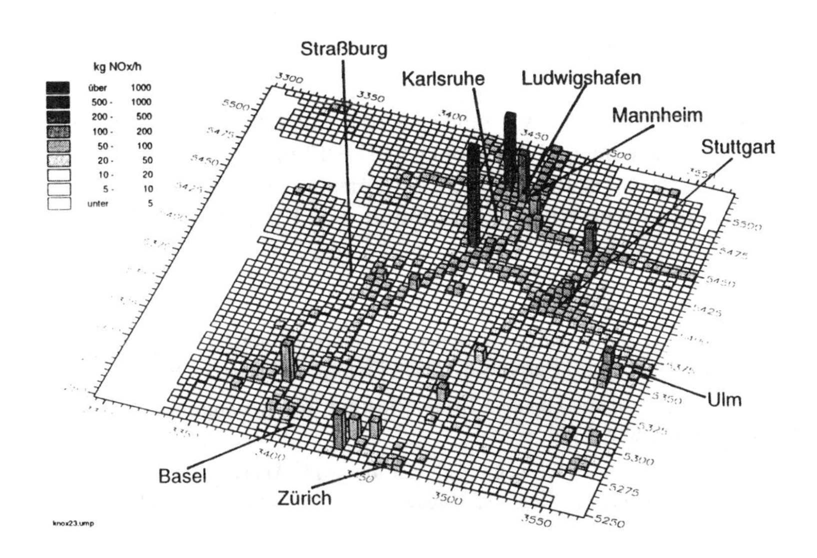

Fig. 6.1.5: Spatial distribution of hourly emissions of NO_x on Wednesday, 16th September between 23:00 and 24:00.

Conclusions

A mesoscale emission inventory for ozone precursors and other major air pollutants was prepared for the period of the TRACT measuring campaign. A special objective of this project was to ensure a high accuracy of emission data in order to enable the evaluation of available air quality models.

To comply with this task a special survey among numerous operators of emission relevant plants was carried out, asking for hourly emission data during the TRACT campaign. For a large number of other installations data on annual emissions were provided from official authorities. The data were included in the data base and converted into hourly values. To calculate hourly emissions of line and area sources in a high spatial resolution sophisticated simulation models were developed further. The models use a large variety of statistical data, road network data and many other information. They represent the current state of the art of regional scale emission modelling. Input data as well as results have been subject of extensive plausibility checks.

The emission inventory prepared within this contribution to TRACT has been made available for the participants of the TRACT and for other interested parties. The data base shows a lot of details, which can not be discussed here completely. However, it can be stated that the emissions of all pollutants show strong variations in time and space. In order to be able to identify clearly events and locations with peak concentrations of ozone and other pollutants, it seems to be of major importance that emission and air quality models operate with a high spatial and temporal resolution.

Acknowledgements

The research described here was funded by the Bundesministerium für Bildung, Wissenschaft, Forschung und Technologie (BMBF) under contract 07EU806. We also appreciate the assistance of the GSF-Forschungszentrum für Umwelt und Gesundheit, Projektträgerschaften für Umwelt- und Klimaforschung.

References

1. Kunz, S., Th. Künzle, B. Rihm, A. Schneider, K. Schläpfer; *TRACT Emissionsmodell Schweiz.* Meteotest, Bern; Carbotech, Basel. Final report, March 1995.
2. Schneider, Ch., *private communication,* Institut für Technische Thermodynamik, Universität Karlsruhe, 1994.
3. Middleton, P., W.R. Stockwell; Aggregation and Analysis of Volatile Organic Compound Emissions for Regional Modelling. *Atmos. Environ.* **24A** (1990) 1107–1133.

6.2 TRACT Data Quality Assessment

V. Mohnen, H.-J. Kanter and F. Slemr

Fraunhofer Institut für Atmosphärische-Umweltforschung, Kreuzeckbahnstr. 19, D-82467 Garmisch-Partenkirchen, Germany

The 1992 EUROTRAC TRACT field experiment was executed under a comprehensive quality assurance plan which covered all primary data obtained from ground based and air borne platforms. The second field intensive on Sept.16/17 1992 has been selected by "TFS Teilprojekt Modellierung" for evaluating on a very preliminary basis the CT-models deployed in TFS and for helping define the Model Quality Objectives (MQO's) and the Evaluation Criteria that ultimately will be used for the validation of models in TFS.

It is important to use only experimental data of known quality for the evaluation process. Data Quality Objectives (DQO's) for measurements are usually defined by:

accuracy and precision;
completeness;
comparability;
representativeness.

While accuracy, precision and completeness can be directly quantified from the experimental results through procedures such as calibration, instrument intercomparison and data review, it is difficult to determine comparability and even more difficult to assess the three-dimensional representativeness of a data set. (These problems will be addressed separately in appendix 1.)

The following tables provide guidance to the modelling community interested in the TRACT primary data products and recommendations for necessary future activities to utilise fully the TRACT data set for its intended application in TFS.

Table 6.2.1: Radiosonde measurements (total of 19 sites).

	accuracy	precision	completeness
wind speed	1.0 m/sec	0.5 m/sec	75 to 90 % of possible data
wind direction	5 °	3 °	"
temperature	1 °C	0.5 °C	"
rel. humidity	5 %	3 %	"
pressure	1.5 hpa	1.0 hpa	"

Table 6.2.2: Tethered balloon measurements (total of 7 sites).

	accuracy	**precision**	**completeness**
wind speed	0.5 m/sec	0.3 m/sec	90 % of possible data
wind direction	3 °	3 °	"
temperature	0.5 ° C	0.5 ° C	"
rel. humidity	3 %	3 %	"
pressure	3 hpa	3 hpa	"

Table 6.2.3: Sodar measurements (total of 6 sites).

	accuracy	**precision**	**completeness**
wind speed	1 m/sec	1 m/sec	90 % (2 <u<10 m/sec)
wind direction	25 °	25 °	"
vert. resolution	50 m	10 m	"

Constant level balloon measurements at 3 sites had the same accuracy, precision and completeness as the radio sonde measurements.

Ground based meteorological stations

Ground based meteorological stations are located in France (total of 18 sites), in Germany (136 sites) and in Switzerland (29 sites).

These stations are operated by METEO France, DWD, LfU's, ANETZ/ENET, RECLIP, ASPA, and BASF. These stations follow their own quality assurance protocol. The data quality is in principle unknown. (It is assumed, however, that the data quality is equal to that established for the temporary TRACT sites.)

Table 6.2.4: Ground based meteorological stations (total of 16 temporary TRACT sites).

	accuracy	**precision**	**completeness**
temperature	0.3 ° C	0.2 ° C	> 90 % of possible data
rel. humidity	2 %	1.5 %	"
wind speed	0.5 m/sec	0.5 m/sec	"
wind direction	5 °	5 °	"
pressure	0.5 hpa	0.5 hpa	"

Ground based chemical measurements

Ground based chemical measurements were made at a total of 20 sites in France, 131 in Germany and 11 in Switzerland.

These stations are operated by ASPA (Strbg), LFU's (Baden-Württemberg, Hessen, Bayern and Rheinland Pfalz), BASF and EMPA/BUWAL and follow their own quality assurance programme. The data quality is in principle unknown. It is assumed, however, that the data quality is equal to that established for the temporary TRACT sites.

Table 6.2.5: Ground based chemical measurements (total of 9 temporary TRACT sites):

	accuracy	**precision**	**completeness**
surface ozone (at 8 sites)	10 % or 2 ppb	5 %	< 90 % of possible data or 2 ppb
NO, NO_2, NO_x	30 %	30 %	"
NO_y (at 1 site)	30 % or 2 ppb	30 % or 2 ppb	"
CO (at 5 sites)	10 % or 15 ppb	10 % or 10 ppb	" "
NMHC C_2-C_7 (at 2 sites)	25 % > 10 ppt C	10 %	(not continuous)
SO_2 (at 3 sites)	10 % or 2 ppb	10 % or 2 ppb	"

Only 3 sites measure O_3 and NO_2 concurrently (for O_x)

Concurrent measurements of O_3 and NO_2 (to determine O_X, which is an essential minimum prerequisit for assessing the comparability of ozone measurements in lightly and heavily polluted areas) is only available at 3 stations in France, 79 in southern Germany and 11 in Switzerland.

Aircraft measurements

A total of 5 aircraft were used for measurements over the main area with the main focus on chemistry.

Table 6.2.6

	accuracy	precision	completeness
ozone	10 % or 2 ppb	5 % or 2 ppb	> 80 % of possible data
NO_2 (2 aircraft)	30 %	30 %	70 %
NO	30 %	30 %	80 %
NO_y (2 aircraft)	30 %	30 %	80 %
SO_2 (4 aircraft)	25 %	20 %	> 80 %
H_2O_2 (1 aircraft)	35 %	35 %	50 %
VOC (2 aircraft)	25 %	20 %	90 %
temperature *	1.5 °C	1.5 °C	90 %
rel. humidity*	5 %	5 %	90 %

*Bias between aircraft have been corrected.
Wind speed and direction have not been quality assured: data quality unknown.

Aircraft measurements over the nested area using a total of 5 aircraft had a main focus on meteorology.

Table 6.2.7

	accuracy	precision	completeness
ozone	10 % or 2 ppb	10 % or 2 ppb	> 80 %
NO_2 (1 aircraft)	20 % or 0.5 ppb	20 % or 0.5 ppb	> 80 %
specific humidity*	0.5 g/kg	0.5 g/kg	90 %
temperature*	0.5 °C	0.5 °C	"
pressure*	2 hpa	2 hpa	"

*Bias between aircraft have been corrected.
** Special calibration program during TRACT
Wind speed and direction : data quality unknown.

Ozone sonde measurements

Ozone sonde measurements were performed over the nested area at 1 station with a tethered balloon.

Table 6.2.8

	accuracy	precision	completeness
ozone concentration	10 % or 3 ppb	10 % or 2 ppb	12 profiles a day
altitude resolution (up to 1000m)	10 m		

Emission inventory

The data quality of the emission inventory is unknown. Defining the uncertainties for emission data deserves the highest priority.

Comparability and representativeness

a. Comparability

All quality assured meteorological TRACT data are comparable within their stated level of uncertainty. This was achieved through adherence to a comprehensive quality assurance plan which included calibration, instrument-/measurement system- intercomparison. In addition, some data sets have been modified to account for measurements biases and the stated uncertainties have been harmonised to assure network wide data comparability. TRACT was the first field campaign in Europe which implemented a harmonised and comprehensive quality assurance program.

b. Representativeness

A data set must be of known quality and adequate for its intended use. The TFS model evaluation study will focus primarily on "meteorology/transport" which requires "adequate" knowledge of the flow field for Sept. 16th/17th 1992 over the entire TRACT domain (275 by 375 km). A total of 17 radio sondes (see Table 6.2.1), 7 tethered balloons (see Table 6.2.2), 6 Sodar stations (see Table 6.2.3), 14 temporary ground based meteorological stations and all DWD, SMA and METEO France routine observation stations provided a harmonised three-dimensional meteorological data base with temporal resolution of better than 3 hours and a geographic resolution of better than 1 ° for the entire area and 1/3 ° for the Rhine valley. In addition, a total of 10 aircraft (see Tables 6.2.6 and 6.2.7) supported the TRACT mission, thus further refining the temporal and geographic resolution. Five aircraft provided primarily meteorological data for the Rhine valley and the nested area and five aircraft covered the entire TRACT area providing additional chemical and meteorological measurements.

It can be concluded that the harmonised meteorological data set is representative for the 60 km grid scale over the entire TRACT area and for the 20 km grid scale over the Rhine valley and the nested area, without further statistical treatment of the meteorological data set such as establishing auto-correlations or subjecting the data to various methods of "kriging". It has already been demonstrated by Löffler-Mang, Kuntze, Nester, Adrian and Fiedler that the observed wind-field for September 16th, 1992 can be simulated with KAMM over the entire TRACT area , whereby the model reproduces the typical orographic effects such as channelling of the flow in the Rhine valley, the flow into the Kraichgau and the flow around the northern part of the Black Forest to Stuttgart.

The TFS model study (16th/17th .September 1992) will likely make only limited use of the available chemical data set, which eliminates the need for a detailed quality assessment. However, a general remark is appropriate at this time. The determination of "data representativeness" (for example ozone) over the entire TRACT area would require further statistical treatment such as outlined above and in the appendix prepared on behalf of the QS-Abteilung by Dr. W. Stockwell. It can be concluded that while the chemical data are undoubtedly of great value for diagnostic case studies within limited regions they are nevertheless not yet at a level that that would allow assigning "representative" values of chemical parameters (*e.g.* ozone, oxides of nitrogen, carbon monoxide, VOCs) for the alpha- or beta- scale. To reach this higher data quality level would require considerable further efforts. The representativeness of the chemical data set is therefore unknown at this time.

Chapter 7

Modelling within TRACT

7.1 Simulation of the Episode of the TRACT Measurement Campaign with the EURAD Model

A. Ebel, H. Hass, H.J. Jakobs, M. Memmesheimer, J. Tippke
and H. Feldmann

EURAD-Project, University of Cologne, Institute for Geophysics and Meteorology,
Aachenerstraße 201-209, D-50931 Köln, Germany

The EURAD modelling system has been used to simulate transport, chemical transformation and deposition of air pollutants in Europe [1–3]. The physical and chemical processes to describe the formation of photo-oxidants and the production of acidity are included in the modelling system. Meteorological fields are calculated by application of the NCAR/Pennstate mesocscale model MM5 [4], Emission rates are generated by using the EURAD Emission Model EEM [5]. Chemical calculations are performed with the EURAD-CTM [6]. Within TRACT the EURAD model is applied to the episode of the TRACT measurement campaign with emphasis on the second intensive measurement phase (2nd IMP).

Aims of the research

The EURAD modelling system is applied to an episode within the TRACT measurement campaign. The episode simulated covers the time span of the second intensive measurement campaign (16th/17th September, 1992) and is extended to the 22nd September. One purpose is to provide the mesoscale models which are used in TRACT (*e.g.* DRAIS/KAMM) with the European scale information generated within the EURAD system (boundary values for chemistry and meteorology). The model results will be compared with measured data for evaluation purposes.

Principal scientific results

The interface between the European scale model EURAD and the DRAIS/KAMM model has been developed within EUMAC and was tested for July 15th, 1986, which is one day within the EUMAC joint dry case. This interface is used within TRACT to couple the European scale results from the EURAD modelling system to the smaller scale calculations performed with the DRAIS/KAMM model.

The meteorological simulation with the EURAD modelling system has been performed for the episode from September 14th–September 22nd, 1992. Emission data has been generated with the EURAD Emission model EEM based on annual data provided by EMEP for the reference year 1992. It is planned to use also information based on the GENEMIS data. These data are currently prepared by GENEMIS for the TRACT episode.

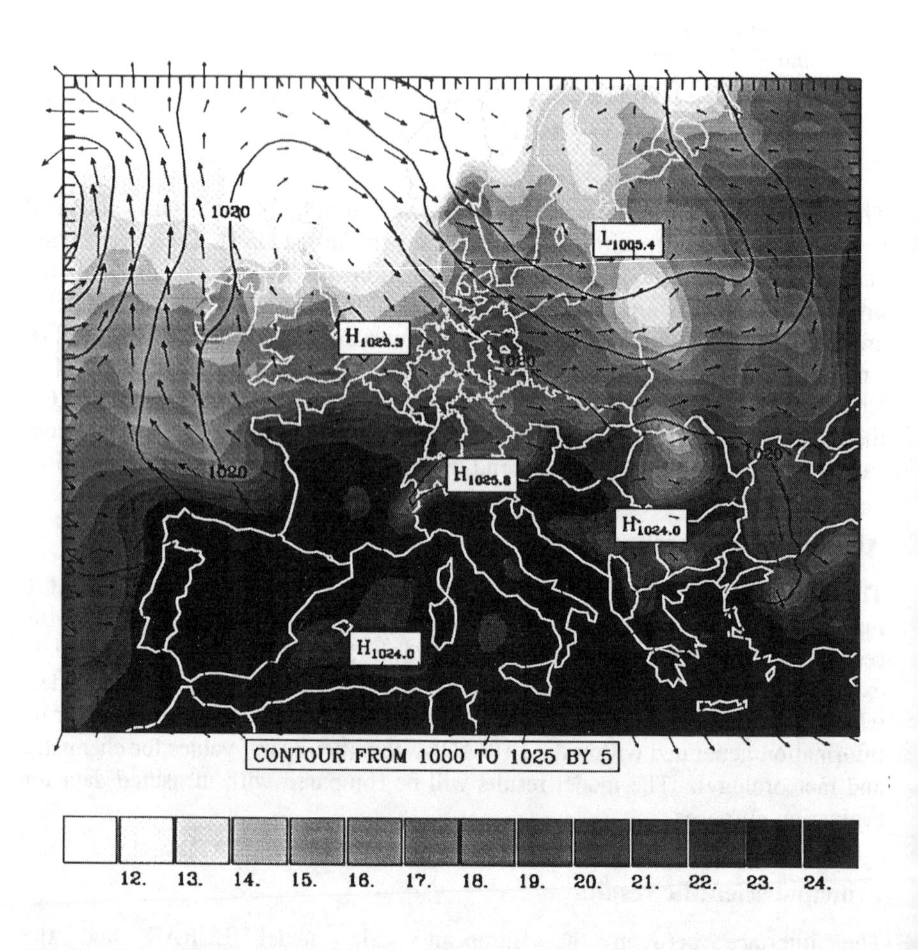

Fig. 7.1.1a: Sea-level pressure, temperature and horizontal wind for the 16th of September 1992, 12:00 UTC (2nd IMP of TRACT).

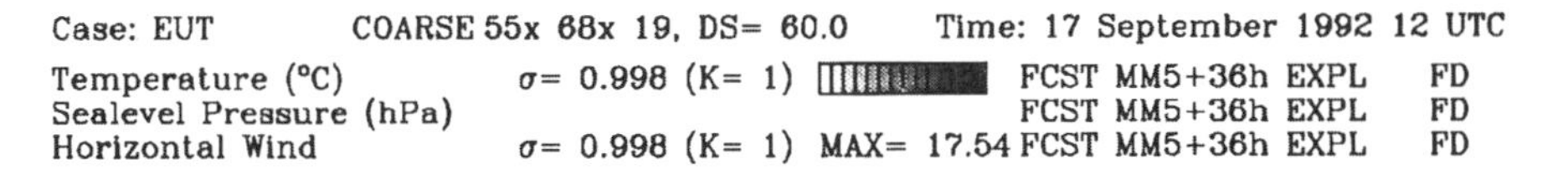

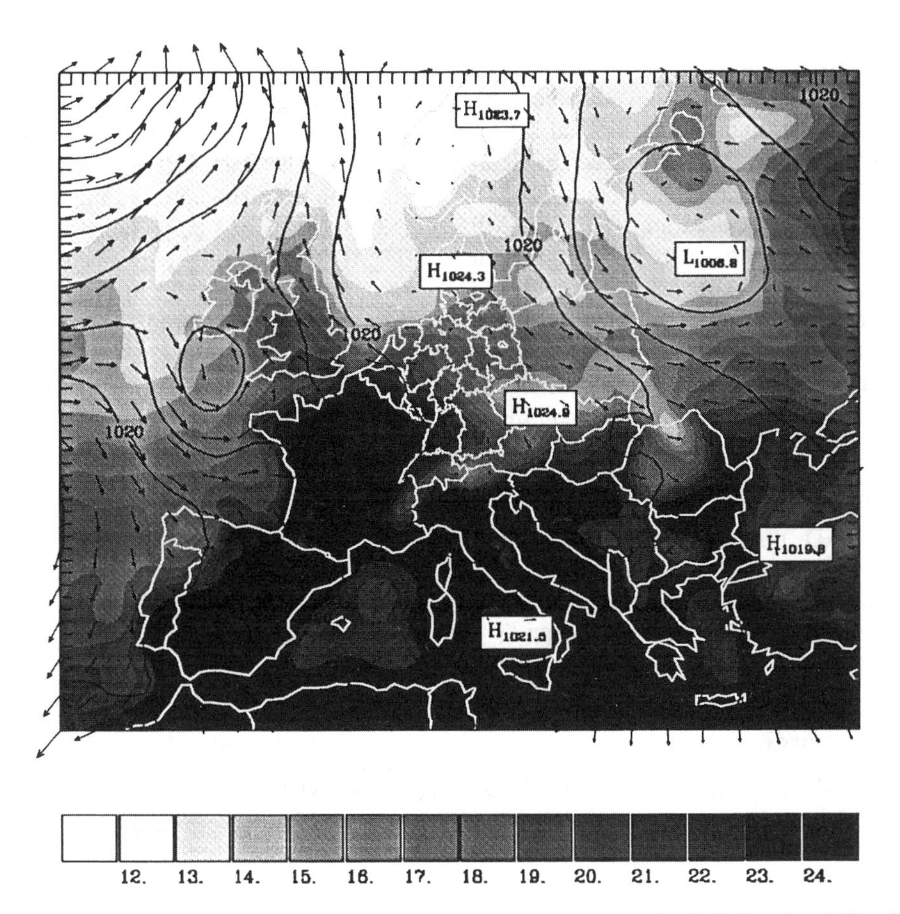

Fig. 7.1.1b: Sea-level pressure, temperature and horizontal wind for the 17th of September 1992, 12:00 UTC (2nd IMP of TRACT).

The horizontal grid size for this simulation is 60 km, in the vertical the model is extended up to 100 hPa with 19 layers. Compared to the standard design which has been used *e.g.* for the EUMAC joint cases the vertical resolution in the planetary boundary layer has been improved. The thickness of the lowest layer in the EURAD-TRACT case is about 38 m compared to about 76 m in the EUMAC joint cases. The model domain covers the whole of Europe except the northern part of Scandinavia.

As an example results for the meteorological simulation are shown in Fig. 7.1.1 for the 16th and 17th September. The meteorological situation over central Europe is characterised by a developing anticyclone with weak pressure gradients and low wind speeds. In the following days the anticyclone was moving to southern Scandinavia, wind speeds over Europe remain low with a predominant southeasterly flow.

Conclusions

The first simulations with EURAD (meteorological part) for the episode of TRACT measurement campaign has been completed. The episode simulated covers the time span of the 2nd IMP of TRACT (16th/17th of September, 1992) and the 3rd IMP (21th/22th of September, 1992). The data will be used to provide the small-scale models (KAMM/DRAIS) with boundary values. Emission data currently used are based on EMEP. GENEMIS data will be included in a step by step procedure. First runs with the EURAD-CTM will be performed till June 1995. The model results will be analysed in detail with respect to dynamic and chemical processes. The combination of chemical and meteorological data sampled during the intensive measurement phases allows for a detailed comparison between model simulations and observations, in particular in the vertical (radiosonde data, balloon soundings, aircraft flights).

Acknowledgements

EURAD development and simulations are done within the framework of EUMAC (see EUMAC annual report 1993). EURAD is funded by the BMFT and the state of NRW (Minister of Science and Research). No special funding is obtained for the contribution to TRACT. EURAD acknowledges the support from EMEP, GENEMIS, KFA Jülich and DWD.

References

1. Ebel, A., H. Hass, H.J. Jakobs, M. Memmesheimer, M. Laube, A. Oberreuter, H. Geiß, Y.-H. Kuo: Simulation of ozone intrusion caused by a tropopause fold and cut-off low. *Atmos. Environ.* **25** (1991) 2131-2144.
2. Hass, H., A. Ebel, H. Feldmann, H.J. Jakobs, M. Memmesheimer, Evaluation studies with a regional chemical transport model (EURAD) using air quality data from the EMEP mo nitoring network. *Atmos. Environ.* **27A** (1993) 867-887.
3. Jakobs, H.J., H. Feldmann, H. Hass, M. Memmesheimer: The use of nested models for air pollution studies: an application of the EURAD model to a SANA episode, *J. Appl. Met.* **34** (1995) 1301–1319.
4. Grell, G.A., J. Dudhia, D.R. Stauffer: A description of the fifth-generation Penn State/NCAR mesoscale model (MM5). *NCAR Technical Note. NCAR/TN-398+IA*, 1993.
5. Memmesheimer, M., J. Tippke, A. Ebel, H. Hass, H.J. Jakobs, M. Laube: On the use of EMEP emission inventories for European scale air pollution modeling with the EURAD model. *Proc. EMEP workshop on Photo-oxidant Modelling for Long-Range Transport in Relation to Abatement Strategies*, Berlin 1991, pp. 307-324.
6. Hass, H: Description of the EURAD Chemistry-Transport-Model Version 2 (CTM2), Mitteilungen aus dem Institut für Geophysik u. Meteorologie der Universität zu Köln, 1991.

7.2 Modeling of Transport, Mixing and Deposition of Air Pollutants over Inhomogeneous Terrain including Chemical Reactions

K. Kuntze, M. Löffler-Mang, K. Nester, G. Adrian and F. Fiedler

Institut für Meteorologie und Klimaforschung, Forschungszentrum Karlsruhe
Universität Karlsruhe, Postfach 3640, D-76021 Karlsruhe, Germany

A simulation of the air pollution for 16th September 1992 was carried out with the model system KAMM/DRAIS. This is a day of the second intensive measuring period of the TRACT experiment. It was selected, because a complex meteorological field was observed at this day. Especially orographically influenced flow systems developed during that day. It is shown that these flows strongly influence the diurnal cycle of the air pollutants. As examples the effects of the valley wind system at Freiburg, the stagnating air masses in parts of the Upper Rhine Valley and the sea breeze flow around Lake Constance are presented. Most of the observed features of the flow field and the species concentration distributions are simulated by the model.

Aims of the research

The aim of the project is the numerical simulation of the diurnal development of the atmospheric boundary layer over complex terrain. In particular the influence of this development on the transport, turbulent mixing, the chemical reactions, the dry deposition and the hand over of air pollutants into the free troposphere is considered. Of special interest are the topographically influenced circulations observed in valleys at mountain slopes and over lakes and their effects on air pollution in this areas.

Principal findings during the project

To study the processes influencing the air pollution in a mesoscale area model simulations are necessary. Only with model simulations it is possible to describe the meteorological and air pollution conditions in an almost complete and area covering manner. In the frame of the TRACT project model simulations are carried out for 16th September 1992, a day of the second intensive measuring period. This day was selected, because wind systems like slope winds, valley winds, land sea breeze and regions of calm winds were observed during this day. All these processes are caused by the complex topography and influence considerably the air pollution in the TRACT area.

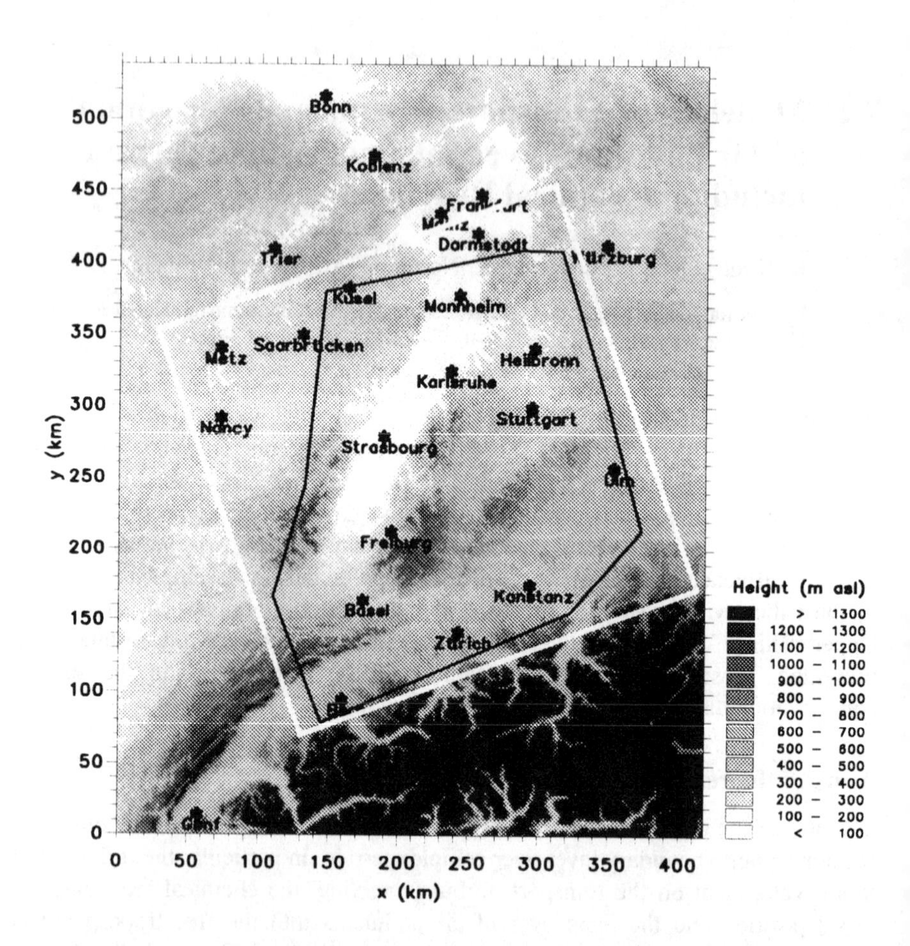

Fig. 7.2.1: The TRACT area and the model area

The models used for this study are the 'Karlsruhe Atmospheric Mesoscale Model' KAMM [1] and the dispersion model DRAIS [2] including the chemical code of the 'Regional Acid Deposition Model' RADM2 [3]. The model domain is a quadratic area with 300 km side length and encloses the south-western part of Germany, a part of Alsace and the northern part of Switzerland. The area is rotated by an angle of 20 degrees against north (Fig. 7.2.1). The terrain is rather inhomogeneous with grid-points on the one hand below 100 m above m.s.l. in the northern part of the Rhine Valley and on the other hand more than 1200 m at the edge of the Alps in the Black Forest and the Vosges Mountains. The model domain is divided into 60 times 60 horizontal grid points. The land use is classified into 11 classes to describe the different interactions between the vegetation and the atmosphere. For the whole area the soil is assumed to be sandy loam, the dominant soil texture in the TRACT area.

The external forcing for the KAMM model is the hydrostatic, geostrophic and non-stationary basic state, which was determined by an objective analysis [4] on the basis of the radio soundings carried out during the TRACT experiment. The geostrophic wind blows from north-west with an increasing speed from about 2 m/s in the south-west corner to about 10 m/s in the north-east corner of the model domain. The KAMM simulation state at 21:00 CET on 15th September, 19992 and the DRAIS Simulation three hours later.

Wind fields

The simulated wind fields show a lot of features which are affected by the complex terrain of the area. These features are in good agreement with the measurements. Figs. 7.2.2 and 7.2.3 show the simulated and measured wind fields at 9:00 CET and 15:00 CET. The measured winds are represented by black-arrow-heads, while the simulated winds are represented by white arrow-heads.

Although the wind speeds are low the channelling of the flow along the Upper Rhine Valley is clearly recognisable in the morning. The Burgundische Pforte acts like a nuzzle which accelerates the flow in this area. Due to intensive vertical mixing the winds in general are stronger in the afternoon.

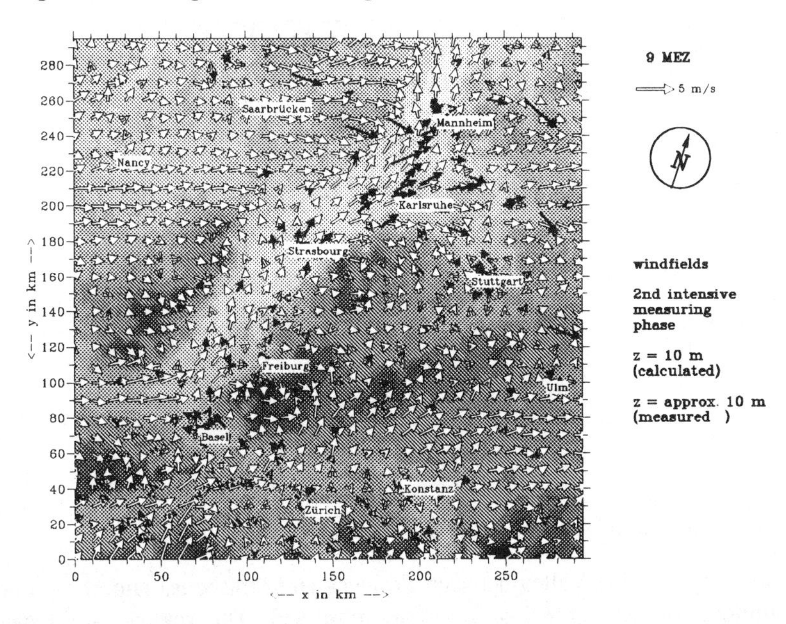

Fig. 7.2.2: Orography and wind fields (z = 10 m), 16th Sept, 1992, 9:00CET.

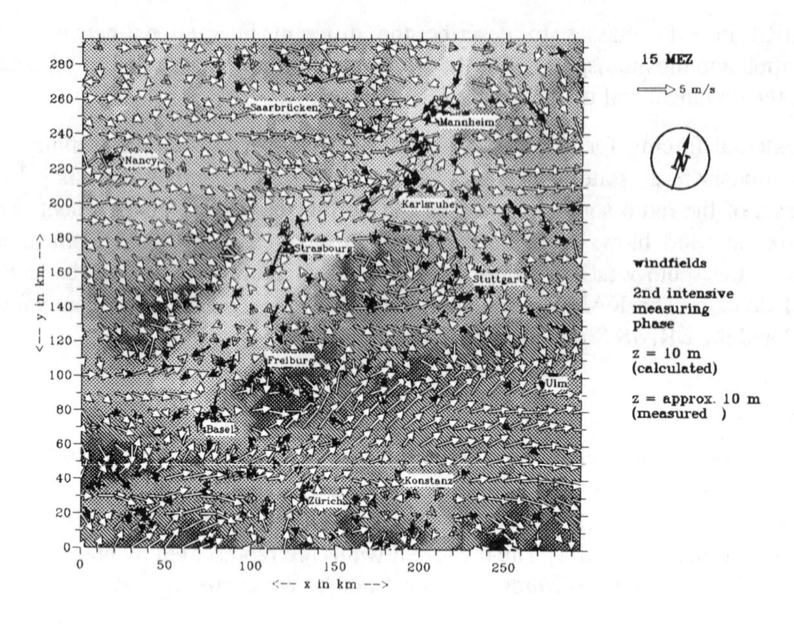

Fig. 7.2.3: Orography and wind fields (z = 10 m), 16th Sept, 1992, 15:00CET.

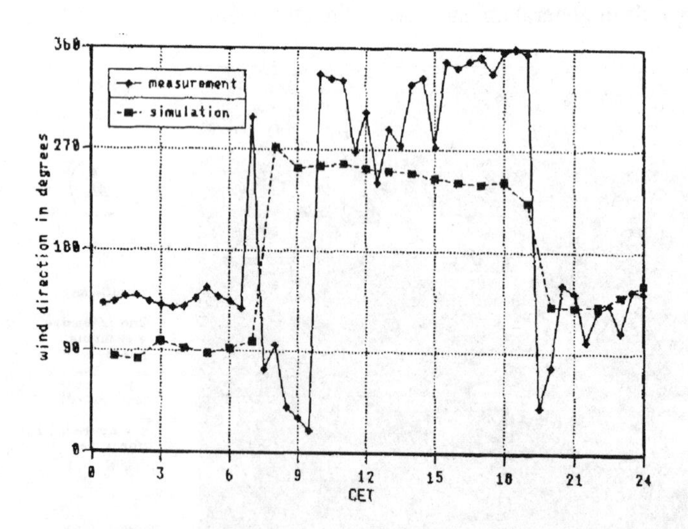

Fig. 7.2.4: Measured and simulated wind direction in Freiburg/Mitte.

But in the Upper Rhine Valley between Freiburg and Strasbourg and in the north of Mannheim regions of calm winds are observed. The regions are located between Vosges Mountains/Black Forest and Pfälzer Wald/Odenwald, respectively. The general flow from north-west reduces the large scale pressure gradient in these two areas causing a stagnation of the air flow. Because of these calm winds pronounced slope-winds occur in these regions.

The measurements show during the whole day an orographically influenced flow from Karlsruhe into the region of Stuttgart around the northern part of the Black Forest, which is reproduced also well by the simulation.

An example for a diurnally varying valley wind system is observed at Freiburg. Fig. 7.2.4 shows a comparison of the wind direction – measured and simulated – at this location. The influence of the Dreisam Valley, oriented perpendicular to the Upper Rhine Valley can be clearly seen in the changing wind direction In the early morning there is an easterly flow down the valley. After sunrise the wind is blowing into the valley and after sunset at 18:00 CET it changes back to an easterly flow.

Another interesting wind system is found around Lake Constance. During daytime the wind directions around the lake are evidently directed from the lake to the surrounding hills (Fig. 7.2.5). This thermal wind system is formed by a combination of land-sea-breeze and uphill-downhill flows. Although the resolution in the model is only 5 km there is really a good agreement of the simulated and measured wind around the lake.

Temperature fields

The temperature distribution (Fig. 7.2.6) reflects the orography in the domain, with low temperatures in the elevated regions (area of Feldberg, Hornisgrinde and Swiss Jura) and high temperatures in the valleys and low regions. The highest temperatures in the model area are found in the Upper Rhine Valley.

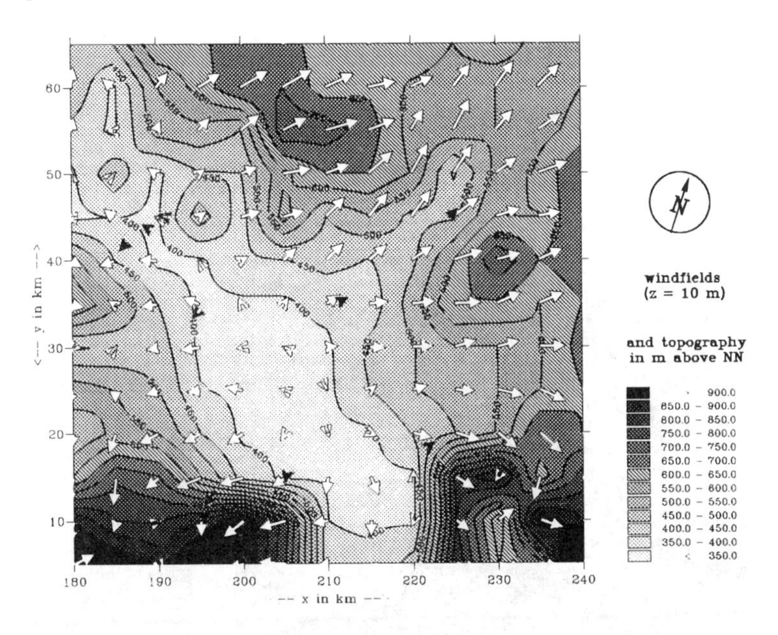

Fig. 7.2.5: Measured and simulated winds at Lake Constance, 16th Sept, 1992, 12:00CET.

One region is located in the lee of the Vosges Mountains and the other in the lee of the Pfälzer Wald. In both regions the calm winds cause the high temperatures. Between this regions the air is transported through the Zaberner Senke to the Kraichgau. This leads to a better ventilation and therefore to lower temperatures in these areas.

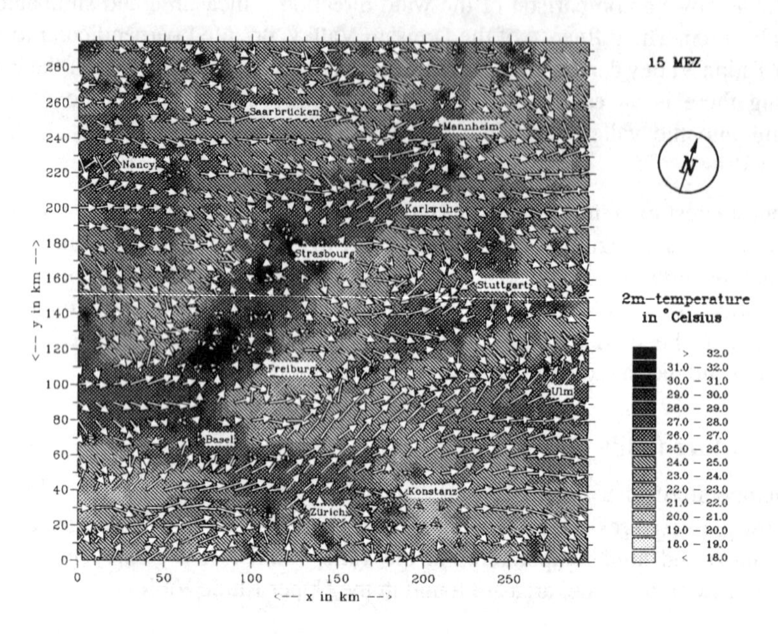

Fig. 7.2.6: Temperature, 2 m above ground, 16th Sept, 1992, 15:00 CET.

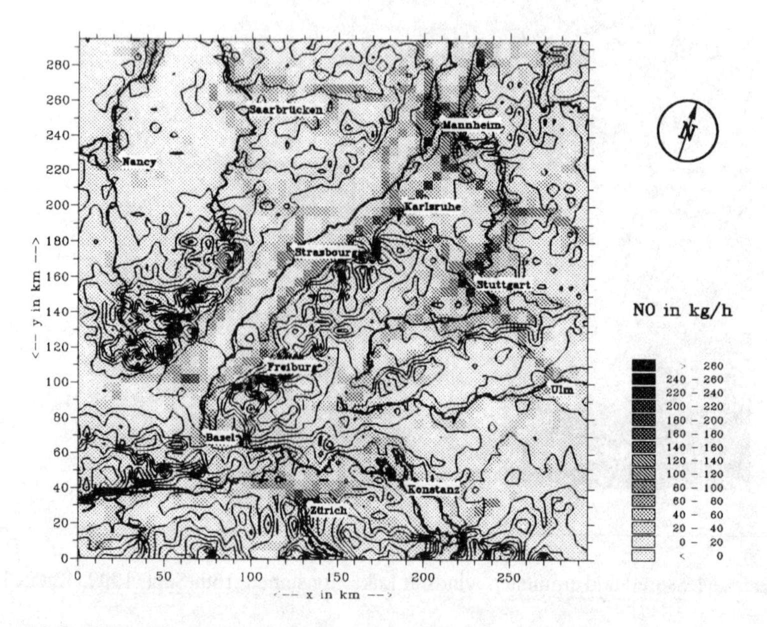

Fig. 7.2.7: NO emissions, 16th Sept, 1992, 9:00 CET.

Concentration of air pollutants

To calculate the concentrations of the different species their background concentrations and additionally the emission data are necessary. The start profiles for the background concentrations were derived from the flight measurements in the morning of 16th September. The final emission data from IER Stuttgart [5] are not yet available, therefore preliminary emission data are taken. Fig. 7.2.7 shows the preliminary emissions of NO at 9:00 CET, which are used in the model simulation. The motorways and the main city areas of Mannheim/Ludwigshafen, Stuttgart and Karlsruhe ca be identified.

Fig. 7.2.8 shows the NO concentrations, calculated with the model. The distribution of the emissions is recognisable, but the calculated concentrations are influenced also by the wind system. There are regions with calm winds like at the eastern edge of the Vosges Mountains or around Strasbourg, where the NO is accumulating although the emissions aren't as high as in other better-ventilated regions.

Another area of high NO concentrations is simulated over Lake Constance, where low winds and stable atmospheric stratification avoid the transport and dilution of the concentrations. Because of the limited mixing layer at that time the emissions stay in the lower parts of the atmosphere and do not reach the elevated regions in the Black Forest or in the Vosges Mountains.

The spatial distribution of NO reflects in the distribution of ozone (Fig. 7.2.9). The lowest ozone concentrations are detected where the NO concentration has its highest values. Because the high NO concentrations are restricted to the lower layers of the atmosphere the highest ozone concentrations are simulated in the elevated regions.

With the increasing mixing during the morning air with higher ozone concentrations is transported into the lower layers where the concentrations rise. The intensive mixing over the whole area in the afternoon provides a more homogeneous distribution of ozone than in the morning hours (Fig. 7.2.10). But due to the ozone destruction caused by NO there are areas of low ozone concentrations, like some parts of the motorways or the region around Stuttgart.

Ozone concentrations higher than the values of the free atmosphere prove the formation of ozone during this day. The highest ozone concentrations are simulated in the lee of the Vosges Mountains and the Pfälzer Wald and near Strasbourg. In these areas the ozone formation is strongest because of the stagnating air masses.

In dependence on the constant water temperature the stratification at Lake Constance stays stable for the whole day. Therefore there is almost no vertical mixing over the lake and consequently a reduced transport of air with higher ozone concentrations down to the lower parts of the atmosphere. Hence the ozone concentrations at Lake Constance stay low during the whole day.

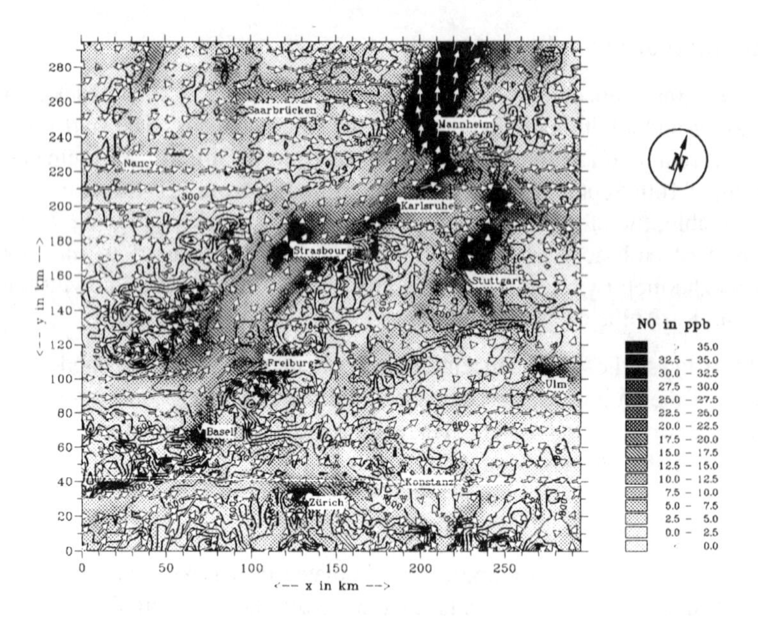

Fig. 7.2.8: NO and orography, 16th Sept, 1992, 9:00 CET.

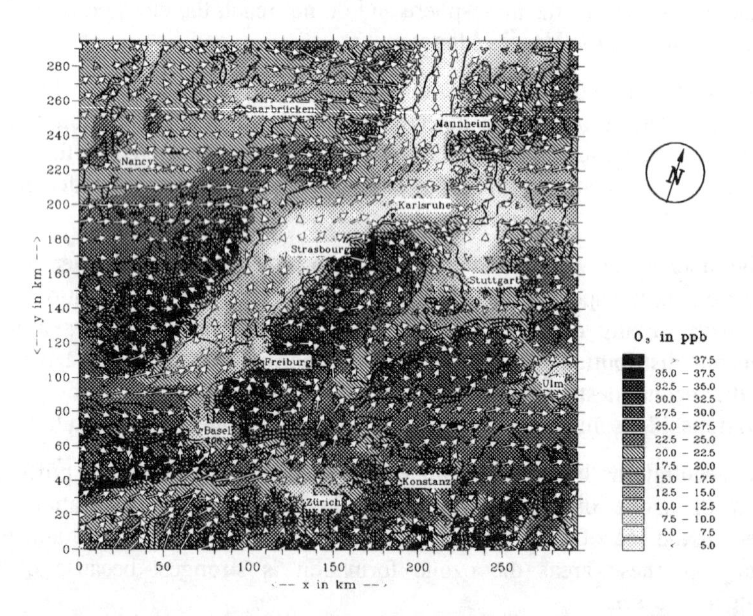

Fig. 7.2.9: Ozone and orography, 16th Sept, 1992, 9:00 CET.

The characteristic change in the wind direction at Freiburg from up valley directions during the day to down valley directions at night influences the ozone concentration at Freiburg. The diurnal cycle of ozone in Freiburg (Fig. 7.2.11) shows higher concentrations in the night as compared to the early morning and

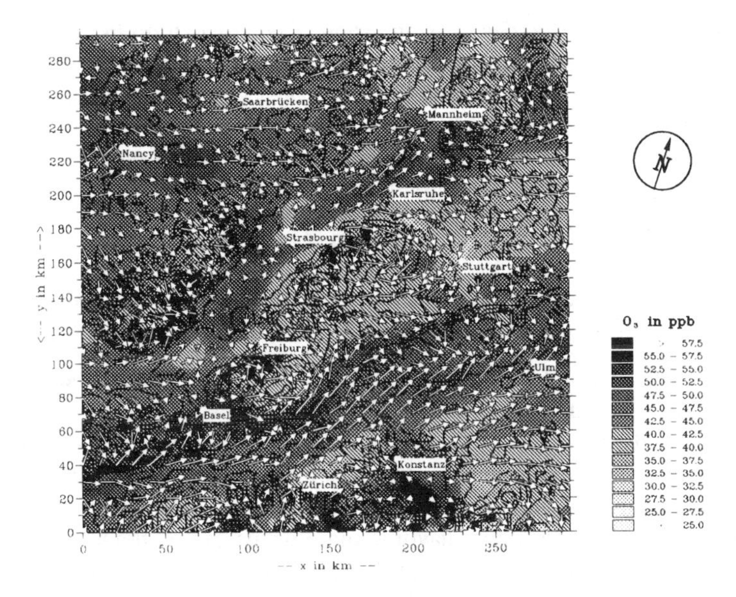

Fig. 7.2.10: Ozone and orography, 16th Sept, 1992, 15:00 CET.

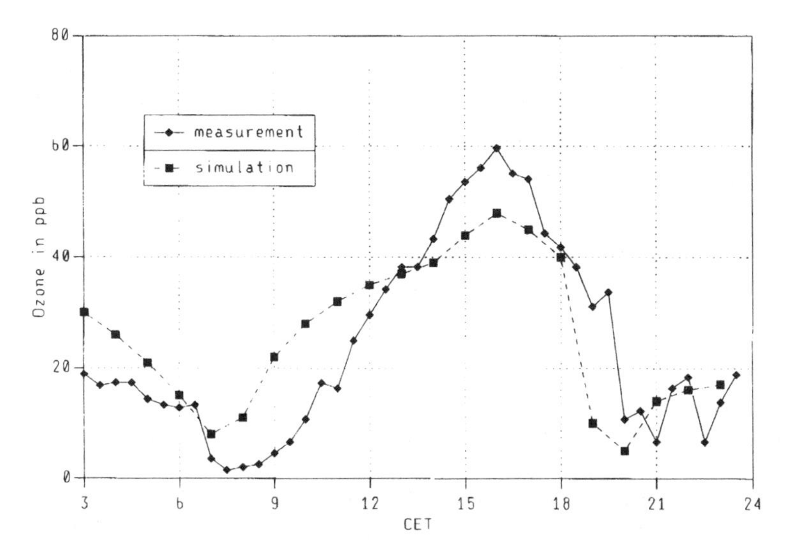

Fig. 7.2.11: Measured and simulated ozone in Freiburg (centre).

evening hours. This increase of the concentration is caused by the advection of ozone from the valley to the city in the night. The simulated diurnal cycle is in good agreement with the measurements.

The deposition of the different species is strongly influenced by the physical, chemical and biological surface properties. In the DRAIS model a big leaf multiple resistance model is applied [6]to determine the deposition velocities. It calculates the uptake of trace gases in dependence of the different land use types.

Therefore the deposition velocities of ozone show a great variation in the model area. Over Lake Constance the deposition velocity of ozone is lowest with about 0.8 mm/s because of its low water solubility. Also over city areas the deposition velocity is small. Here the deposition velocity varies between 0.1 cm/s and 0.5 cm/s. Highest values are simulated over forests, where the strongest diurnal variation of the deposition velocity occurs. During the afternoon ozone deposition velocities of about 1 cm/s can be reached.

An assessment of the achievement of the contribution in the light of its original aims and those of the subproject

The simulation with the model system KAMM/DRAIS for 16th September 1992 clearly shows that orographically influenced flow systems like valley winds, mountain slope winds, channelling effects and regions of stagnating air determine mainly the diurnal variation of the air pollution in these regions. Especially in the latter case accumulations of the primary air pollutants occur in the morning hours. If the period of stagnation lasts over daytime more ozone can be produced there than in other regions. Over greater water areas like Lake Constance a sea breeze circulation develops. The stable stratification reduces the vertical mixing. Therefore the diurnal cycle of the concentrations of air pollutants over the lake is quite different from that over land.

The comparison of the simulations with the measurements shows that most of the observed features of the diurnal cycle of the meteorological and air pollutant distributions can be reproduced by the model KAMM/DRAIS.

Acknowledgements

The EUROTRAC subproject TRACT is supported by the Bundesministerium für Bildung und Forschung (BMBF), Germany.

References

1. Adrian, G., Fiedler, F.: Simulation of unstationary wind and temperature fields over complex terrain and comparison with observations. *Beitr. Phys. Atmos.* **64** (1991) 27–48.
2. Nester, K., Fiedler, F.: Modelling of the diurnal variation of air pollutants in a mesoscale area, proceedings of the 9th Clean Air Congress, 1992, paper No. IU-16C.02.
3. Stockwell, W.R., P. Middleton, J.S. Chang, X. Tang: The second generation regional acid deposition model chemical mechanism for air quality modelling, *J. Geophys. Res.* **95** (1990) 16343–16368.
4. Adrian, G.: Zur Dynamik des Windfeldes über orographisch gegliedertem Gelände, *Berichte des Deutschen Wetterdienstes*, Nr.188, 1994.
5. Friedrich, R., Berner, P., John, C., Obermeier, A., Seier, J.: Generation of Emssion Data Base for TRACT. *EUROTRAC, Annual Report 1993, part 10, TRACT*, EUROTRAC ISS, Garmisch-Partenkirchen 1994, pp. 118–123
6. Bär, M., Nester, K.: Parametrization of trace gas dry deposition velocities for a regional mesoscale diffusion model, *Ann. Geophysicae*, **10** (1993) 912–923.

7.3 Measurements and Model Simulations: A Contribution by the Italian Participants

D. Anfossi[1], E. Ferrero[1,4], D. Sacchetti[1], S. Trini Castelli[1], S. Finardi[2], A. Marzorati[3], S. Bistacchi[3], G. Bocchiola[3], G. Brusasca[3], P. Marcacci[3], G. Morselli[3], U. Pellegrini[3] and G. Tinarelli[3]

[1]C.N.R. - Istituto di Cosmogeofisica, Corso Fiume 4, 10133 Torino, Italy
[2]ENEL SpA, Environment and Material Research Centre, via Rubattino 54, 20134 Milano, Italy
[3]CISE SpA, Environment Department, via R.Emilia 39, 20090, Segrate, Milano, Italy
[4]DIP. Scienze e Tec. Av., Univ. Torino, C. Borsalino 54, 15100 Alessandria, Italy

Our joint ENEL–SpA and CNR teams participated in the experimental activity of TRACT in September 1992 at Memmingen-Baden Württemberg, Germany, and in the 3 TRANSALP exercises (1989, 1990 and 1991) in Switzerland. Furthermore it performed the simulation of the TRANSALP I (1989) experiment: the transport and diffusion of tracer release inside an Alpine valley has been simulated through the use of a mass-consistent model (for the wind field reconstruction) and of a Lagrangian particle model.

Aims of the research

The main scientific objective of our activities was to define and measure meteorological variables and turbulent parameters of the atmosphere in complex terrain. Firstly we performed experimental activities during the three TRANSALP and TRACT campaigns using traditional and Remote Sensing instrumentation. In particular, the TRANSALP objective was to improve the knowledge of the pollutant transport across the Alpine barrier. This objective has been pursued by means of tracer experiments at different scales from 40 to 200 km and in different meteorological conditions: valley breezes, synoptic forcing, PBL evolution. The TRACT objective was to study the transport diffusion, chemical transformation and deposition of air pollutants in a mesoscale domain (400 × 600 km).

Due to the participation of various European teams, an almost complete data set was built up in each campaign and could be used as a benchmark for modelling applications. The final goal was to simulate the transport and diffusion of pollutants in complex terrain and in a mesoscale area using three dimensional numerical modelling.

Principal scientific results

Experimental activities

The experimental equipment used during the campaigns by our team were:

* 3-D Doppler SODAR for monitoring wind profiles in the PBL, pilot balloons and radio soundings; and
* complete ground-based meteorological station.

Only for TRACT campaign:

* a radio theodolite for wind, temperature and humidity profiles up to 10 km height;
* 3-axis ultrasonic anemometer; and
* a Radio Acoustic Sounding System (RASS) for temperature profiles in the PBL

The data have been acquired in real-time by means of the Mobile Intelligent Node and stored in an appropriate data-base. In the TRANSALP campaigns our instruments were located in the Leventina valley [1–3] while in the TRACT experiment [4] they were located at the SE point of the investigated mesoscale domain.

Different studies were performed by us to investigate the dispersive characteristics of the atmosphere. The reliability of Sodar data in complex terrain was assessed (Fig. 7.3.1).

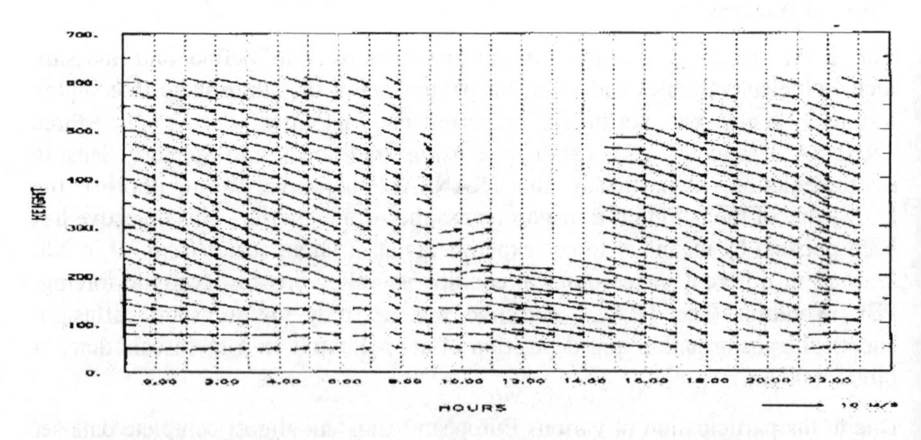

Fig. 7.3.1: Wind evolution at DALPE (TRANSALP 1991) means by Sodar data

This allowed us to depict wind evolution inside the alpine valley [5] and the evolution of stability classes calculated by means of 50 m σ_W (Fig. 7.3.2). The availability of ground measurements, vertical profiles into the PBL and the upper level, allowed a well defined evolution of the PBL vertical structure to be made (Fig. 7.3.3). Particular attention was paid to elaborate sonic anemometer data.

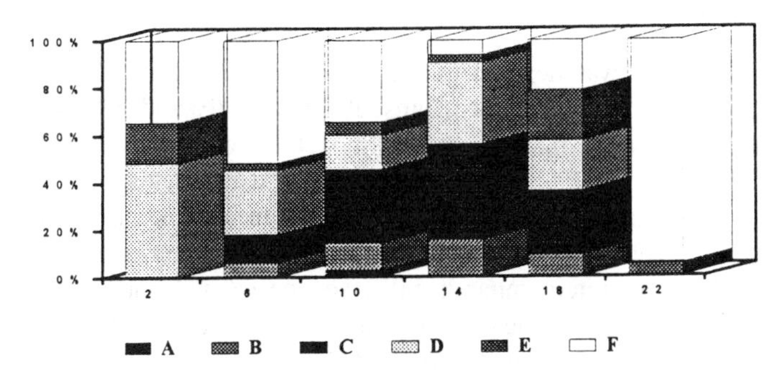

Fig. 7.3.2: Stability evolution at Airolo (TRANSALP 1990) means by 50m σ_w Sodar data.

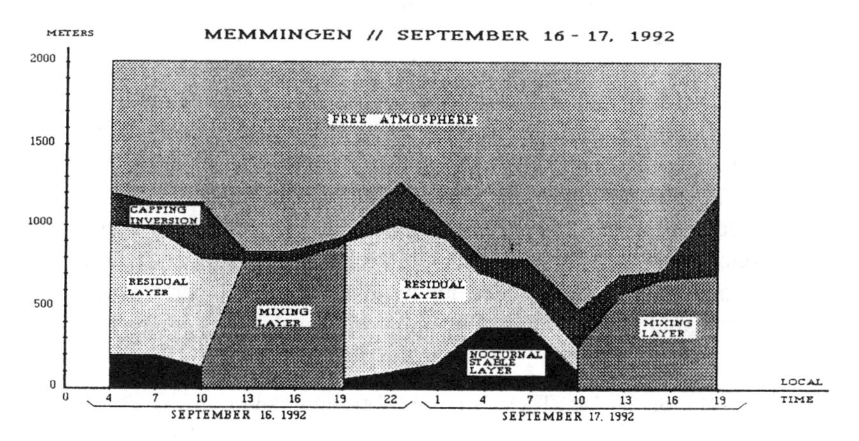

Fig. 7.3.3: PBL evolution in the TRACT campaign

A software package [6] was built to solve the misalignment problem and to estimate sensible and latent heat fluxes (Fig. 7.3.4).

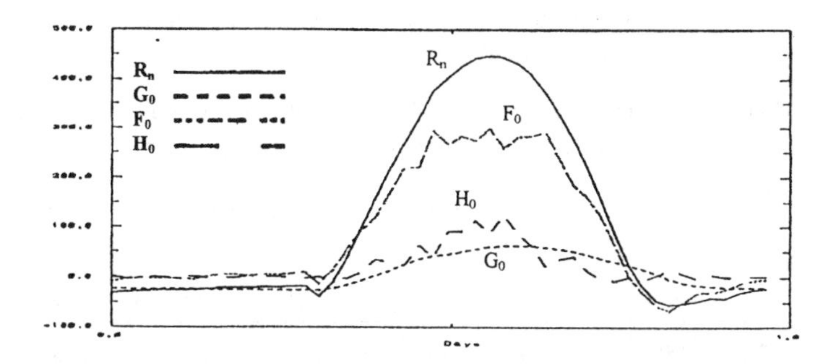

Fig. 7.3.4: Net radiation (R_n), latent heat flux (F_0), sensible heat flux (H_0), heat flux into the ground (G_0) in the TRACT campaign.

Modelling activities

Our modelling activity was devoted to the simulation of the tracer experiment that took place on October 19, by means of MINERVE and SPRAY codes. MINERVE is a diagnostic mass-consistent model, developed by Electricité de France [7], that can quickly generate 3-D mean wind fields satisfying the continuity equation, starting from the available wind measurements. SPRAY is a 3-D Lagrangian particle dispersion model developed by our group [8]. Our simulation results were intercompared with the one carried out by the ENEA group from Rome [9], who applied a different mass-consistent model, CONDOR [10], and a Lagrangian particle model ARCO [11].

MINERVE produced a sequence of 13 wind fields, one every half a hour, on a 49 × 49 computational mesh, with an horizontal resolution of 500 metres [12]. The vertical domain had a depth of 7500 metres filled by 21 vertical sigma levels, stretched in order to obtain a higher resolution near the ground. The wind stations used to initialise the models have been chosen on the base of their representativeness of the main flow characteristics. The models were run with the surface observations and SODAR profiles measured at Iragna, near the release point. Due to their position in the southern part of the valley in a relatively homogeneous area, these measurements describe the upwind flow oncoming on the release point and on the valley bifurcation. The surface stations of Somprei and Comprovasco were added to improve the model performances inside the Leventina and Blenio valleys. The input information was completed by the use of the 500 HPa geostrophic wind at the top of the computational domain.

The results of the MINERVE run at 12:00 are shown in Fig. 7.3.5 by the wind field cross section at the 10 m sigma level, where the channelling effect of the topography is evident. A statistical evaluation of MINERVE at some stations, not used as input data, is included in Tables 7.3.1 and 7.3.2, where the results are compared with those obtained by CONDOR. The comparison between observed and computed winds shows that the RMSE are below 1 m/s for wind speed, and below 40 degrees for wind direction. Starting from a few input data (1 vertical wind profile and 3 ground stations) MINERVE was able to simulate the main characteristics of the flow in the studied valleys.

Table 7.3.1: Model evaluation of MINERVE (M) and CONDOR (C) runs. Comparison between observed (o) and computed (c) wind direction. Mean values, standard deviations, Bias and RMSE are indicated.

	$\langle d_o \rangle$	$\langle d_c \rangle$		σ_0	σ_c		Bias		RMSE	
		C	M		C	M	C	M	C	M
Semione	230	180	164	70	24	23	−50	−66	73	84
Pizzoerra	193	184	187	9	16	13	−9	−6	25	22
Camperio	194	186	160	25	31	19	−8	−34	25	39

The general structure of the wind field produced by the two different mass-consistent models was quite similar showing only small local differences. This result is confirmed by the comparison of the statistical results obtained by MINERVE and CONDOR.

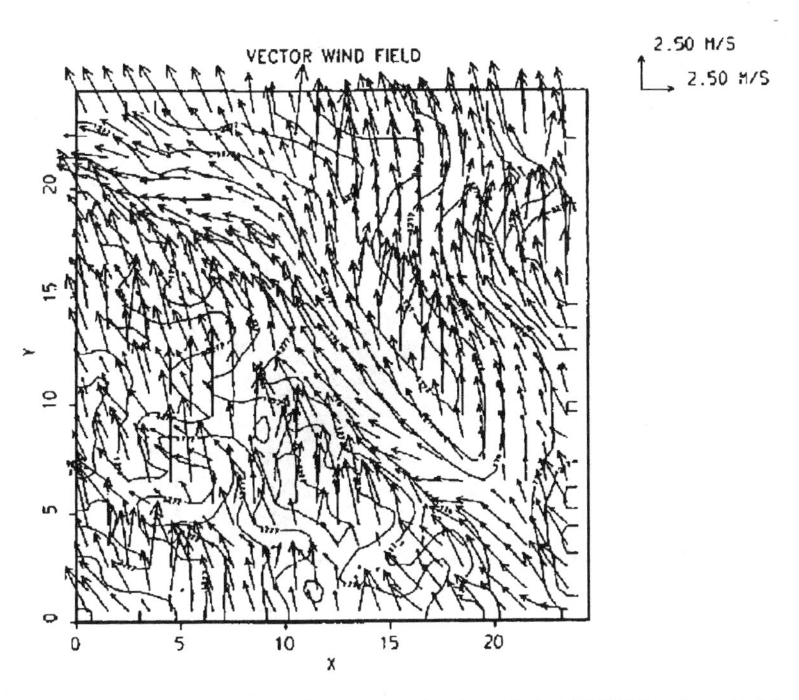

Fig. 7.3.5: Wind field at 10m sigma level simulated by MINERVE code at 12:00.

Table 7.3.2: Model evaluation of MINERVE (M) and CONDOR (C) runs. Comparison between observed (o) and computed (c) speed. Mean values, standard deviations, Bias and RMSE are indicated.

	$<v_o>$	$<v_c>$		σ_0	σ_c		Bias		RMSE	
		C	M		C	M	C	M	C	M
Semione	1.44	1.84	1.78	0.40	0.43	0.40	0.40	0.27	0.65	0.63
Pizzoerra	3.48	2.78	3.05	0.51	0.42	0.69	–0.70	–0.43	0.91	0.68
Camperio	1.62	1.34	1.71	0.17	0.30	0.52	–0.28	0.09	0.49	0.65

SPRAY used the 3-D wind fields produced by MINERVE and 3-D turbulence fields produced by a parametrisation scheme developed by Hanna [13] and based on the data measured by the JRC team during the campaign, to evaluate the surface layer and boundary layer variables. The main results of the dispersion

The plume moved at an averaged speed of about 2 m/s towards the valley bifurcation. The ratio of tracer mass entering the two valleys (4 to 1) was correctly reproduced. An example of the plume simulated is shown in Fig. 7.3.6 with respect to the actual profiles within the two valleys and the difficulty to reproduce local circulation characteristics by the mass-consistent models.

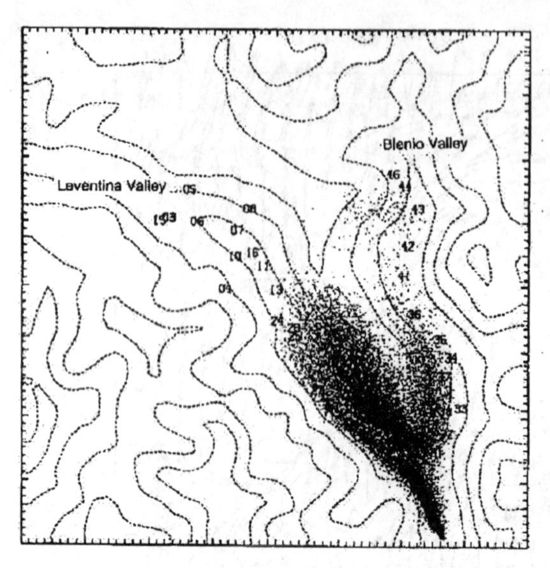

Fig. 7.3.6: Particle plume simulated by SPRAY code at 12:00. Numbers represent sampler positions.

To estimate the accuracy of simulations, a model evaluation, based on a point-to-point comparison of measured and predicted values, was performed on 193 pairs of half-hour concentration values measured and modelled by ARCO and SPRAY to calculate the following statistical indexes: mean values, standard deviations, maximum value, fractional bias (FB), normalised mean square error of the normalised ratios distribution (NNR) [14] and correlation coefficient (CORRE). The results are summarised in Table 7.3.3 which shows that SPRAY correctly captures mean and maxima values while ARCO tends to overestimate.

Table 7.3.3: Statistical indexes relative to ARCO and SPRAY simulation (entire sample)*

	Mean (ng/m^3)	St.dev. (ng/m^3)	Max (ng/m^3)	FB	NNR	CORRE
Observed	113.6	216.8	1502.3			
SPRAY	108.4	214.4	1553.3	–0.048	1.545	0.74
ARCO	138.4	313.9	2221.1	–0.197	1.682	0.84

*Reprinted from *Atmos. Environ.* **31** (1997) with kind permission from Elsevier Science Ltd.

On the other hand ARCO obtains a better correlation coefficient. Tables 7.3.4 and 7.3.5 report the same indexes but for the time integrated concentration distribution and for concentration maxima at each sampler respectively. They confirm the results indicated in Table 7.3.3, that ARCO shows a better correlation coefficient and a general overestimation, while SPRAY doesn't indicate an appreciable bias. Finally, ARCO performs better with respect to the NNR index.

Table 7.3.4: Statistical indexes for the time integrated concentration distribution*.

	Mean (ng/m^3)	St.dev. (ng/m^3)	Max (ng/m^3)	FB	NNR	CORRE
Observed	664.6	739.0	3066.5			
SPRAY	633.7	679.3	3452.0	0.048	0.795	0.67
ARCO	809.3	918.2	4672.4	–0.197	0.381	0.87

Table 7.3.5: Statistical indexes for the g.l.c. maxima (at each sampler) distribution*.

	Mean (ng/m^3)	St.dev. (ng/m^3)	Max (ng/m^3)	FB	NNR	CORRE
Observed	303.7	347.5	1502.3			
SPRAY	303.8	298.8	1553.3	0.0002	0.855	0.70
ARCO	458.0	480.3	2221.1	–0.405	0.370	0.88

Taking into account both the limited available meteorological information and the extreme complexity of the site, the present results indicate an overall capacity of the two models to capture the ensemble characteristics of the tracer plume distribution inside the two valleys. Except during the last hour, the concentration field is quite correctly simulated even if some disagreements appear for particular samplers. Furthermore it resulted that, even in this particularly complicated topography, the two models though being different in many respects yield g.l.c. fields with the same degree of accuracy and describe the 3-D structure of the dispersion phenomena in a reasonable way.

Conclusions

A very intense experimental activity, performed by our group jointly with other TRACT participants, allowed us to obtain 4 different data sets to investigate the behaviour of atmospheric characteristics in complex terrain at mesoscale. These data sets could be also used to investigate the ability of numerical models to simulate the transport and diffusion of pollutants in these severe conditions. A first test was conducted by our teams in the TRANSALP 89 experiment giving encouraging results. Considering the greater computational power nowadays available on workstations, we are able to show that 3-D models are powerful tools for studying the airborne dispersion in highly complex terrain configurations and in this respect could be applicable for practical purposes.

Acknowledgements

No support from any funding agencies was given to our teams.

References

1. Ambrosetti P., D. Anfossi, S. Cieslik, P. Gaglione, R. Lamprecht, A. Marzorati, S. Sandroni, F. Spinedi, A. Stingele, S. Vogt: The TRANSALP-89 Exercise: a Tracer Release Experiment in a Subalpine Valley. *EUR report 13474 EN*, 1991.
2. Ambrosetti P., D. Anfossi, S. Cieslik, G. Graziani, G. Grippa, R. Lamprecht, A. Marzorati, K. Nodop, A. Stingele, H. Zimmermann: Mesoscale Transport of Atmospheric Trace Gases Across the Alps: the TRANSALP 1991 Campaign. EUR report 16046 EN, *ENEL/CRAM internal report G12/94/11*, 1994.
3. Ambrosetti P., D. Anfossi, S. Cieslik, G. Graziani, G. Grippa, R. Lamprecht, A. Marzorati, A. Stingele, H. Zimmermann: The TRANSALP-90 Campaign: the Second Tracer Release Experiment in a Sub-Alpine Valley. *EUR report 15952 EN, ENEL/CRAM internal report G12/94/10*, 1994.
4. Mohnen V., Slemr F., Kanter H.J., Fiedler F., Corsmeier U.: Quality assurance/quality control in TRACT. *EUROTRAC Newsletter* **12**, EUROTRAC ISS, Garmisch-Partenkirchen 1993, pp. 18–23.
5. Marzorati A., D. Anfossi: Doppler SODAR Measurements and Evaluations in Complex Terrain. *Il Nuovo Cimento* **16 C** (1993) 141–154.
6. Anfossi D., C. Cassardo, E. Ferrero, A. Longhetto, D. Sacchetti, G. Brusasca, V. Colombo, A. Marzorati, M.G. Morselli, F. Rocchetti, G. Tinarelli: On the Analysis of Ultrasonic Anemometer Data. *ENEL/CRAM internal report E1/93/04, ICG/CNR report 282/93*, 1993.
7. Geai P.: Methode d'interpolation et de reconstitution tridimensionelle d'un champ de vent: le code d'analyse objective MINERVE. E.d.F. Chatou, France, *Report DER/HE/34-87.03*, 1987.
8. Tinarelli G., D. Anfossi, G. Brusasca, E. Ferrero, U. Giostra , M.G. Morselli, J. Moussafir, F. Tampieri, F. Trombetti: Lagrangian Particle Simulation of Tracer Dispersion in the Lee of a schematic Two-Dimensional Hill. *J. App. Meteor.* **33** (1994) 744–756.

9. Desiato F.: A dispersion model evaluation study for real-time application in complex terrain. *J. Appl. Meteor.* **30,8** (1991) 1207–1219.
10. Moussiopoulos N., Flassak T., and Knittel G.; A refined diagnostic wind field model. *Environ. Software* **3** (1988) 85–94.
11. Anfossi D., Desiato F., Tinarelli G., Brusasca G., Ferrero E., Sacchetti D.: TRANSALP 1989 experimental campaign - Part II: Simulation of a tracer experiment with Lagrangian particle models. *Atmos. Environ.* **32** (1998) 1157–1166.
12. Desiato F., Finardi S., Brusasca G., Morselli M.G.: TRANSALP 1989 experimental campaign - part I: Simulation of 3-D flow with diagnostic wind field models. *Atmos. Environ.* **32** (1998) 1141–1156.
13. Hanna S.R.: Applications in air pollution modeling. In: F.T.Nieuwstadt, H.Van Dop (eds), *Atmospheric Turbulence and Air Pollution Modelling*, Reidel, Dordrecht 1982.
14. Poli A., Cirillo M.: On the use of the normalized mean square error in evaluating dispersion model performance. *Atmos. Environ.* **27A** (1993) 2427–2434.

Chapter 8

Atmospheric Processes

8.1 Transport Processes over Complex Terrain

A.M. Jochum, C. Strodl, H. Willeke, N. Entstrasser and H. Schlager

Institut für Physik der Atmosphäre, DLR Oberpfaffenhofen, Germany

The transport of energy, moisture and ozone in the atmospheric boundary layer over heterogeneous terrain was investigated in the context of the field experiment TRACT. An area of 12 km by 10 km in the upper Rhine valley was selected for this study of small scale inhomogeneities. Three powered gliders performed measurements to investigate the horizontal variability of near surface fluxes of energy, moisture and ozone, to derive vertical profiles of these fluxes, and to study the effects of land-use and local emissions on these fluxes. The extrapolation of the flux profiles gives estimates of the surface fluxes which are compared with data from surface stations. Another aircraft equipped for atmospheric chemistry measurements was flying in the main area of TRACT to study the vertical and horizontal distribution of pollutants on a larger scale.

Aims of the research

The transport, transformation and deposition of pollutants directly influenced by turbulent processes in the atmospheric boundary layer. Therefore, this contribution was aiming at investigating turbulent transport processes over inhomogeneous terrain. The focus is on two different scale ranges: small scale inhomogeneities within the upper Rhine valley were studied in a nested area of 10 km × 10 km using three turbulence aircraft, whereas larger scale effects over the whole TRACT area of several hundreds of kilometres in diameter were addressed using one (of a total of five) atmospheric chemistry aircraft. It was the aim of the measurements to assess the spatial variability of atmospheric parameters and ozone concentration (of further components on the larger scale), and of heat, moisture, and ozone fluxes (on the smaller scale) and to relate it to inhomogeneities of the underlying terrain, land-use and emissions.

Principal scientific results

The observational strategy for the turbulence aircraft (three motor gliders with identical instrumentation) was designed to give information about the horizontal distribution of mean parameters and fluxes across the nested area ('grid flights') and about the mean vertical profiles of the same quantities over the depth of the whole atmospheric boundary layer ('profile flights'). Fig. 8.1.1 shows a schematic representation of both patterns and the location of the nested area in the upper Rhine valley. Due to the limited size of the nested area most of the flight lines were 10 km long. In order to overcome the difficulties associated with inadequate sampling of turbulence events along short lines, the three aircraft were used in parallel. The resulting sampling errors for the fluxes are of the order of 20–40 %. A total of eight flights were performed during three intensive observation periods (IOPs).

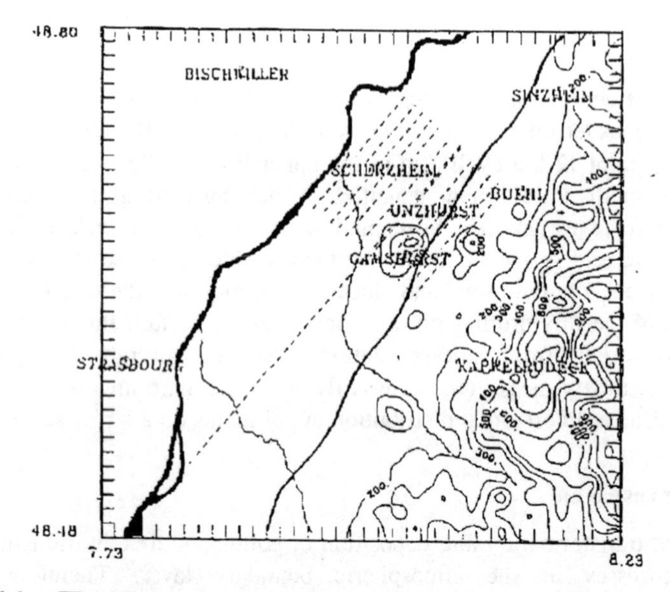

Fig. 8.1.1: TRACT nested area with flight pattern of the motor gliders (dashed lines). The thick line on the left shows the Rhine river, the solid line on the right shows the main highway. X and Y axes in degrees; each tick mark is 1 km..

The grid flights were performed at low level (75–100 m above ground) with the three aircraft flying in parallel at a lateral separation of 800–1000 m. A total of 12 straight lines (legs) were defined to cover the entire nested area. Data along these flight legs were averaged over 500 m to give comparable resolution along and across the individual legs. The resulting distributions show similar features during all IOPs. In order to eliminate random variability the results were averaged over several consecutive mapping patterns. Fig. 8.1.2 shows an example of such a composite pattern from the second IOP. There are gradients of temperature and specific humidity across the valley of 6 K and 2 gkg^{-1}, respectively, indicating the influence of the mountain range. The ozone concentration (Fig. 8.1.2c), however, seems to be fairly homogeneously distributed. The instantaneous covariance of temperature and vertical wind (a measure of sensible heat flux) (Fig. 8.1.2d) does not show any persistent pattern nor any correlation with the underlying terrain.

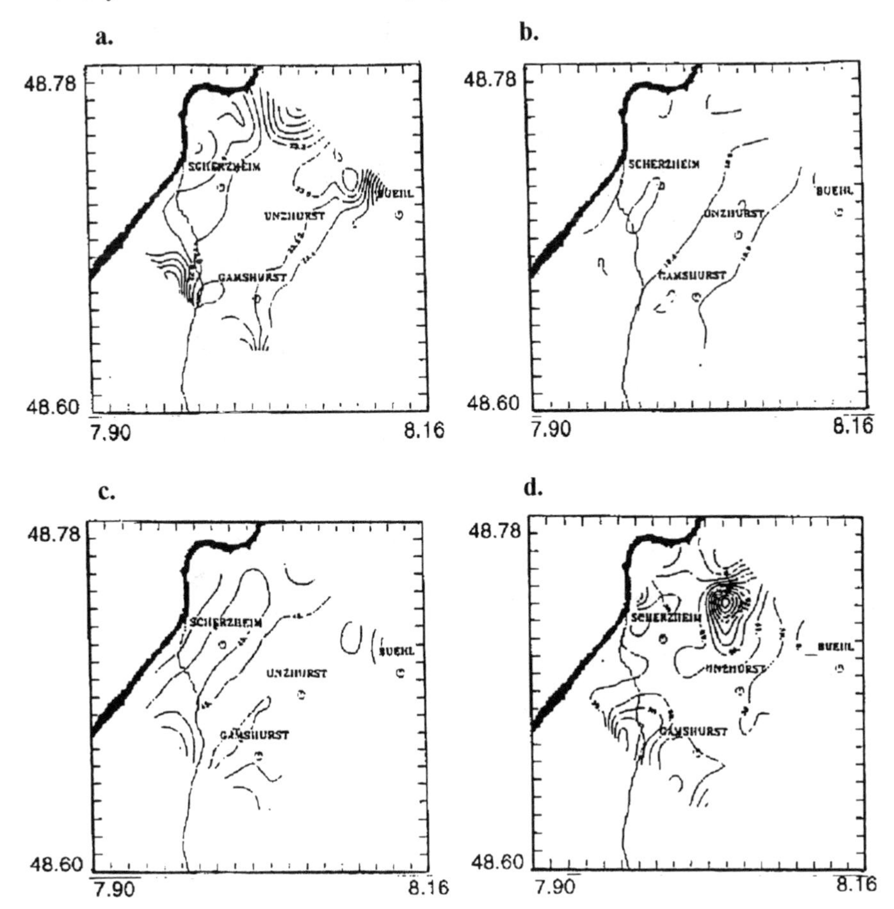

Fig. 8.1.2: Composite distributions of potential temperature (a), specific humidity (b), ozone concentration (c), and an instantaneous covariance of temperature and vertical wind (d) on 17th September 1992. X and Y axes in degrees; each tick mark is 1 km.

In order to investigate the horizontal variability of near-surface fluxes a grid structure of the nested area was defined and pixel values of fluxes were computed from a twofold averaging process: firstly, the time series were averaged over segments of 4 km overlapping by 2 km, and secondly, three laterally adjacent segment mean values were averaged.

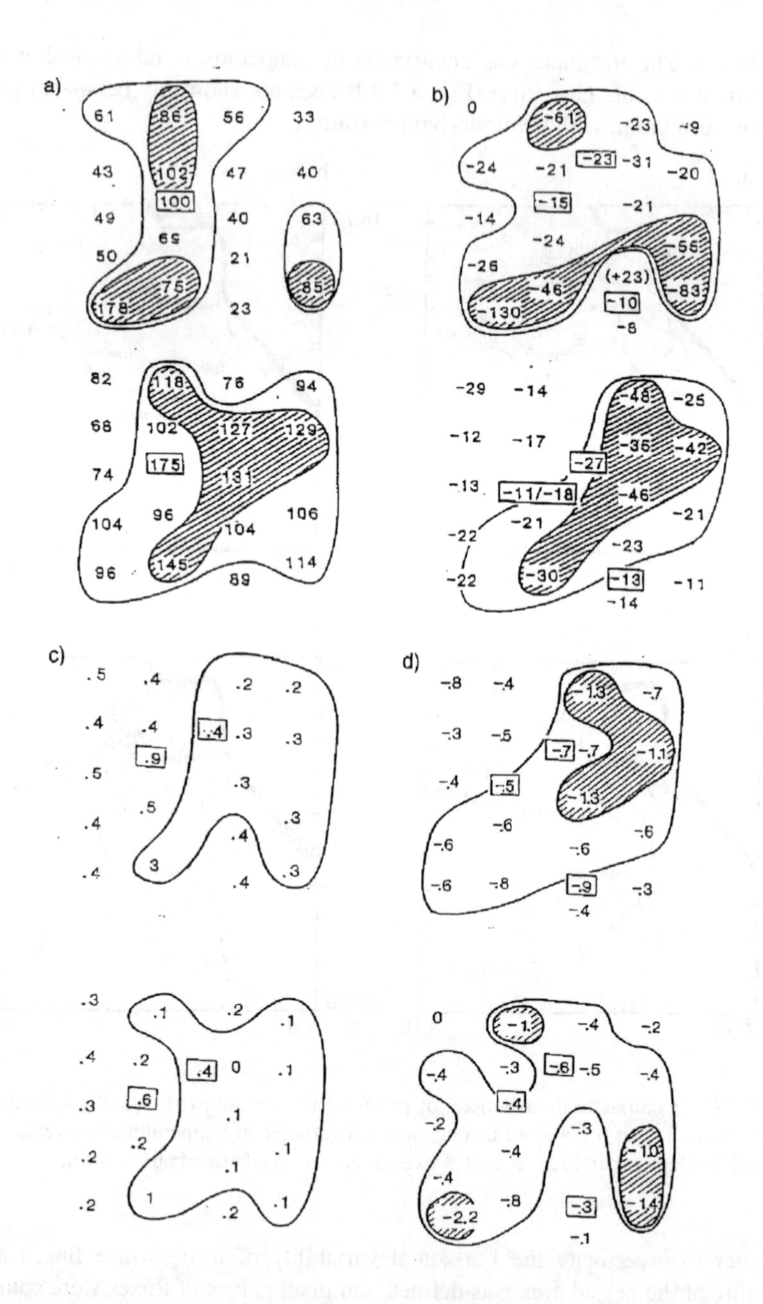

Fig. 8.1.3: Low-level maps of moisture flux in Wm^{-2} (a), ozone flux in ppb ms^{-1} (b), Bowen ratio (c), and deposition velocity in cm s^{-1} (d) on 17th September 1992 at 13:00 (top) and 15:00 local time (bottom). The vertical axis of the plots is oriented along the Rhine valley.

Fig. 8.1.3 shows the resulting flux maps at two different times in the afternoon of 17th September 1992. The structure of the moisture (3a), heat (not shown) and ozone flux (3b) change considerably from 13:00 to 15:00. This also applies to the ozone deposition velocity (3d). The Bowen ratio (3c), however, shows some persistent features, with higher values towards the centre of the valley. The values in the boxes in Fig. 8.1.3 represent data from surface stations (Dlugi, Güsten, and Jensen, personal communication). In general there seems to be fairly good agreement with the values measured from aircraft at 80 m above ground (the numbers give pixel values), possibly indicating that the vertical flux divergence in the surface layer is small in this case.

The profile flights consisted of repeated passes along the same leg in the centre of the grid pattern (on some days extending further south, see Fig. 8.1.1). All three aircraft were flying at 3 to 4 different altitudes ranging from 75 m above ground to 1.2 km (z_i = mixed layer depth). Fig. 8.1.4 gives a typical example. The symbols represent mean values over individual flight legs. The vertical profiles of sensible and latent heat flux are linear and become negative both around 450 m above ground, indicating strong entrainment of warm, moist air. Extrapolation of the regression lines to the ground gives surface fluxes of 65 (163) Wm^{-2} at 12 noon and of 20 (87) Wm^{-2} at 14:00 for sensible (latent) heat, respectively. The low values in the early afternoon are due to the onset of low cloud cover.

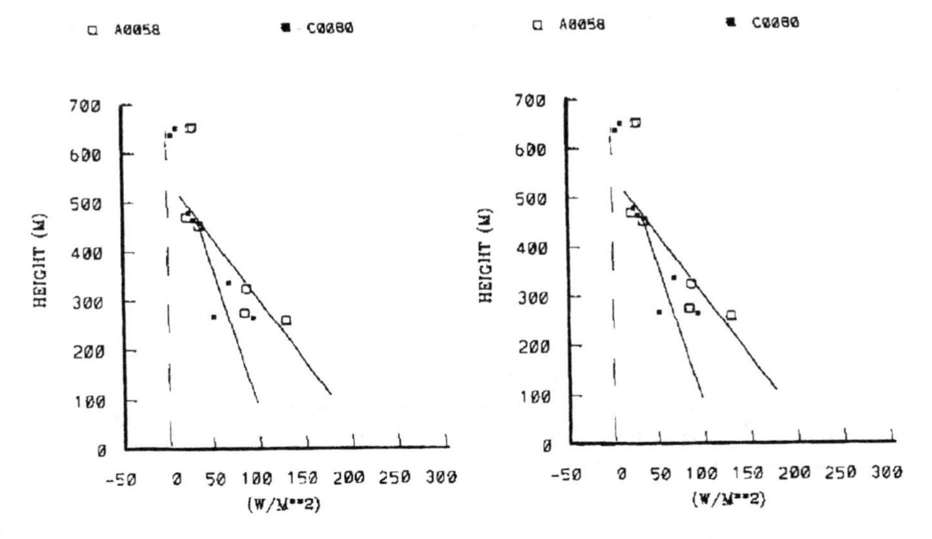

Fig. 8.1.4: Vertical profiles of sensible (a) and latent (b) heat flux on 17th September 1992. The horizontal dashed lines indicate the mixed layer depth at 12:00–14:00 local time.

Data from other IOPs [1] exhibit similar features. Ozone fluxes (not shown here, [1]) are often positive in the upper part of the atmospheric boundary layer, indicating that the ozone-rich reservoir layer above the boundary layer had already disappeared due to entrainment into and photochemical production of ozone within the mixed layer.

The data set of the DO 228-D-CFFU measurements includes one flight during the first intensive measuring period (IMP) and three flights during the second and third IMP. All flights were performed in the east box (first flight) or the south box (all other flights). An inspection of the data reveals that the height of the boundary layer can be easily determined for most of the flight sequences, *e.g.* from the measured relative humidity or concentration of nitrogen oxides. In some regions relatively polluted air masses were found in the lower boundary layer. In general the observed concentrations of pollutants were higher in the western part of the measuring area than in the eastern part.

Conclusions

Gradients of temperature and moisture across the Rhine valley were observed regularly. In contrast, the ozone concentration was distributed more uniformly in most cases. The spatial variations of turbulent fluxes and structures were not found to be systematic and did not show any correlation with the underlying terrain. The vertical profiles of the turbulent fluxes are essentially linear, exhibiting a rather classical shape and showing no evidence of any significant modification by terrain inhomogeneities. The transport of ozone seems to be equally influenced by turbulent and chemical processes. In summary, this suggest that the turbulent transport processes within the nested area in the upper Rhine Valley can well be described by parameterisation methods for homogeneous terrain.

Measurements on the larger scale of the entire TRACT area suggest that there are gradients of pollutant concentration across the whole area.

Acknowledgements

This work was partially funded by BMFT.

References

1. Jochum, A.M., C.S. Strodl, H. Willeke, N. Entstrasser; The transport of heat, moisture and ozone in the atmospheric boundary layer in the Upper Rhine Valley. in: P.M. Borrell, P. Borrell, T. Cvitaš, W. Seiler (eds), *Proc. EUROTRAC Symp. '94,* SPB Academic Publishing bv, The Hague 1994, pp. 766–769.

8.2 Balloon Air Turbulence Measurements at Hegeney

G. Schayes and P. Hakizimfura

Institut d'Astronomie et de Géophysique G.Lemaître,
Université catholique de Louvain, 2, ch. du cyclotron;
B-1348 Louvain-la-Neuve, Belgium

Turbulence measurements made at Hegeney (Alsace, France) have been evaluated in terms of the heat fluxes and other statistical characteristics. The data have also been compared to the Oberbronn soundings and to the surface data of the Hegeney REKLIP station.

Aims of the research

Obtain a view of the boundary layer structure, turbulent fluxes and characteristics in the boundary layer at the west side of the Rhine valley.

Principal scientific results

Introduction

A balloon borne sounding system has been used to measure the air turbulence with a 3 Hz scan speed of 6 fast response sensors. It has been used to acquire more some 20 hours of turbulence data during the SOP II and III campaigns [1, 2].

Comparison with Oberbronn soundings and Hegeney REKLIP data

Our profile data have been compared with the nearby soundings from Oberbronn. For the temperature, the agreement between the two data sets is very good in the upper part of the soundings (up to 500 m max height of our measurements), the difference being generally less than 1 K. A larger difference exists sometimes in the lower part of the temperature profiles, which expectedly reflects more a local effect of the ground surface. For the humidity data, the agreement is not so good, even in the upper part, the Oberbronn data frequently showed a relative humidity 10 to 15 % higher than the Hegeney values, which may be due to instrumental errors. Wind directions are very close. The lower wind speed shows the largest discrepancies but this also reflects the strongest terrain effects. Figs. 8.2.1a and 8.2.1b show examples of profile comparisons.

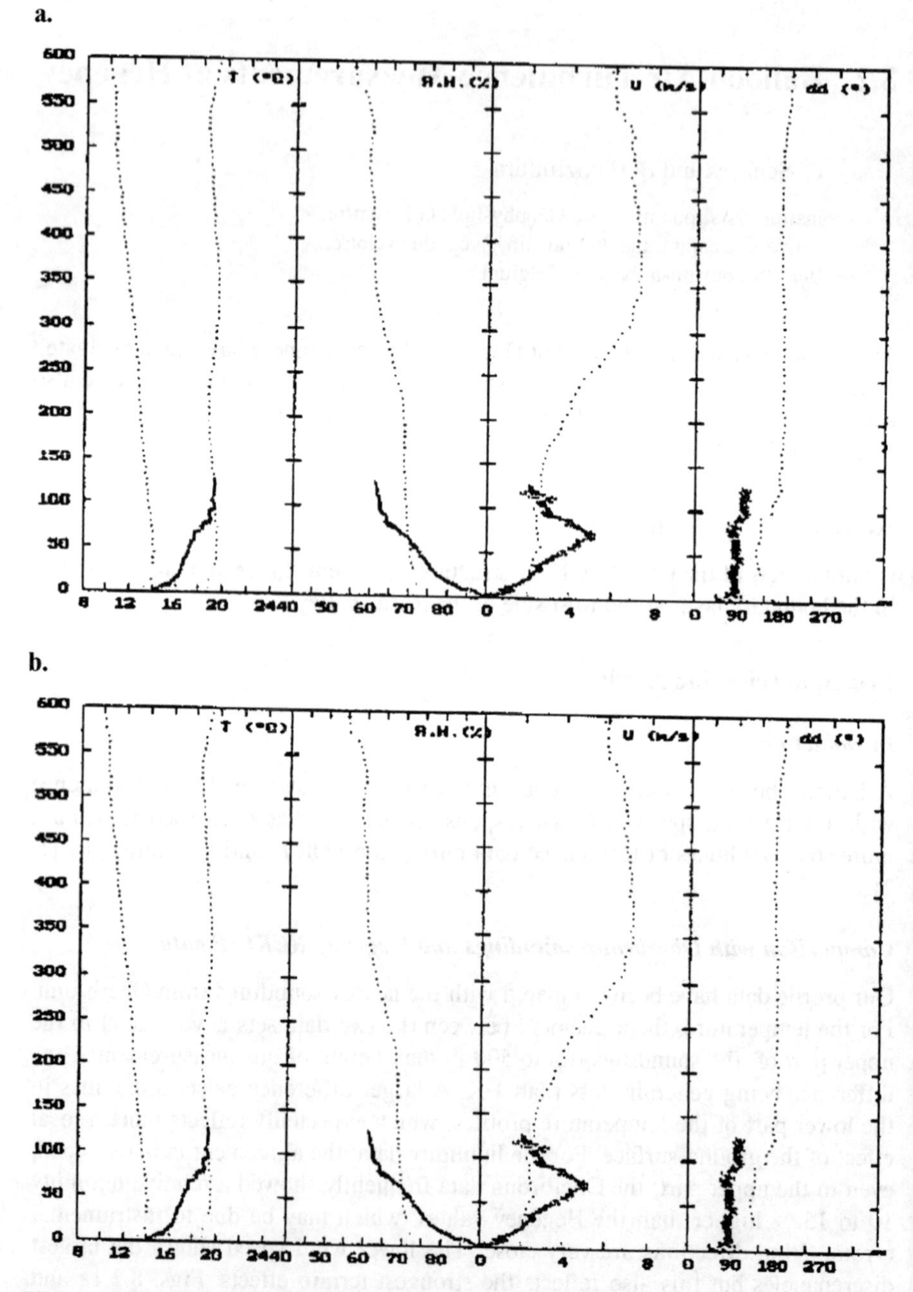

Figs. 8.2.1a and 8.2.1b: Comparison between the tethered balloon soundings at Hegeney (full lines) and the radiosoundings from Oberbronn (dots). From right to left, the graphs show temperature profiles (with wet and dry bulb for Oberbronn), relative humidity, horizontal wind speed and direction. a. September 16th, profile at 590 m; b. September 22nd, profile at 602 m.

As our balloon system was located less than 100 m away from the Hegeney REKLIP station, our lower soundings have been compared with the surface data. Fig. 8.2.2 shows the temperature and the wind speed data for 16th–17th September 1994. The temperature agreement is good (within less than 1 K) for most of the cases. For the wind speed, the comparison shows similar values, but the averaging process in both measurements is too different to allow further comparisons.

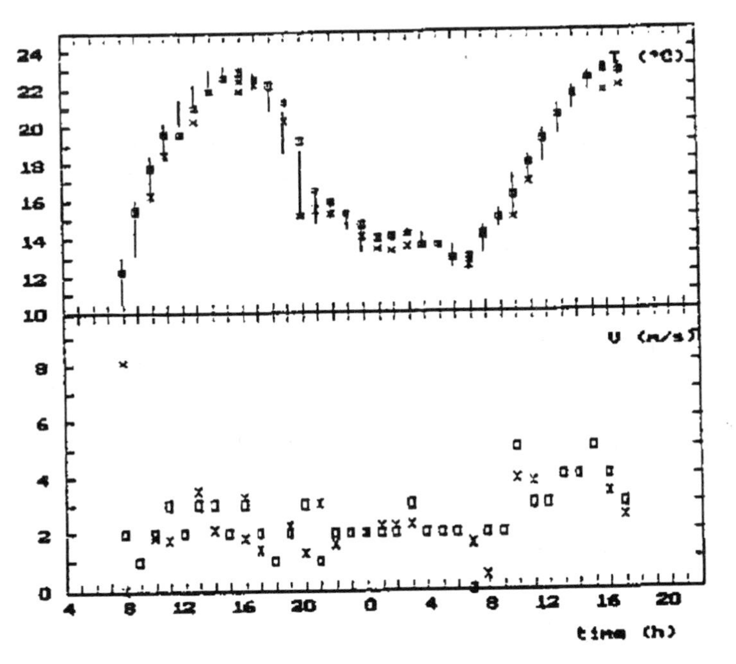

Fig. 8.2.2: Comparison of temperature and horizontal wind speed between the lower soundings and the data from the REKLIP Hegeney station; crosses denote the balloon data. Vertical bars join the min and max data for each hour from the Hegeney station and squares denote their mean values.

In conclusion, our balloon data have been found to be consistent with the other TRACT measurements and may be used for other intercomparisons.

Turbulence statistics and flux data.

The data obtained at a constant level above ground is used to compute turbulence statistics and heat fluxes. The duration of all data sets is close to 30 min. A preliminary detrending of the data had to be done in order to obtain better flux evaluations. Table 8.2.1 presents the results obtained for all data for the mean values and the main turbulence characteristics.

The sigmas represent the standard deviation of temperature, T, specific humidity, q, and the 3 wind components u, v and w in the mean wind reference frame. The quantity u^* is the local friction velocity of the turbulence. Finally, H and E are the sensible and latent heat flux determined by eddy correlation.

Table 8.2.1: Statistical data obtained for all turbulence measurements. Data columns are : actual height above ground, mean values of temperature, specific humidity, wind speed and direction, standard deviation for the same quantities (u, v and w denote the wind components in mean wind), u^* is the local friction velocity, H and E are the sensible and latent heat fluxes.

Data set	Time	Level	T	q	U	Dir	sig. T	sig. q	sig. u	sig. v	sig. w	u^*	H	E
		m	deg C	g/kg	m/s	deg.	deg C	g/kg	m/s	m/s	m/s	m/s	W/m^2	W/m^2
heg 1608p	08:54-09:07	199	8.7	5.0	3.97	259	0.19	0.11	0.23	0.25	0.06	0.06	–1.9	–0.7
heg 1610p	10:30-11: 01	92	15.3	7.7	2.80	250	0.24	0.16	0.62	0.62	0.30	0.30	5.4	49.0
heg 1612p	11:51-12:26	327	15.11	7.9	2.68	286	0.24	0.17	0.88	0.79	0.42	0.19	–35.9	83.6
heg 1613p	13:12–13:40	187	18.3	7.9	3.00	294	0.29	0.16	1.01	0.79	0.51	0.37	–21.6	30.8
heg 1613q	13:45–14:15	95	19.8	8.0	3.45	281	0.30	0.18	0.71	0.68	0.38	0.18	26.1	53.8
heg 1614p	14:35–15:05	42	20.9	8.2	2.91	277	0.40	0.26	1.25	0.84	0.42	0.30	61.4	106.2
heg 1617p	17:00–17:29	79	20.9	8.1	1.93	325	0.31	0.17	0.93	0.55	0.29	0.34	28.5	77.6
heg 1618p	18:01–18:31	305	18.4	8.1	1.27	337	0.26	0.16	0.71	0.60	0.18	0.15	–2.5	14.0
heg 1619p	19:20–19:48	177	18.8	7.6	1.73	342	0.27	0.15	0.12	0.13	0.02	0.04	–0.1	–0.4
heg 1619q	19:52–20:23	91	19.2	7.3	2.55	13	0.28	0.13	1.42	1.05	0.07	0.11	–3.2	–10.4
heg 1621p	20:55–21:29	26	18.8	7.2	3.99	28	0.28	0.12	0.11	0.35	0.04	0.03	1.3	–2.8
heg 1622p	22:18–22:38	162	16.6	7.6	7.35	74	0.28	0.13	0.14	0.46	0.05	0.04	0.0	–0.2
heg 1622q	22:39–22:49	85	16.4	7.6	6.11	66	0.46	0.21	0.33	0.48	0.19	0.17	–1.6	–4.7
heg 1700p	00:29–00:52	35	15.1	7.3	3.68	72	0.30	0.12	0.39	0.41	0.17	0.11	–6.6	–4.3
heg 1701p	01:32–01:55	100	14.6	7.4	6.44	66	0.27	0.13	0.50	0.64	0.29	0.25	–1.5	1.2
heg 1708p	08:09–08:29	84	12.3	6.8	4.46	70	0.22	0.15	0.54	0.59	0.29	0.10	–7.1	–8.5
heg 1708q	08:31–08:54	153	11.9	6.8	6.00	73	0.23	0.12	0.57	0.77	0.40	0.33	–4.2	1.3
heg 1710p	10:20–10:48	93	13.6	6.4	4.14	91	0.21	0.18	0.53	0.55	0.32	0.20	4.0	11.1

Data set	Time	Level	T	q	U	Dir	sig. T	sig. q	sig. *u*	sig. *v*	sig. *w*	*u**	*H*	*E*
		m	deg C	g/kg	m/s	deg.	deg C	g/kg	m/s	m/s	m/s	m/s	W/m^2	W/m^2
heg 1710q	10:51–11:20	195	13.4	7.6	3.24	89	0.22	0.12	0.76	0.50	0.38	0.14	22.5	31.9
heg 1712p	12:05–12:34	461	16.0	7.5	3.51	99	0.31	0.15	0.31	0.31	1.10	0.06	3.1	–7.6
heg 1712q	12:38–12:51	341	14.5	8.2	3.83	96	0.33	0.19	0.89	0.50	0.38	0.28	1.2	16.7
heg 1716p	16:11–16:32	175	19.9	8.4	4.56	83	0.34	0.16	0.88	0.67	0.43	0.38	–9.5	17.1
heg 1716q	16:35–16:57	93	21.0	8.5	4.23	77	0.35	0.17	0.74	0.61	0.35	0.22	18.2	55.2
heg 2111p	11:34–12:02	175	12.2	8.4	1.48	125	0.19	0.14	0.48	0.41	0.19	0.17	2.1	7.3
heg 2111q	12:06–12:36	70	13.6	8.6	1.43	155	0.39	0.15	0.44	0.46	0.19	0.05	5.5	10.0
heg 2113p	13:20–13:50	461	15.6	8.3	5.42	197	0.33	0.12	0.37	0.69	0.13	0.17	14.1	–4.6
heg 2113q	1 3:52–14:22	321	15.0	8.8	2.19	170	0.65	0.21	0.93	0.87	0.28	0.33	42.4	–45.1
heg 2115p	14:55–15:25	45	18.9	9.5	1.68	100	0.30	0.20	0.55	0.40	0.24	0.11	8.6	46.6
heg 2116p	16:00–16:30	197	19.2	9.2	1.69	113	0.29	0.19	0.35	0.41	0.20	0.20	–12.7	9.4
heg 2116q	16:32–17:03	97	20.5	9.5	1.90	103	0.29	0.16	0.43	0.32	0.19	0.12	0.9	22.4
heg 2117p	17:46–18:13	282	19.3	8.6	3.54	95	0.30	0.15	0.53	0.38	0.17	0.22	2.6	–13.7
heg 2120p	20:39–20:52	33	19.3	8.6	3.15	68	0.47	0.20	0.45	0.38	0.16	0.12	–7.4	–13.4
heg 2122p	22:32–23:02	26	18.5	8.9	2.14	80	0.34	0.21	0.43	0.26	0.08	0.09	–5.2	–9.3
heg 2124p	23:56–00:26	26	19.0	8.9	2.03	68	0.38	0.14	0.21	0.26	0.04	0.03	2.2	–2.4
heg 2201p	01:41–02:11	75	19.6	8.6	2.26	96	0.28	0.12	0.36	0.30	0.09	0.02	–0.8	–1.0
heg 1710p	10:20–10:48	93	13.6–	6.4	4.14	91	0.21	0.18	0.53	0.55	0.32	0.20	4.0	11.1
heg 1710q	10:51–11:20	195	13.4	7.6	3.24	89	0.22	0.12	0.76	0.50	0.38	0.14	22.5	31.9
heg 1712p	12:05–12:34	461	16.0	7.5	3.51	99	0.31	0.15	0.31	0.31	0.10	0.06	3.1	–7.6
heg 1712q	12:38–12:51	341	14.5	8.2	3.83	96	0.33	0.19	0.89	0.50	0.38	0.28	1.2	16.7

Data set	Time	Level	T	q	U	Dir	sig. T	sig. q	sig. *u*	sig. *v*	sig. *w*	*u**	*H*	*E*
		m	deg C	g/kg	m/s	deg.	deg C	g/kg	m/s	m/s	m/s	m/s	W/m^2	W/m^2
heg 1716p	16:11–16:32	175	19.9	8.4	4.56	83	0.34	0.16	0.88	0.67	0.43	0.38	–9.5	17.1
heg 1716q	16:35–16:57	93	21.0	8.5	4.23	77	0.35	0.17	0.74	0.61	0.35	0.22	18.2	55.2
heg 2111p	11:34–12:02	175	12.2	8.4	1.48	125	0.19	0.14	0.48	0.41	0.19	0.17	2.11	7.3
heg 2111q	12:06–12:36	70	13.6	8.6	1.43	155	0.39	0.15	0.44	0.46	0.19	0.05	5.5	10.0
heg 2113p	13:20–13:50	446	15.6	8.3	5.42	197	0.33	0.12	0.37	0.69	0.13	0.17	14.1	–4.6
heg 2113q	13:52–14:22	321	15.0	8.8	2.19	170	0.65	0.21	0.93	0.87	0.28	0.33	42.4	–45.1
heg 2115p	14:55–15:25	45	18.9	9.5	1.68	100	0.30	0.20	0.55	0.40	0.24	0.11	8.6	46.6
heg 2116p	16:00–16:30	197	19.2	9.2	1.69	113	0.29	0.19	0.35	0.41	0.20	0.20	–12.7	9.4
heg 2116q	16:32–17:03	97	20.5	9.5	1.90	103	0.29	0.16	0.43	0.32	0.19	0.12	0.9	22.4
heg 2117p	17:46–18:13	282	19.3	8.6	3.54	95	0.30	0.15	0.53	0.38	0.17	0.22	2.6	–13.7
heg 2120p	20:39–20:52	33	19.3	8.6	3.15	68	0.47	0.20	0.45	0.38	0.16	0.12	–7.4	–13.4
heg 2122p	22:32–23:02	26	18.5	8.9	2.14	80	0.34	0.21	0.43	0.26	0.08	0.09	–5.21	–9.3
heg 2124p	23:56–00:20	26	19.0	8.9	2.03	68	0.38	0.14	0.21	0.26	0.04	0.03	2.2	–2.4
heg 2201p	01:41–02:11	75	19.6	8.6	2.26	96	0.28	0.12	0.36	0.30	0.09	0.02	–0.8	–1.0
heg 2202p	02:45–03:15	135	19.3	8.5	2.86	101	0.28	0.12	0.42	0.41	0.08	0.06	–0.3	–2.2
heg 2204p	04:00–04:31	38	16.4	8.6	2.52	83	0.33	0.13	0.29	0.33	0.04	0.07	0.2	–0.7
heg 2209p	09:37–10:08	155	18.1	7.8	4.48	68	0.44	0.13	0.44	0.42	0.15	0.16	40.3	–19.5
heg 2209q	10:11–10:40	69	16.8	8.9	1.90	54	0.36	0.27	0.61	0.51	0.13	0.08	–5.5	11.1
heg 2211p	11:20–11:44	296	18.3	7.8	2.98	59	0.35	0.17	0.45	0.34	0.14	0.23	–17.81	23.1
heg 2212p	11:44–12:05	296	17.2	9 1	3.84	284	0.38	0.39	0.67	0.92	0.13	0.28	12.4	–87.1
heg 2212p	12:58–13:28	176	20.5	10.0	0.45	491	0.41	0.31	0.85	0.58	0.27	0.23	1.9	22.3

Note the surprising negative fluxes sometimes measured around noon at heights above 200 m (see Fig. 8.2.3). Soundings showed surprising low altitudes for the boundary layer height. So these data were taken close to the BL top where fluxes become small or even negative. Note that a correct interpretation will need additional soundings or aircraft data.

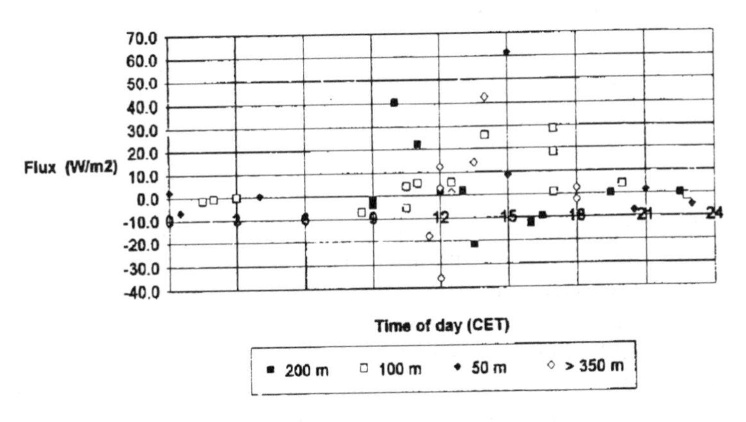

Fig. 8.2.3: Sensible heat fluxes measured at various heights during the SOP II and III campaigns.

Conclusions

The assessment of the measurements has been done and the data set may now be considered as valid. Our data must still be compared with those taken by other teams, *e.g.* the Do-128 (D-IBUF) aircraft data.

Acknowledgements

The continuous support of the University of Louvain-la-Neuve and the Institut d'Astronomie et de Géophysique has been greatly appreciated.

References

1. Schayes, G., P. Hakizimfura; Measurement of atmospheric boundary layer turbulence with a captive balloon during the TRACT experiment (preliminary report). Progress Report 1993/1, Institut d'Astronomie et de Géophysique G. Lemaître, UCL, Louvain-la-Neuve 1993.
2. Schayes, G., P. Hakizimfura; Balloon air turbulence measurements at Hegeney. *EUROTRAC Annual Report 1993, part 10, TRACT,* EUROTRAC ISS, Garmisch-Partenkirchen 1994, pp.18–23.

8.3 Surface Flux Variability in Relation to the Mesoscale

Niels Otto Jensen[1], Charlotte Bay Hasager[1], Poul Hummelshøj[1], Kim Pilegaard[2] and Rebecca Barthelmie[1]

[1]Meteorology and Wind Energy Department, Risø National Laboratory, DK-4000 Roskilde, Denmark

[2]Environmental Science and Technology Department, Risø National Laboratory, DK-4000 Roskilde, Denmark

This contribution to EUROTRAC deals with the variation that takes place in the surface fluxes (for example in the dry deposition of the various air pollutants) when the boundary layer flow encounters changes in the land-use type. The project has had three components: participation in the TRACT field campaign with ground based flux measurements; development of a model for integration of the flux variation over the variety of surface types in a given land-use pattern from scales of 25 m up to 10 km or more (aggregation model) and to apply it to the TRACT area; and to develop an interpretation techniques that will allow remote sensing (satellite) products to be used for estimating the necessary boundary conditions for the aggregation model. The following gives the background for the work and an overview of the results obtained.

Aims of the research

The background for this work is that large scale models are very much needed for rational policy decisions. The subject could be the effect of local cut backs in certain emissions to the atmosphere. Such a discussion is only meaningful if long range transport of pollutants is taken into account. Due to a relatively limited computer capacity such large scale models have to be fed with parameterised forms of the boundary conditions, because the resolution is typically no better than tens of kilometres. The wealth of variability in the surface conditions below this scale must somehow be represented with a single value (for each parameter) for each of these grid cells.

Due to the non-linear nature of the relevant physics it is not correct to obtain this by any type of simple averaging procedure, especially not of the surface roughness. The problem is well known. The more frequent the surface change the worse the problem is. The air has no possibility of "forgetting the past" and adjust to the local conditions. The method developed here tries to go a step forward in objectivity as compared with currently used methods (for example the method of the blending height suggested by Mason [1]).

Principal scientific results

During the TRACT field campaign, Risø operated one continuous flux station near Scherzheim in the so-called nested area, plus one extra station exposed at different places in the neighbourhood in order to check the representativeness of the fluxes. Measurements of fluxes of momentum, heat and water vapour, as well as NO_2 and O_3 were taken along with other micrometeorological variables. The chemical flux measurements were formally a BIATEX contribution to TRACT [2]. During most periods of the so-called intensive measurement phases the fetch conditions were less than ideal and the measured momentum fluxes were thus affected by local conditions. However, the conclusion so far is that it is possible to determine meaningful fluxes of scalar quantities also during times with less ideal wind conditions [3]. The experimental setup and some flux results are given in Pilegaard *et al.* [4].

In addition, Risø operated a tethered balloon at Hornenberg, measuring pressure (height), wind speed and direction, dry and wet bulb temperature (air temperature and humidity), in regular intervals up to 600 m above the local terrain. The balloon ascents were synchronised with ascents at other sites in the area and contributed to the understanding of the mesoscale flow patterns during the measurement periods. This part of the work is not reported here.

Surface layer data measured during the TRACT campaign, by Risø as well as all other groups that did such observations (21 stations), have been collected in one data base. The data base includes detailed descriptions of the various sites. This work is described in Barthelmie and Jensen [5, 6]. The data can be accessed via ftp. Please email n.o.jensen@risoe.dk for further information.

The basic ideas or principles of the aggregation model were presented by Jensen [7]. It is of the same type as the model by Walmsley *et al.* [8] but simpler. A more formal description is given in Hasager and Jensen [9]. The model is based on a linearised and much simplified version of the equation of motion that can be solved by Fourier techniques. This allows a very large number of cases to be calculated with reasonable use of computer time. An analytical solution for a periodically varying surface roughness has been developed. This serves as a test case for the next step, a numerical (Fast Fourier Transform) solution for the same case. Another basic test is to compare the model to the literature examples of the classical 1-D abrupt roughness change (see for example Jensen [10]). Examples are given in Astrup *et al.* [11]. The final step is extension to 2-D. Fig. 8.3.1 shows some results from the model calculations for an artificial 2-D land-use pattern.

This further allows a comparison with fields of data measured in "real" terrain. For this, the model needs an input of the actual surface conditions (roughness map). The rational way of obtaining this is through remote sensing. The use of satellite images for outlining the land-use pattern is a unique opportunity that cannot be matched by using "normal" maps.

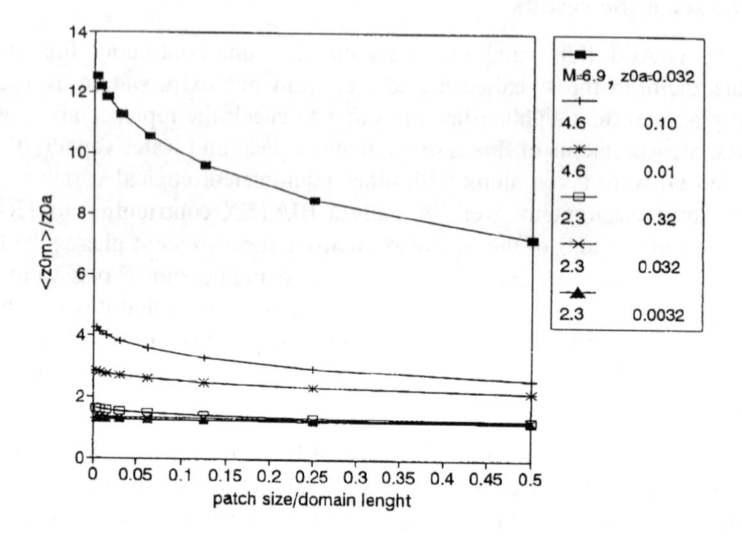

Fig. 8.3.1: The response of the aggregation model to a simplified land-use pattern consisting of a checkerboard of two different surface roughnesses. z_{0a} is the geometric mean of the two roughnesses, while M is the logarithm of the ratio of the two roughnesses (the magnitude of the variations). The ordinate shows the effective roughness $<z_{0m}>$ relative to the average roughness. Its excess value is seen to be quite sensitive to M, which means that if there is a large difference between the two roughnesses, the model weighs the rougher parts the most. The dependence on z_{0a} is much less. The abcissa shows the dependence of the excess value on the relative patch sice of the checkerboard. The more frequent the surface conditions change the larger is the excess value everything else being equal.

One of the classical applications of remote sensing is to calculate surface fluxes by various approximate methods. In the present context of flux aggregation, the main need is not so much the ability to estimate fluxes directly from remote sensing but rather to be able to precisely outline the land-use pattern and be able to estimate the aerodynamic roughness. The Landsat TM (optical) is suitable for that since it has a sufficient spatial resolution (30 m × 30 m). The disadvantage being that cloudless scenes of a particular land area are relatively rare. In theory the coverage is once per 16 days, but since the areas are often covered by clouds the images are often of no use. It is therefore attractive to use the ERS-1 SAR (radar) satellite with all-weather capability and a comparable resolution.

Satellite images in digital form have been retrieved from the upper Rhine valley covering part of Germany/Switzerland/France. Three Landsat TM scenes from 1993 are geometrically corrected and classified into 20 land-cover classes by Konrad *et al.* [12]. The classified image form the basis for empirical roughness length mapping on a spatial resolution of 30 m × 30 m. With this as a boundary condition we have calculated the map of surface stresses shown in Fig. 8.3.2.

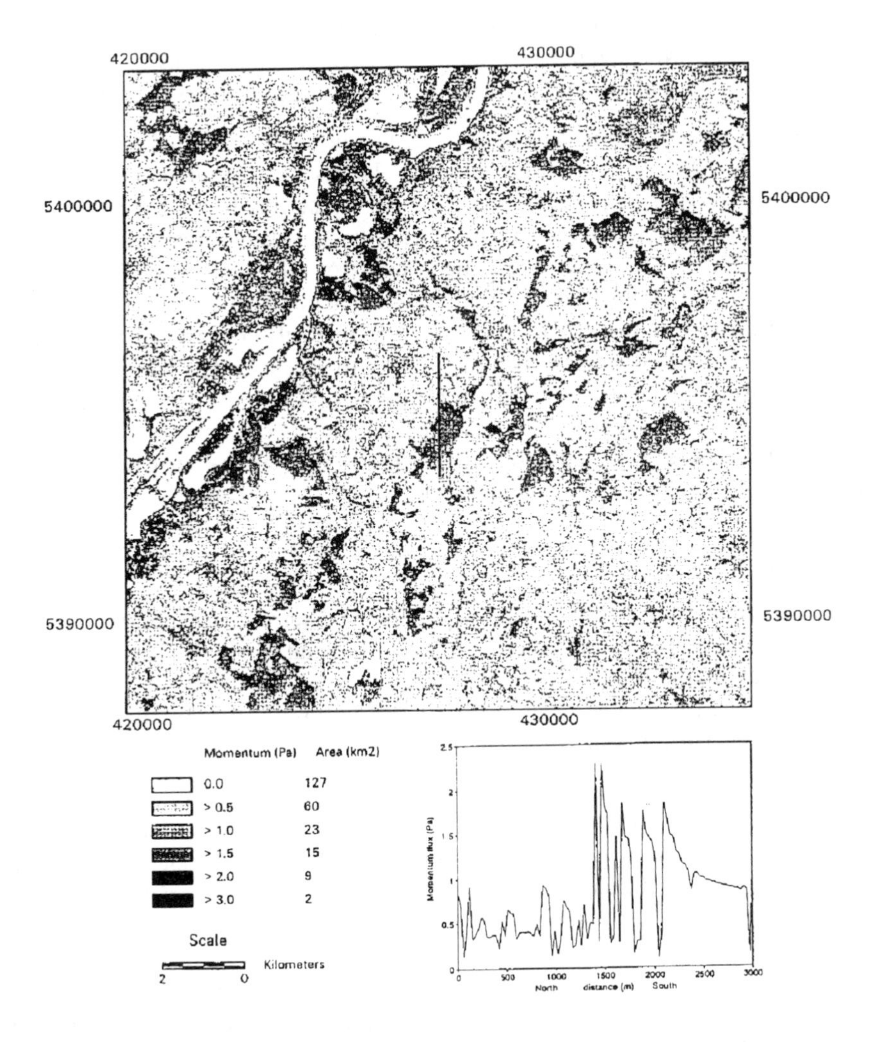

Fig. 8.3.2: The map shows the calculated distribution of momentum flux for a 15 km × 15 km area in the Rhine valley as a response to a 5 m/s wind from the north (at the height of 4 m). Our station, Scherzheim, is in the middle of the map. The characteristic bend in the river is easily identified on a regular map. The legend states the flux levels and the areas in km^2 of the particular levels. The value of z_{0a} for the terrain is 0.13 m while $\langle z_{0m} \rangle$ is 0.29 m according to this model (see caption of Fig. 8.3.1 for definitions). The resulting average surface stress over this area is 66 % higher than one would anticipate from an equilibrium calculation. This is due to the dynamic flow effects with undershoots in rough to smooth transitions and overshoots in smooth to rough transitions. The response is visualised in the insert below the map. The modelled fluxes are shown along the line indicated on the map. The first part of the traverse (from north to south) is above mixed agricultural land. Then at about 1400m the forest edge is encountered. The forest has several clearings within the first 600 m. In the remaining part the flux relaxes over a relatively homogeneous forest with a gradual decline towards equilibrium conditions corresponding to extensive forest.

The average of these stresses is 66 % higher than that one would get from a terrain with an even roughness equal to the area weighted sums of the various patches (on a logarithmic basis, otherwise the excess would be even higher) [13].

ERS-1 SAR images from four dates have also been retrieved and geometrically corrected image-to-image to the Landsat TM scenes mentioned above. The ERS-1 SAR processing analysis focused on direct roughness estimation. The results so far are not promising. Therefore it is considered of greater interest to use classified SAR data. Classified data of the so-called "Radarkarte Deutschland" by Markwitz *et al.* [14] are used as input to a roughness map for 1992 with a spatial resolution of 25 m $\times$ 25 m. Real terrain results of friction velocity and stresses compare within $\pm$ 50 % to the field data from September 1992 measured by eddy correlation techniques [13] and also compare well spatially for the two years. Work on direct roughness estimation based on fully polarised airborne SAR data from Denmark is in progress [15] but far from finished.

Conclusions

The main conclusion so far is that it is possible to determine meaningful fluxes of scalar quantities also during times with less ideal wind conditions. This statement is supported by the observations of a relatively small scatter in for example the canopy resistance $r_c(O_3)$ under different conditions. The point needs further investigation. The analysis of the TRACT data is still in progress.

So far the development of the aggregation model has concentrated on the momentum flux. Aggregation of sensible heat flux, evaporation, evapo-transpiration as well as other scalar fluxes are in principle analogous. However, there are additional difficulties in obtaining the surface values. Another difficulty is that the spatial resolution of relevant satellite images is too coarse. Thus NOAA AVHRR and ERS-ATSR digital satellite images for surface radiation temperature, T_s, are on a spatial resolution of 1 km $\times$ 1 km. The Russian satellite Resurs-O1 gives 600 m $\times$ 600 m pixels. Only Landsat TM of 120 m $\times$ 120 m is adequate but only infrequently available. Resolution in T_s remains a problem.

Some basic work has been done on the correction to T_s to a temperature that can be used in equations (bulk formulations) for calculating the sensible heat flux. We are pursuing this problem beyond the end of this project.

Another piece of work that has been carried out is the analysis of the motor-glider data from the TRACT campaign. It was hoped that they would provide validation for the flux aggregation model described here. However, it was found that the motor-glider flux results were not significant statistically due to an insufficient number of repeat-flight tracks. Another upcoming possibility is the NCAR-Electra flights during the recent BOREAS experiment.

The method of dealing with the "up-scaling" problem as suggested here can also be used for the opposite problem, the "down-scaling". By using the exact same aggregation model, but in reverse, it is possible to calculate the deposition of a given compound to a specific ecosystem somewhere in a grid cell of, say, the European Monitoring and Evaluation Project (EMEP) model. This will allow a much more accurate estimate on whether the critical load for that ecosystem is exceeded or not. This application has not yet been made.

Acknowledgements

In addition to direct funding from Risø National Laboratory, the work was supported by the National Agency of Industry and Trade; the Danish Strategic Environmental Research Programme; and the Commission of the European Communities (DGXII, EV5V-CT93-0276).

References

1. Mason, P.J.; The formation of areally-averaged roughness lengths. *Quart. J. Roy. Met. Soc.* **114** (1988) 399–420.
2. Pilegaard, K., P. Hummelshøj, N.O. Jensen, L. Kristensen; Measurement of fluxes of particles, ozone, nitrogen oxides and carbon dioxide to coniferous and decidous forests, in: S. Slanina (ed), *Biosphere-Atmosphere Exchange of Pollutants and Trace Substances*, Springer Verlag, Heidelberg 1996, pp. 391–396.
3. Hummelshøj, P., K. Pilegaard, N.O. Jensen; Flux measurements of O_3 and NO_2 over a regrown wheat field. in: P.M. Borrell, P. Borrell, T. Cvitaš, W. Seiler (eds), *Proc. EUROTRAC Symp. '94*, SPB Academic Publishing bv, The Hague 1994, pp. 544-548.
4. Pilegaard, K., P. Hummelshoj, N.O. Jensen; Fluxes of ozone and nitrogen dioxide measured by eddy correlation over a harvested wheat field, *Atmos. Environ.* **32** (1998) 1167–1177.
5. Barthelmie, R.J., N.O. Jensen; TRACT Database No. 4: surface measurements, station and data description. Risø-R-816, 1995, 54pp.
6. Barthelmie, R.J., N.O. Jensen; Comparison of surface meteorological measurements in TRACT Database 4. *Annales Geophysicae* **14** (1996) 574-583.
7. Jensen, N.O.; Flux-aggregation modelling, *European Geophys. Soc.* XIX General Assembly, Grenoble, France 1994.
8. Walmsley, J.L., P.A. Taylor, T. Keith; A simple model of neutrally stratified boundary-layer flow over complex terrain with surface roughness modulations (MS3DJH/3R). *Boundary-Layer Meteorol.* **36** (1986) 157-186.
9. Hasager, C.B., N.O. Jensen; Surface flux-aggregation in heterogeneous terrain, *Quart. J. Roy. Meteorol. Soc.* **125** (1999) 2075–2102.
10. Jensen, N.O.; Change of surface roughness and the planetary boundary layer. *Quart. J.R. Met. Soc.* **104** (1976) 351-356.
11. Astrup, P., N.O. Jensen, T. Mikkelsen; Surface roughness model for LINCOM. Risø-R-900, 1996, 30pp.
12. Konrad C. K. Segl, J. Wiesel; Multitemporal land-use classification of the Upper Rhine Valley and adjacent areas with Landsat TM image data. in: F.Fiedler (ed), *Heat, Moisture and Mass Exchange Processes on a Regional Scale in a Non Homogeneous Terrain, Final report (Contract No. EV5V-CT93-0276),* Institut für Meteorologie, University of Karlsruhe 1994, a 1–13.

13. Hasager, C.B.; Surface fluxes in heterogeneous landscape. Ph. D. thesis, Risø R-922, 1996, 177pp.

14. Markwitz W., T. Winter, D. Kosmann, M. Wiggenhagen, W. Hagg, M. Sties; Radarkarte Deutschland. *Zeitschrift für Photogrammetrie und Fernerkundung* **4** (1995) 150-159.

15. Søgaard, H., S.N. Madsen, C.B. Hasager; Estimation of aerodynamic roughness and soil moisture in agricultural areas using SAR data.Final Report to ESA, contract AO2.DK104, Institute of Geography, University of Copenhagen 1998, 6pp.

8.4 Investigation of Mixing Processes in the Lower Troposphere over Orographically Structured Terrain

R. Vögtlin, M. Koßmann, H.-J. Binder, F. Fiedler, N. Kalthoff, U. Corsmeier and H. Zimmermann

Institut für Meteorologie und Klimaforschung, Universität Karlsruhe / Forschungszentrum Karlsruhe GmbH, Postfach 6980, 76128 Karlsruhe, Germany

Meteorological and chemical measurements on the β - γ mesoscale, made during the TRACT campaign, show a strong influence of the orography on the vertical mixing and horizontal transport of air masses in the lower troposphere under convective conditions.

The exchange of trace gases between the planetary boundary layer and the free troposphere (handover) over hilly terrain is caused substantially by steps in boundary layer height and by gaps in boundary layer inversion. Over rather large mountain crests convective bursts of trace gases can be observed through inversion gaps contributing to enhanced vertical entrainment of trace gases into the free troposphere. Handover of air pollutants by the mean horizontal wind can also be accomplished at orographically induced inversion steps. A comparison of the transport efficiency of both processes mentioned (gaps and steps) indicates a clear preponderance by inversion steps.

Prognostic equations of the boundary layer depth for homogeneous terrain over mountain ridges lead to an overestimation compared to observations. Valuations of parameterisations of boundary layer development are addressed which can lead to a better description of the boundary layer depth over orographically structured terrain. It is shown that in hilly terrain secondary circulation systems have to be taken into account which may cause a reduction of the boundary layer depth by generating an inversion at the top of the slope wind layer.

Aims of the scientific research during the project

The relevant processes determining the structure of the boundary layer over complex terrain (that means variation of orography and land use) and the simultaneously occurring chemical transformations are not sufficiently known to allow the transport and diffusion of air pollutants over orographically structured terrain to be identified and quantified. Especially

* the transport of trace gases from the boundary layer into the free troposphere or vice versa at boundary layer steps induced by changing terrain elevation,

* the mass transport efficiency of convective cells to handover processes at boundary layer gaps, and
* the description of the daily variation of the mixing layer depth over complex terrain

have to be investigated. Therefore, in September 1992, the TRACT (Transport of Air Pollutants over Complex Terrain) field campaign was conducted in southwestern Germany, eastern France and the northern part of Switzerland [1]. Within this main area the transition zone between the Rhine valley and the northern part of the Black Forest was chosen for detailed investigations with higher resolution in space and time.

All measured data have been filed in different data bases by types (*i.e.* aerological data, aircraft data, ground based measurement data) to have them readily available. In a further step the different data bases will be merged to one common data base.

Principal experimental findings during the project

In the TRACT field campaign three special observation periods (SOPs) with durations of 25, 37, and 31 hours were accomplished [2]. Because of the scale of the processes of interest a lot of measurement systems were necessary to get sufficient spatial resolution: 10 aircraft, 18 radiosonde stations, 7 tethered balloon stations, 6 sodar stations, 2 RASS and 1 lidar station, and 25 ground based meteorological and chemical stations were operated. In addition, data are available from about 350 (operationally working) network stations (meteorological and/or chemical measurements). To achieve good comparability of the measurements and to get a data set of well known accuracy, a comprehensive quality assurance/quality control (QA/QC) programme was designed and executed for the TRACT research aircraft [3, 4] as well as for radiosonde systems [5] and for ground based meteorological and chemical measurements [6, 7].

The main area

Since turbulent diffusion of air pollutants is most effective in the atmospheric boundary layer, the boundary layer depth z_i is one key parameter for the mixing ratios of trace gases in the troposphere. In addition, knowledge of z_i is of fundamental importance to investigating the handover efficiency of air pollutants between the boundary layer and the free troposphere caused by boundary layer steps and gaps. The boundary layer inversion constitutes the upper boundary condition (sink or source) for budget studies of air pollutants in the boundary layer. Handover processes also determine the amount of pollutants taking part in long-range transport and, hence, contribute to the background concentrations on which photochemical episodes are based [8].

Under convective conditions, the boundary layer depth z_i is normally given by the base height of the lowest inversion above ground level (a.g.l.). The boundary layer height $h = z_i + z_s$ is given by the base height of the lowest inversion above sea level (a.s.l.) with z_s representing the terrain height above sea level.

In the afternoon of September 16th, 1992 a nearly terrain-following capping inversion was observed in the (southern) TRACT area (Fig. 8.4.1). But a closer look reveals that the differences with respect to a boundary layer of constant thickness are significant. Low-turbulent heat fluxes over the water surface of Lake Constance and compensating subsidence caused by the divergence in the horizontal wind field due to lake breezes and slope winds [9] result in a thin boundary layer of only 300 m above the water surface. East of the Black Forest (at the end of the flight pattern) the highest values of z_i of about 1400 m a.g.l. were detected. Furthermore, above some mountain crests like the Black Forest the boundary layer was markedly thinner than in the surroundings, while above other crests this phenomenon was not observed. This behaviour of the boundary layer depth was also proved by analysis of data obtained by vertical sounding systems [10]. At the Jura the horizontal wind vector of 3–4 m s^{-1} was nearly perpendicular to the mountain crest. Therefore, the boundary layer inversion was displaced to the leeward side (see Fig. 8.4.1 to the left part of the Jura). This can also be seen in the up-current plumes of water vapour and NO_2.

Above all rather large mountain crests these bursts of water vapour, NO_2 and O_3 can be observed which penetrate up to heights of about 1800 m a.s.l., due to upslope movement of air from the valley bottoms induced by solar heating [11, 12]. So, through orographically induced effects air masses (pollutants) are brought up to much greater heights compared to the valley regions. It can be expected that during the following night, when the boundary layer height drops, parts of these bursts are handed over to the free troposphere which increases the possibility of the pollutants to be transported over great distances. The distribution of O_3 exhibits a very complex behaviour on that day (Fig. 8.4.1). Higher O_3 concentrations in the boundary layer and in the upper part (1500 m–2000 m a.s.l.) of the interpolated cross-section were separated by a local minimum in O_3 concentrations between 1100 m and 1500 m a.s.l. Over the crests the O_3 bursts connect the two layers with higher O_3 concentrations.

During two of the three special observation periods fog was generated in many parts of the TRACT area in the early morning hours. Depending on the orography, the thickness and the duration of the fog were very different. This affects markedly the daily cycle of O_3 which is illustrated by data from the two stations, Karlsruhe Nordwest and Kehl Hafen, in the Rhine valley (Fig. 8.4.2).

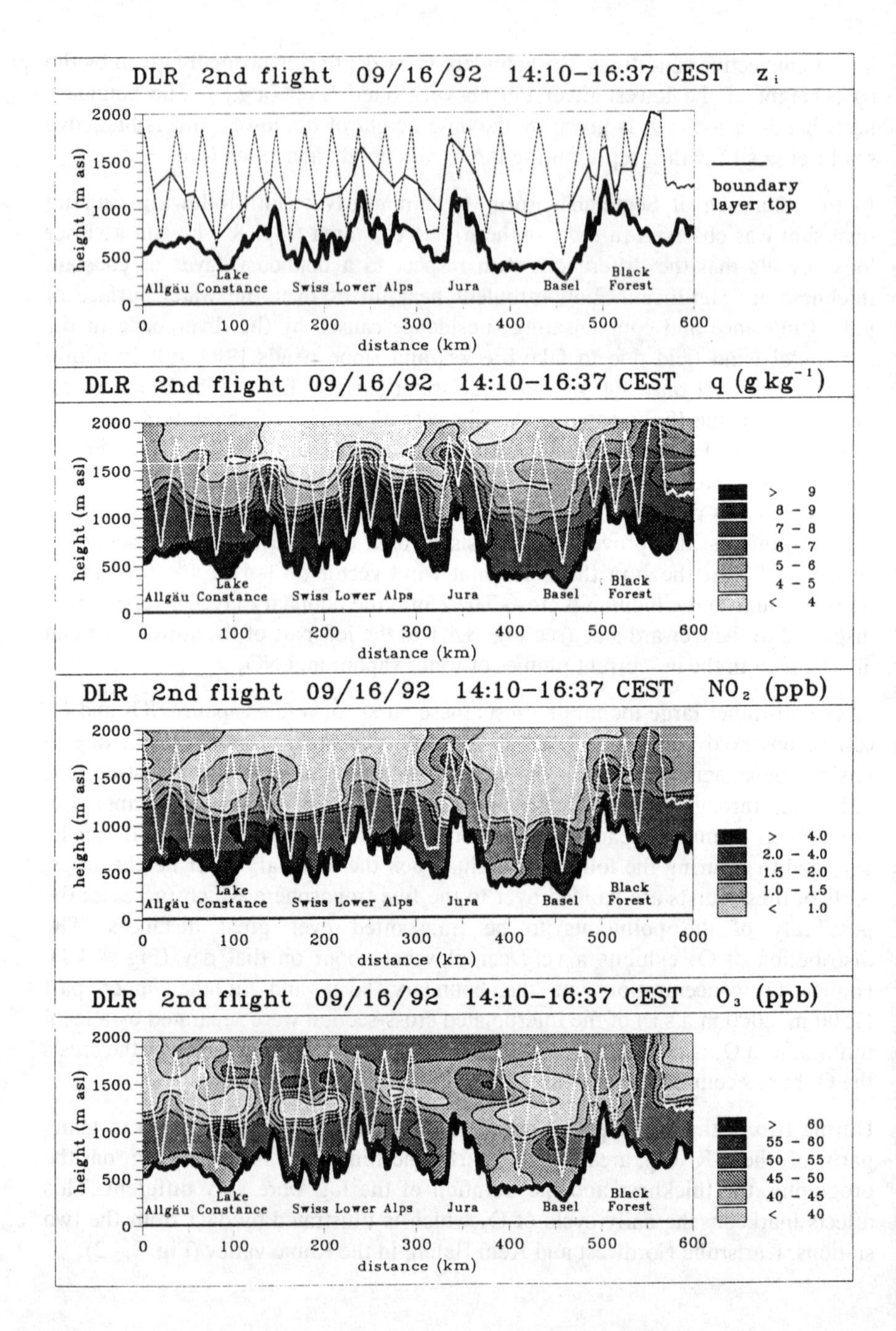

Fig. 8.4.1: Height of the boundary layer top and interpolated cross-sections of specific humidity *q*, NO_2 and O_3, measured by the DLR aircraft in the southern TRACT area in the afternoon of September 16th, 1992. The zig-zag line indicates the flight pattern and the thick solid line the orography.

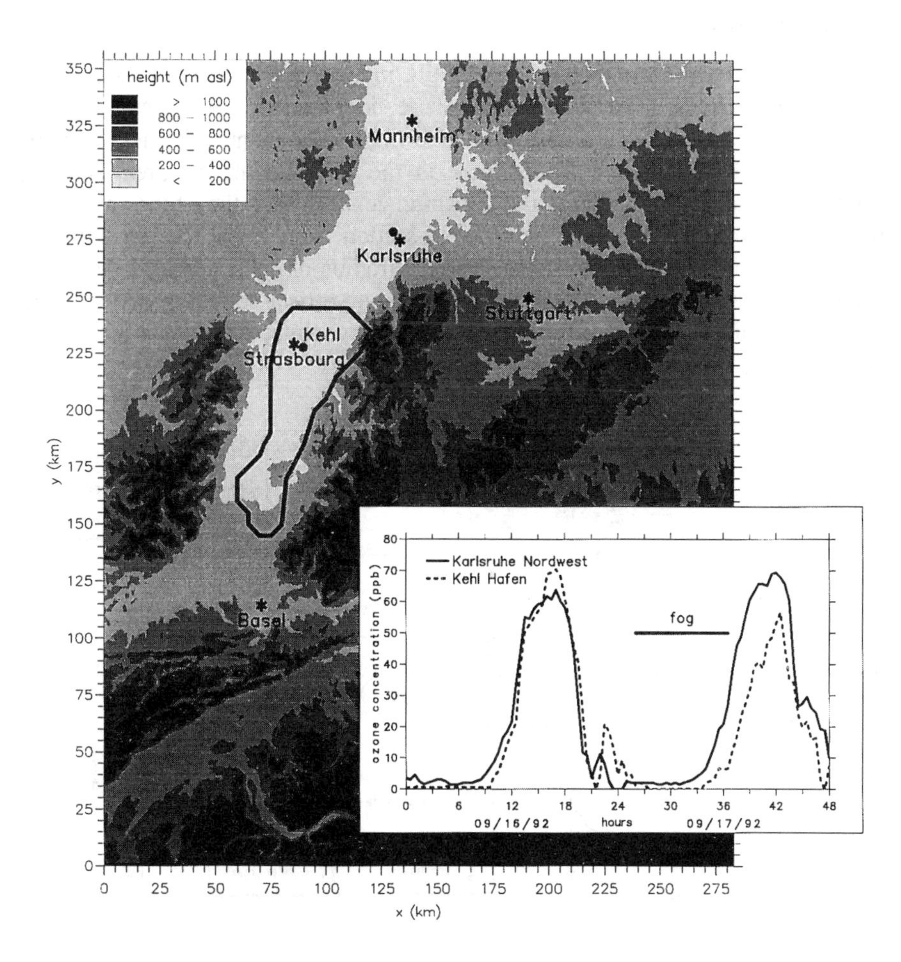

Fig. 8.4.2: Location and time series of O_3 for two stations in the Rhine valley during September 16th and 17th, 1992. The thick solid line shows the area where fog was still present on September 17th, 1992, 11:00 CEST. In the time series plot the thick solid line marks the fog period at the Kehl Hafen station on September 17th.

On September 16th, a day without fog, both stations showed nearly the same daily cycle with almost the same O_3 concentrations. This changed on September 17. Under easterly flow conditions, fog was observed on the leeside of the Black Forest until noon (Fig. 8.4.2), while at the station near Karlsruhe the fog dissolved quickly in the early morning hours. Hence, the daily cycle was quite different. At the station near Karlsruhe the development of O_3 concentrations was similar to that on the previous day, even reached higher values in the afternoon. However, due to the persistence of fog and the accompanying reduced solar radiation in the southern Upper Rhine valley, the increase in the O_3 concentrations at the Kehl station was delayed and did not reach the values of the stations without fog. So, orographically modified meteorological conditions also influence the daily cycle and the maximum values of the O_3 concentrations.

The nested area

One characteristic feature of orographically induced phenomena are secondary circulation systems. To investigate these systems measurements on the appropriate meso γ scale are necessary. For these reasons, measurement systems were installed in a denser arrangement in the so-called 'nested area' between Strasbourg and Baden-Baden in the transition zone between the Rhine valley and the Black Forest [1, 2]. Along this cross-section from the Rhine river to the eastern edge of the Black Forest, the spatial variation of the boundary layer depth z_i and the boundary layer height h derived from vertical sounding systems and aircraft measurements is given in Fig. 8.4.3 for three time intervals on September 16th.

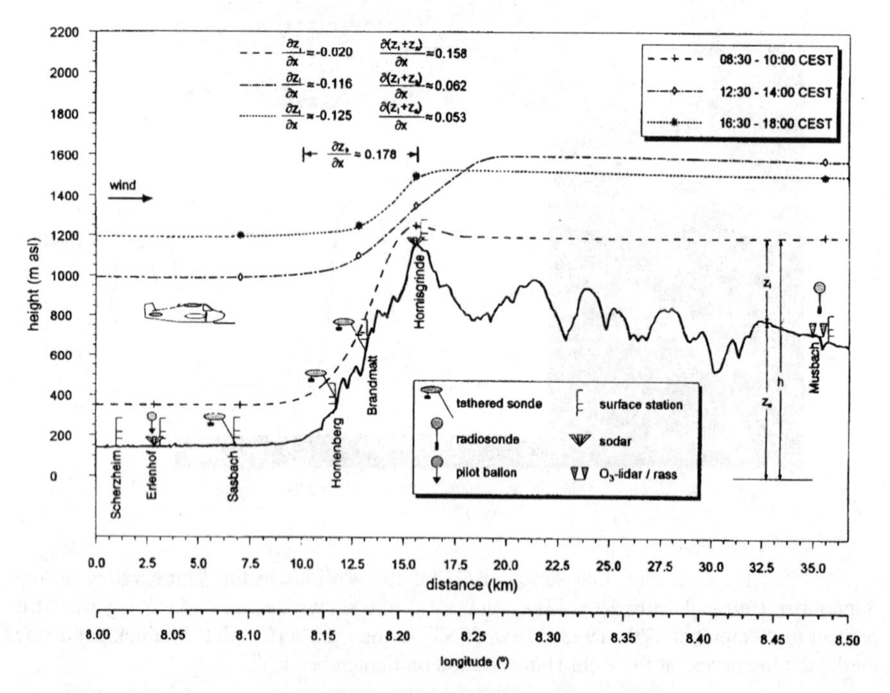

Fig. 8.4.3: Measuring sites in the nested area along the cross-section from the Rhine valley to the northern Black Forest and observed boundary layer depths z_i and boundary layer heights $h = z_i + z_s$ for three time periods during September 16th, 1992. Additionally, the horizontal gradients in the boundary layer depth $\partial z_i / \partial x$ and in the boundary layer height $\partial(z_i + z_s) / \partial x$ between the foot of the Black Forest and the top of the Hornisgrinde mountain are given.

Again, at noon and in the afternoon a nearly terrain following capping inversion was detected, except over the Hornisgrinde mountain crest, where z_i remained small during the whole day. Predictions of z_i for the Sasbach and the Hornisgrinde sites were calculated with the formula of Deardorff [13] which was derived for applications on homogeneous terrain:

$$\frac{\partial z_i}{\partial t} = \frac{1.8\left(w_*^3 + 1.1u_*^3 - 3.3u_*^2 f z_i\right)}{\frac{g}{\theta_0} z_i^2 \gamma + 9w_*^2 + 7.2u_*^2} = w_e, \qquad (1)$$

where [illegible] is the friction velocity, f the Coriolis parameter, g the gravity acceleration, θ_0 the potential temperature near the surface, γ the vertical gradient of potential temperature just above zi, $w_* = \left(g / \theta_0 z_i \overline{w'\theta'_0}\right)^{1/3}$ the convective vertical scaling velocity, and $\overline{w'\theta_0}$ the vertical kinematic sensible heat flux in the surface layer. w_e is called the entrainment velocity. The calculations significantly overestimate the boundary layer depth at the Hornisgrinde, while good agreement with observations is found for Sasbach in the Rhine valley (Fig. 8.4.4). It is assumed that advective processes are the main reasons for this disagreement between calculated and observed boundary layer depths over the Hornisgrinde.

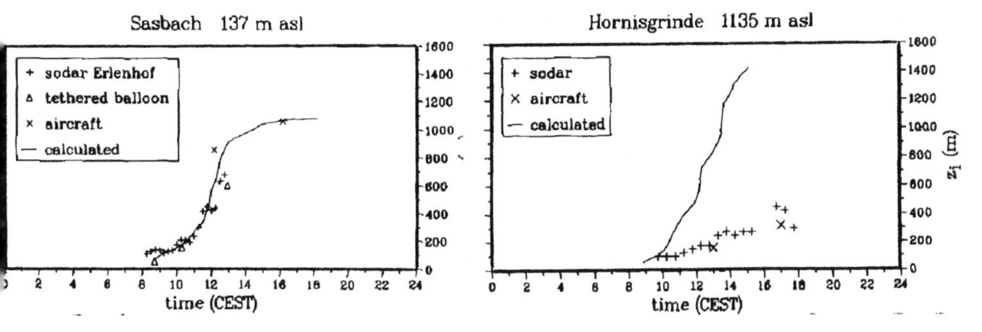

Fig. 8.4.4: Observed and calculated boundary layer depths z_i at the Sasbach station (left) and the Hornisgrinde mountain station (right) on September 16th, 1992. The calculations were made with formula (1).

To include advective processes in calculations of z_i the following formulation was proposed by Deardorff [14]:

$$\frac{\partial z_i}{\partial t} = w_e - u_{z_i} \frac{\partial (z_i + z_s)}{\partial x} - v_{z_i} \frac{\partial (z_i + z_s)}{\partial y} + w_{z_i}, \qquad (2)$$

where u_{z_i}, v_{z_i} and w_{z_i} are the mean wind velocity components at the height z_i. Due to westerly winds on September 16th (*e.g.* [12]), we neglect the contribution by the v component in a first estimate. Between the Rhine valley and the Hornisgrinde station $\partial z_i / \partial x$ was negative and $\partial (z_i + z_s) / \partial x$ was positive during the day (Fig. 8.4.3). In connection with the positive u component it is obvious that the effect of horizontal advection is able to reduce the overestimation of z_i resulting from $\partial z_i / \partial t = w_e$ (eq. 1) at the Hornisgrinde. This means that the boundary layer capping inversion is forced by the mean horizontal wind towards the western slope of the northern Black Forest. This process counteracts the CBL growth due to w_e.

A further possibility of parameterising the phenomenon of thin convective boundary layers above mountain crests was elaborated by de Wekker [15].

Calculations of z_i for the Hornisgrinde site, based on prognostic formulae, were fitted to observations by using reduced surface heat fluxes. This method takes into account that the mountain top area has only a small spatial extension. So, if a decrease of turbulent vertical heat fluxes with the height is assumed, an area averaged turbulent heat flux through a horizontal plane placed at the mountain top covering also surrounding areas will be lower than the local heat flux at the mountain top.

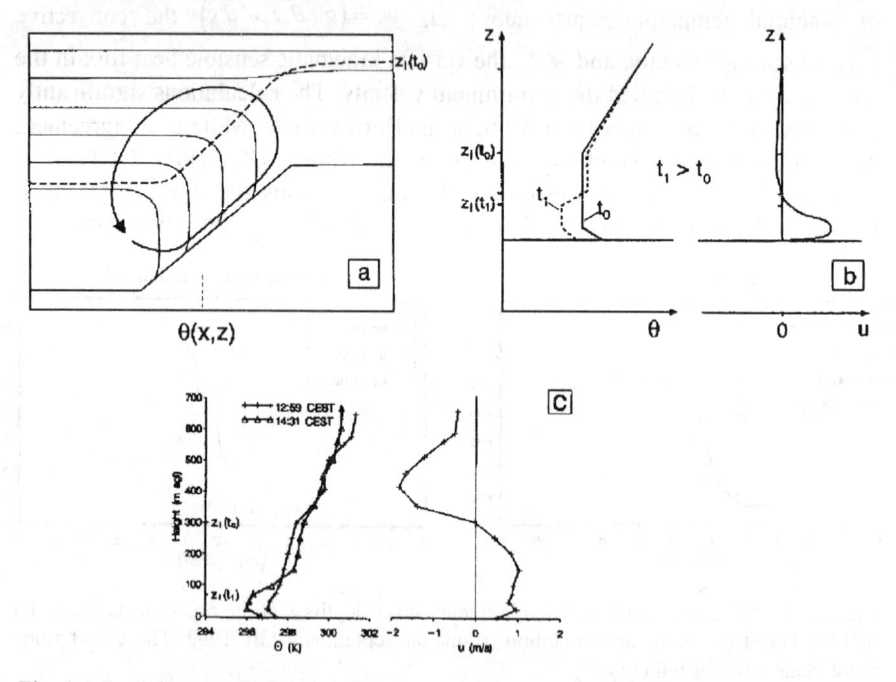

Fig. 8.4.5: Influence of slope wind systems on the boundary layer depth.. a) and b): schematic views (large scale flow is neglected). c): tethered balloon measurements at the Brandmatt station on the western slope of the Black Forest (see Fig. 8.4.3) on September 17th, 1992. The generation of a low-level stable layer due to upslope advection of cold air near the ground leads to a decrease of the boundary layer depth z_i, as sketched in Fig. 8.4.5b, which represents the vertical structure of potential temperature and horizontal wind velocity at the slope site, marked in Fig. 8.4.5a by a short dotted line.

However, in the case of non-uniform boundary layer heights combined with height-dependent wind directions as given by secondary circulation systems, the advective processes can not be described correctly by the methods mentioned. These secondary circulation systems significantly affect the heat budget and, hence, the growth of the boundary layer height. This was observed at the Brandmatt tethered balloon station on the western slope of the Black Forest during the 2nd SOP on September 16th and 17th [11, 16]. The advection of cold air in the upslope wind layer generated an inversion at the top of the slope wind layer and hence a decrease of z_i (Fig. 8.4.5). On September 17th, there were strong temperature gradients between the Rhine valley and the Brandmatt station. They were caused by the fog in the Rhine valley on that day (see Fig. 8.4.2).

Therefore, the upslope wind system led to more than 1 K cooling in the lowest layers at the slope station (Fig. 8.4.5c). In the measurements the return flow above the slope wind layer was superimposed by an easterly mean flow. Therefore, higher negative values in the u component were observed than entered in Fig. 8.4.5b, where large-scale flow is neglected.

The higher spatial resolution of the measurements in the nested area can also be used for investigating handover processes acting on this scale. Fig. 8.4.6 shows aircraft measurements of potential temperature, specific humidity, and horizontal wind vectors performed during the afternoon of September 17th, 1992. On that day, there was fog in the Rhine valley until about 12:30 CEST (see Fig. 8.4.2), while the mountainous area was fog-free. It can be seen from the potential temperature field that in the Rhine valley the base of the capping inversion was at 600 m a.s.l. while above the Black Forest the inversion base was at 1800 m a.s.l. This means that in the transition zone between the valley and mountainous area the top of the boundary layer was characterised by a step of about 1200 m with respect to mean sea level. At the heights between about 1200 m and 1600 m a.s.l. easterly winds transported humid air masses from the boundary layer of the mountain area to the free troposphere above the Rhine valley. Additionally, humid air from the Rhine valley bottom was brought towards the mountain ridge by the slope wind system [11, 16], which constitutes a second source of moisture. This air merged with the westward moving air from the mountain boundary layer. Beside the non-horizontal boundary layer top, this handover of water vapour by advective venting requires that the three-dimensional wind vector at the boundary layer top is not parallel to the boundary layer inversion [16]. Therefore, a portion of air was able to leave the boundary layer in horizontal direction.

The step in the boundary layer height enables the transport of humidity from the boundary layer to the free troposphere and is one important process, which cannot be found under horizontal homogeneous conditions.

In order to calculate the accompanying mass transport of moisture for that day, we apply the characteristic parameters, *i.e.* we assume a layer of 400 m thickness, which is accompanied by easterly winds, a width of the step area of about 50 km (from Offenburg to Baden-Baden), a mean easterly wind of 3 ms^{-1}, an air density of 1 kg m^{-3}, and a specific humidity of 7.5 g kg^{-1} in the boundary layer above the mountain area. Assuming a duration of 6 hours for the time scale of the process leads to a total moisture transport of 9.7×10^6 t d^{-1}. This corresponds to a normalised flux of 8×10^4 t h^{-1} for an area of 1 km^2. This is an enormous amount of mass transport through a boundary layer step and the data will be compared below to mass transport by a boundary layer gap.

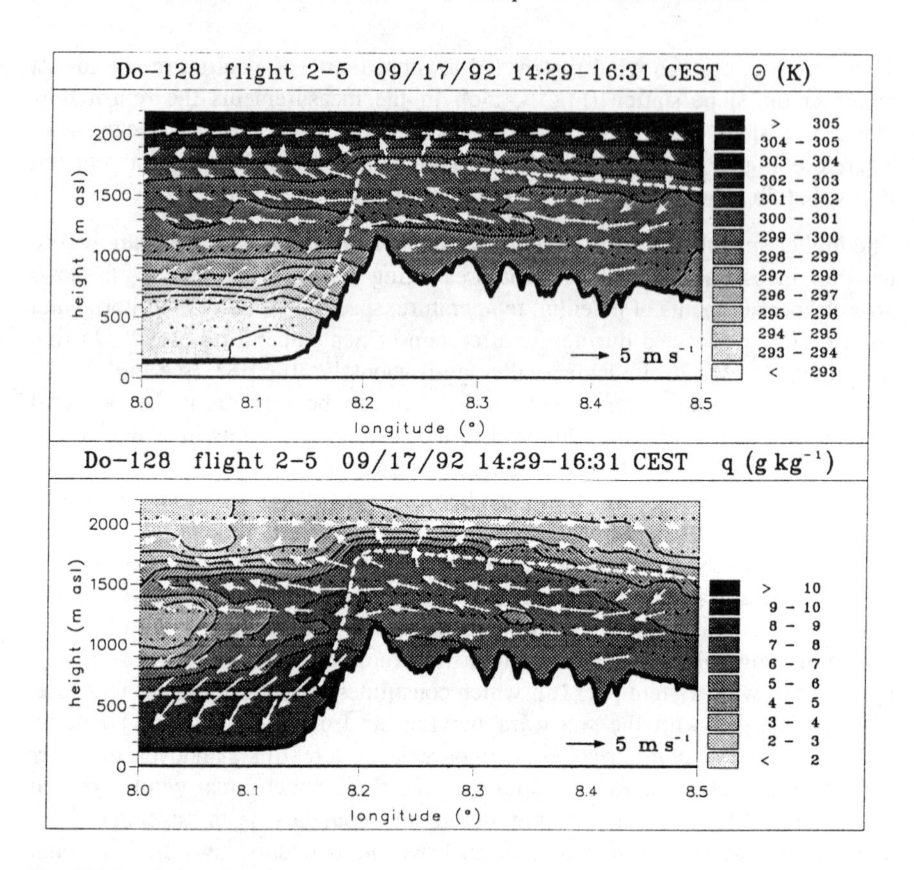

Fig. 8.4.6: Interpolated cross-sections of potential temperature θ, specific humidity q, and horizontal wind vectors from the Rhine valley to the northern Black Forest, measured by the Do-128 aircraft in the afternoon of September 17th, 1992. The dotted lines indicate the flight pattern, the dashed line marks the boundary layer top. The horizontal range is about 36 km.

At a specific site, the change in moisture with time is given by divergence of turbulent flux and/or by advection according to the budget equation. In order to be able to compare the efficiency of the turbulent flux to the advection process on that day, we assume a turbulent latent heat flux of 200 W m^{-2} at the surface, an inversion height of 500 m above ground in the Rhine valley where the turbulent latent heat flux is zero, a wind speed of 3 m s^{-1}, and a horizontal gradient in specific humidity of 0.2 g kg^{-1} km^{-1}. This results in a relation which gives a 4 times higher value for the advection process in comparison to the transport by the turbulent flux, and again reveals the importance of the processes arising under non-homogeneous conditions.

Another effective handover process was observed in the afternoon of September 21st, 1992 (Fig. 8.4.7): a mountain-induced convective cell pumping water vapour from the boundary layer through an inversion gap to the free troposphere [17].

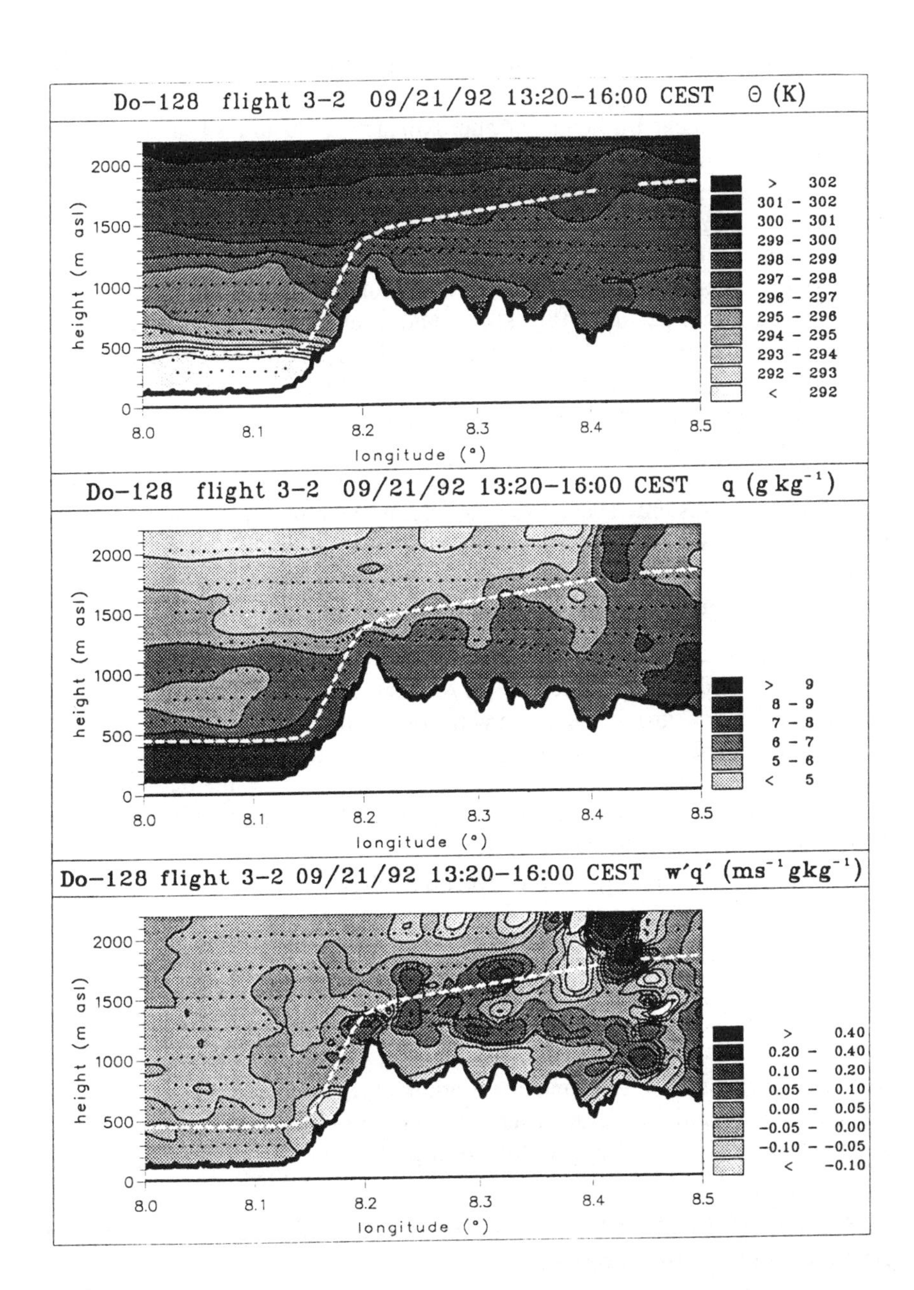

Fig. 8.4.7: Interpolated cross-sections of potential temperature θ, specific humidity q, and $w'q'$ from the Rhine valley to the northern Black Forest measured in the afternoon of September 21st, 1992 by the Do-128 aircraft. The dotted lines indicate the flight pattern, the dashed lines the boundary layer top. The horizontal range is about 36 km. Between 8.4° and 8.5° a convective cell with vertical wind speeds up to 2 m s^{-1} was measured, leading to an effective transport of humidity into the free troposphere as shown by the highly positive values of $w'q'$ in the cell.

In case of moderate horizontal winds a wedge-shaped boundary layer developed over the mountains, beginning at the slope. About 15 km downstream, the boundary layer reached a height of approximately 1900 m a.s.l. and one single convective cell was able to penetrate the inversion. For about 1 hour about 5.8×10^4 t water vapour was transported by the cell with a diameter of 1 km, an air density of 1 kg m^{-3}, a specific humidity of 8 g kg^{-1}, and a vertical wind speed of 2 m s^{-1}. The area-averaged evaporation in the TRACT nested area on September 16th, 1992 - a day with meteorological conditions similar to that on September 21st - was 2.5 mm per day [18]. The radiation temperature of the soil, surface and land use data estimated by satellite as well as ground based data from the REKLIP network (Regio-Klima-Projekt; [19]) were used to calculate area-averaged evaporation. Therefore, an area of 25 km^2 was necessary to replace the water vapour transported from the boundary layer to the free troposphere by the convective cell. However, the observed decrease in the specific humidity below the cell at noon was only about 1–2 g kg^{-1} because the real intake area of the cell is enlarged by advection. This means that a small-scale convective plume, forming a gap in the capping inversion, substantially affects the trace gas mass budget of the boundary layer far around and below the cell.

Similar observations were made during all three special observation periods of TRACT. The wedge-shaped increase in the boundary layer height, beginning at the top of the Hornisgrinde mountain with severe convective activity about 15 km downstream, turned out to be a typical feature in this mountainous area. On September 11th, a gap in the inversion occurred just at the same location. But in this case a cumulus cloud was formed between 1700 m and 2200 m a.s.l. On September 16th, convection of warm and moist air was again observed at the same place. But on that day the convection cell was not able to penetrate the inversion layer.

Finally, comparison of the two processes characterising the transport of moisture and hence of air pollutants under non-homogeneous conditions yields the result that the normalised transport through the boundary layer step is by about 25 % greater than the transport through the boundary layer gap (assuming for both processes an area of 1 km^2 and a duration of 1 hour). Taking into account realistic spatial and temporal scales of both processes, the transport by the inversion step is 170 times greater than transport through a single inversion gap.

Assessment of the contribution and its success in the light of its original aims and those of the subproject

The presented results focus on the influence of orography on the convective boundary layer depth and on the handover of mass between the boundary layer and the free troposphere by boundary layer steps and gaps. So, this contribution deals with unsolved questions on transport of air pollutants over complex terrain which are the subject of TRACT. Phenomena were analysed which had not been investigated previously (advection of z_i structures, influence of secondary circulations, handover processes). In a further step modified parameterisations for the daily cycle of the boundary layer depth were discussed (advection term, reduction of surface heat flux, influence of mountain shape) which have still to be tested in numerical model simulations. Further tasks will be

* the parameterisation of the observed handover processes,
* the investigation of the flow field response to the interaction of thermal forcing (due to differences in the heating rate) and mechanical forcing (geostrophic wind) in relation to orographical parameters which is decisive regarding the main transport paths and wind speeds [20], and
* the analysis of the influence exerted by the terrain on diffusion and chemical transformations which leads to very variable depositions in the lee of strong pollutant sources [21].

The results of this TRACT contribution help to better understand the governing mixing processes over orographically structured terrain and in this way improve mesoscale atmospheric flow and dispersion models (see TRACT contribution 'Modelling of transport, mixing and deposition of air pollutants over inhomogeneous terrain including chemical reactions' by Kuntze *et al.*, [8]).

Acknowledgements

This contribution to the EUROTRAC subproject TRACT was funded by the Bundesministerium für Forschung und Technologie (Germany) under contract no. 07EU734 and by the Commission of the European Community (CEC).

References

1. Fiedler, F., TRACT Operational Plan, Institut für Meteorologie und Klimaforschung, Universität Karlsruhe / Forschungszentrum Karlsruhe 1992, 66 pp.
2. Zimmermann, H., Field phase report of the TRACT field measurement campaign, *EUROTRAC Special Publication*, EUROTRAC ISS, Garmisch-Partenkirchen 1995, 196 pp.
3. Mohnen, V., Corsmeier, U., Fiedler, F., Quality assurance project plan - aircraft measurement systems, Institut für Meteorologie und Klimaforschung, Universität Karlsruhe / Forschungszentrum Karlsruhe 1992, 184 pp.
4. Kanter, H.-J., Slemr, F., Mohnen, V., Corsmeier, U., Airborne chemical and meteorological measurements made during the 1992 TRACT experiment: quality control and assessment, *J. Air & Waste Management* **46** (1996) 710-724.
5. Kolle, O., Kalthoff, N., Intercomparison of radiosonde data during the TRACT campaign, summer 1992, *EUROTRAC Annual Report 1992, TRACT*, EUROTRAC ISS, Garmisch-Partenkirchen 1993, pp. 142-150.
6. Kalthoff, N., Kolle, O.: Intercomparison of meteorological surface measurements during the TRACT campaign, summer 1992, *EUROTRAC Annual Report 1992, TRACT*, EUROTRAC ISS, Garmisch-Partenkirchen 1993, pp. 104-119.
7. Heinrich, G., Weppner, J., Güsten, H., Clauss, G., Antz, H., Target, A., Quality assurance for trace gas sensors used in the nested area of TRACT, *EUROTRAC Annual Report 1993, TRACT*, EUROTRAC ISS, Garmisch-Partenkirchen 1994, pp. 103-117.

8. Builtjes, P.J.H., Interaction of planetary boundary layer and free troposphere, in: T. Schneider *et al.* (eds.), *Atmospheric Ozone Research and its Policy Implications*, Elsevier Science Publishers bv, Amsterdam 1989, pp. 605-612.
9. Kuntze, K., Löffler-Mang, M., Nester, K., Adrian, G., Fiedler, F., Modeling of transport, mixing and deposition of air pollutants over inhomogeneous terrain including chemical reactions, this volume.
10. Kalthoff, N., Binder, H.-J., Koßmann, M., Vögtlin, R., Corsmeier U., Fiedler, F., Schlager, H., The temporal evolution and spatial variation of the boundary layer over complex terrain, *Atmos. Environ.* **32** (1998) 1179–1194.
11. Vogel, B., Kolle, O.: Observed temperature structures at the tethered balloon stations Sasbach and Brandmatt during the second intensive measuring phase of TRACT, *EUROTRAC Annual Report 1992, TRACT*, EUROTRAC ISS, Garmisch-Partenkirchen 1993, pp. 216-226.
12. Koßmann, M., Vögtlin, R., Corsmeier, U., Vogel, B., Fiedler, F., Binder, H.-J., Kalthoff, N., Beyrich, F.: Aspects of the convective boundary layer structure over complex terrain, *Atmos. Environ.* **32** (1998) 1323–1348.
13. Deardorff, J.W., Three-dimensional numerical study of the height and mean structure of a heated planetary boundary layer, *Boundary-Layer Meteor.* **7** (1974) 81-106.
14. Deardorff, J.W., Parameterization of the planetary boundary layer for the use in general circulation models, *Monthly Weather Review* **100** (1972) 93-106.
15. de Wekker, S.F.J., The behaviour of the convective boundary layer height over orographically complex terrain. Diplomarbeit (thesis), Institut für Meteorologie und Klimaforschung, Karlsruhe and Department of Meteorology, Wageningen Agricultural University 1995.
16. Koßmann, M., de Wekker, S.F.J., Vögtlin, R., Kalthoff, N., Corsmeier, U., Binder, H.-J., Fiedler, F.: Advective venting - an effective process for the handover of trace gases between the atmospheric boundary layer and the free troposphere over complex terrain, in: P.M. Borrell, P. Borrell, T. Cvitaš, K. Kelly W. Seiler (eds), *Proc. EUROTRAC Symp. '96, Vol. 1*, Computational Mechanics Publications, Southampton 1996, pp. 693–697.
17. Corsmeier, U., Koßmann, M., Vögtlin, R., Fiedler, F.: Water vapour transport by stationary convective cells over hilly terrain, in: P.M. Borrell, P. Borrell, T. Cvitaš, W. Seiler (eds), *Proc. EUROTRAC Symp. '94*, SPB Academic Publishing bv, The Hague 1994, pp. 761-765.
18. Wenzel, A., Kalthoff, N., Fiedler, F., Olesen, F., The evaporation in the Upper Rhine valley. European Geophys. Soc. XIX General Assembly, Grenoble, April 1994.
19. Fiedler, F., Das Regio-Klima-Projekt: Wie regeln die natürlichen Energieumsetzungen das Klima in einer Region, KfK-Nachrichten, Jahrg. 24, 3/92, (1992) 125-131.
20. Löffler-Mang, M., Zimmermann, H., Fiedler, F.: Analysis of ground based operational network data acquired during the September 1992 TRACT campaign, *Atmos. Environ.* **32** (1998) 1229–1240.
21. Vögtlin, R., Koßmann, M., Corsmeier, U.: Aircraft measurements in the main area during TRACT, *EUROTRAC Annual Report 1993, TRACT*, EUROTRAC ISS, Garmisch-Partenkirchen 1994, pp. 38-48.

Chapter 9

TRACT Publications 1991 – 1995

Some later publications have been added.

1991

Ambrosetti, P., D. Anfossi, S. Cieslik, P. Gaglione, R. Lamprecht, A. Marzorati, S. Sandroni, F. Spinedi, A. Stingele, S. Vogt; The TRANSALP-89 Exercise. A tracer release aexperiment in a subalpine valley, *Report* EUR 13474 EN, CEC, Brussels 1991.

Marzorati, A., A. Longhetto, D. Anfossi, G. Elisei; Wind variance spectra obtained by Doppler sodar measurements in different sites, in: *Proc. 7th Symp. on Meteorological Observations and Instrumentation*, American Meteorological Society, New Orleans 1991.

1992

Anfossi, D., A. Marzorati; Complex terrain Doppler sodar measurements, in: *Proc. Int. Symp. of Acoustic Remote Sensing*, Athens 1992.

Brusasca, G., D. Anfossi, E. Ferrero; Modelli per la dispersione degli inquinanti in atmosfera, *Le Scienze* **288** (1992) 38–49.

Fiedler, F.; *TRACT Operational Plan,* Institut für Meteorologie und Klimaforschung, Universität Karlsruhe / Forschungszentrum Karlsruhe 1992, 66 pp.

Mohnen, V., Corsmeier, U., Fiedler, F.; *Quality assurance project plan - aircraft measurement systems*, Institut für Meteorologie und Klimaforschung, Universität Karlsruhe / Forschungszentrum Karlsruhe 1992, 184 pp.

Nodop, K., P. Ambrosetti, D. Anfossi, S. Cieslik, P. Gaglione, R. Lamprecht, A. Marzorati, A. Stingele, H. Zimmermann; Tracer experiment to study air flow across the Alps, in: P.M. Borrell, P. Borrell, T. Cvitaš, W. Seiler (eds), *Proc. EUROTRAC Symp. '92,* SPB Academic Publishing bv, The Hague 1992, pp. 203–207.

Vogel, H., Bär, M., Fiedler, F.: Influence of different land-use types on dry deposition, concentration and mass balances: a case study, in: P.M. Borrell, P. Borrell, T. Cvitaš, W. Seiler (eds.), *Proc. of EUROTRAC Symp. 92,* SPB Academic Publishing bv, The Hague 1992, pp. 559–562.

Vogel, H., F. Fiedler, B. Vogel; The contributions of transport, diffusion and chemical reactions to the diurnal cycles of ozone in the regional scale, in: P.M. Borrell, P. Borrell, T. Cvitaš, W. Seiler (eds), *Proc. EUROTRAC Symp. '92,* SPB Academic Publishing bv, The Hague 1992, pp. 190–192.

Wachs, P., Corsmeier, U., Fiedler, F.; TRACT aircraft operational plan, Institut für Meteorologie und Klimaforschung, Universität Karlsruhe / Forschungszentrum Karlsruhe 1992, 73 pp.

1993

Anfossi, D., C. Cassardo, E. Ferrero, A. Longhetto, D. Sacchetti, G. Brusasca, V. Colombo, A. Marzorati, M. G. Morselli, F. Rocchetti, G. Tinarelli; On the analysis of ultrasonic anemometer data, *Report ICG/CNR* n. 282/93.

Anfossi, D., A. Marzorati, D. Sacchetti; Studio del trasporto transalpino di traccianti, *Bolletino Geofisico* **16** (1993) 13–14.

Cassardo, C., D. Anfossi, D. Sacchetti, M.G. Morselli; Turbulence measurements in the PBL by the use of an ultrasonic anemometer, *Bolletino Geofisico* **16** (1993) 27–31.

Desiato, F.; Preprocessing of meteorological data and modelling the atmospheric dispersion over complex terrain with a particle model, in: *Proc. Workshop Intercomparison of Advanced Practical Short-Range Atmospheric Dispersion Models*, JCR, Ispra 1993, pp. 95–102.

Fiedler, F.: Development of meteorological computer models, *Interdisciplinary Science Reviews* **18** (1993) 192–198.

Kanter, H.-J., Mohnen, V., Slemr, F., Corsmeier, U.; Qualitätskontrolle der Flugzeugmessungen bei der TRACT-Messkampagne, *VDI Berichte Nr. 1059*, VDI Verlag 1993, pp. 395–405.

Minella, M., S. Finardi, M. G. Morselli, G. Brusasca; Simulation of a complex terrain release experiment (the TRANSALP Campaign 1989) through mass consistent (wind field) and Gaussian puff (dispersion) model, in: *Proc. Workshop Intercomparison of Advanced Practical Short-Range Atmospheric Dispersion Models*, JCR, Ispra 1993.

Marzorati, A., D. Anfossi; Doppler SODAR measurements and evaluations in complex terrain, *Il Nuovo Cimento* **16** (1993) 141–154.

1994

Adrian, G., F. Fiedler, F., K. Kuntze, M. Löffler-Mang, K. Nester: Wind Field Simulations over Complex Terrain for an Episode of the TRACT Experiment, *Annales Geophysicae, Suppl. II*, **12**, 1994, p C354.

Ambrosetti, P., D. Anfossi, S. Cieslik, G. Graziani, G. Grippa, R. Lamprecht, A. Marzoratti, A. Stingele, H. Zimmeramann; The TRANSALP-90 campaign. The second tracer release experiment in a subalpine valley, *Report* EUR 15952 EN, EC, Brussels 1994.

Ambrosetti, P., D. Anfossi, S. Cieslik, G. Graziani, G. Grippa, R. Lamprecht, A. Marzoratti, K. Nodop, A. Stingele, H. Zimmeramann; Mesoscale transport of atmospheric trace gases across the Alps: the TRANSALP–1991 campaign, *Report* EUR 15952 EN, EC, Brussels 1994.

Corsmeier, U., M. Koßmann, R. Vögtlin, F. Fiedler; Water vapour transport by stationary convective cells observed during TRACT over hilly terrain, in: P.M. Borrell, P. Borrell, T. Cvitaš, W. Seiler (eds), *Proc. EUROTRAC Symp. '94,* SPB Academic Publishing bv, The Hague 1994, pp. 761–765.

Dlugi, R., L. Kins, D. Müller, G. Roider, H. Schoers, T. Seiler, M. Weber, M. Zelger, T. Foken, B. Wichura, H. Güsten, G. Kramm; The exchange of trace gases and the surface energy budget: An intercomparison of results from field studies, in: P.M. Borrell, P. Borrell, T. Cvitaš, W. Seiler (eds), *Proc. EUROTRAC Symp. '94,* SPB Academic Publishing bv, The Hague 1994, pp. 643.

Fiedler, F.; TRACT field measurements and model simulations: first results, in: P.M. Borrell, P. Borrell, T. Cvitaš, W. Seiler (eds), *Proc. EUROTRAC Symp. '94,* SPB Academic Publishing bv, The Hague 1994, p. 614.

Güsten, H., G. Heinrich, E. Mönnich, D. Sprung; Ozone depletion within the nocturnal inversion layer: Night-time chemistry versus surface deposition, in: P.M. Borrell, P. Borrell, T. Cvitaš, W. Seiler (eds), *Proc. EUROTRAC Symp. '94,* SPB Academic Publishing bv, The Hague 1994, pp. 639–642.

Hummelshøj, P., K. Pilegaard, N.O. Jensen; Flux measurements of O_3 and NO_2 over a regrown wheat field, in: P.M. Borrell, P. Borrell, T. Cvitaš, W. Seiler (eds), *Proc. EUROTRAC Symp. '94,* SPB Academic Publishing bv, The Hague 1994, pp. 544–548.

Jeannet, P., B. Hoegger, P. Viatte, D. Schneiter; Vertical ozone profiles over Payerne, Switzerland, during TRACT and POLLUMET, in: P.M. Borrell, P. Borrell, T. Cvitaš, W. Seiler (eds), *Proc. EUROTRAC Symp. '94,* SPB Academic Publishing bv, The Hague 1994, pp. 362–366.

Jochum, A.M., C.S. Strodl, H. Willeke, N. Entstrasser; The transport of heat, moisture and ozone in the atmospheric boundary layer in the upper Rhine valley, in: P.M. Borrell, P. Borrell, T. Cvitaš, W. Seiler (eds), *Proc. EUROTRAC Symp. '94,* SPB Academic Publishing bv, The Hague 1994, pp. 766–769.

Kalthoff, N., H.-J. Binder, F. Fiedler, M. Koßmann, U. Corsmeier, R. Vögtlin; The boundary layer evolution during TRACT, in: P.M. Borrell, P. Borrell, T. Cvitaš, W. Seiler (eds), *Proc. EUROTRAC Symp. '94,* SPB Academic Publishing bv, The Hague 1994, pp. 748–752.

Koßmann, M., B. Vogel; Comparison between calculated and observed boundary layer heights during TRACT, in: P.M. Borrell, P. Borrell, T. Cvitaš, W. Seiler (eds), *Proc. EUROTRAC Symp. '94,* SPB Academic Publishing bv, The Hague 1994, pp. 753–756.

Kramm, G., R. Dlugi, T. Foken, N. Mölders, H. Muller, K. T. Pau U; On the determination of sublayer Stanton numbers of heat and matter for different types of surfaces, in: P.M. Borrell, P. Borrell, T. Cvitaš, W. Seiler (eds), *Proc. EUROTRAC Symp. '94,* SPB Academic Publishing bv, The Hague 1994, pp. 644.

Lamprecht, R., D. Berlowitz; Performance evaluation of a Lagrangian air quality model for passive tracers by a sensitivity study with field data, in: P.M. Borrell, P. Borrell, T. Cvitaš, W. Seiler (eds), *Proc. EUROTRAC Symp. '94,* SPB Academic Publishing bv, The Hague 1994, pp. 863–868.

Lightman, P., A.R. Marsh; Pollutant concentration profiles over complex terrain: Evidence for flow between valleys, in: P.M. Borrell, P. Borrell, T. Cvitaš, W. Seiler (eds), *Proc. EUROTRAC Symp. '94,* SPB Academic Publishing bv, The Hague 1994, pp. 776–779.

Reuder, J., T. Gori, A. Ruggaber, L. Kins, R. Dlugi; Photolysis frequencies of nitrogen dioxide and ozone: Measurements and model calculations, in: P.M. Borrell, P. Borrell, T. Cvitaš, W. Seiler (eds), *Proc. EUROTRAC Symp. '94,* SPB Academic Publishing bv, The Hague 1994, pp. 180–184.

Stingele, A., U. Corsmeier, K. Nodop, M. Koßmann, R. Voegtlin; Tracer experiment in the nested area of the TRACT campaign, in: P.M. Borrell, P. Borrell, T. Cvitaš, W. Seiler (eds), *Proc. EUROTRAC Symp. '94,* SPB Academic Publishing bv, The Hague 1994, pp. 748–752.

Vogel, H., F. Fiedler, B. Vogel; The sensitivity of modelled ozone distributions in the State of Baden-Württemberg to biogenic VOC emissions, in: P.M. Borrell, P. Borrell, T. Cvitaš, W. Seiler (eds), *Proc. EUROTRAC Symp. '94,* SPB Academic Publishing bv, The Hague 1994, pp. 492–496.

Vogel, H., F. Fiedler, B. Vogel, J.S. Chang, A. Obermeier, C. John, R. Friedrich; Aggregation factors and rate constants derived for the emission conditions in the TRACT area, in: P.M. Borrell, P. Borrell, T. Cvitaš, W. Seiler (eds), *Proc. EUROTRAC Symp. '94,* SPB Academic Publishing bv, The Hague 1994, pp. 757–760.

Vögtlin, R., M. Löffler-Mang, M. Koßmann, U. Corsmeier, F. Fiedler, O. Klemm, H. Schlager; Spatial and temporal distribution of air pollutants over complex terrain during TRACT, in: P.M. Borrell, P. Borrell, T. Cvitaš, W. Seiler (eds), *Proc. EUROTRAC Symp. '94,* SPB Academic Publishing bv, The Hague 1994, pp. 770–775.

1995

Kalthoff, N., Kolle, O.; Intercomparison of meteorological surface measurements and radiosonde data during the TRACT campaign, summer 1992, *KfK Primärbericht 22.01.01P01E*, 1995, 45 pp.

Koßmann, M., F. Fiedler, U. Corsmeier: Untersuchung der Überströmung eines Bergrückens mit einem Meßflugzeug, *Ann. Meteor.* **31** (1995) 398–399.

Koßmann, M., Vögtlin, R., Fiedler, F., de Wekker, S.F.J., Kalthoff, N., Binder, H.-J., Schlager, H., Klemm, O.; Austausch von Spurengasen zwischen der atmosphärischen Grenzschicht und der freien Troposphäre, *Annalen der Meteorologie N.F. 31*, 208–209, Selbstverlag des Deutschen Wetterdienstes, Offenbach 1995.

Obermeier, A., R. Friedrich, C. John, J. Seier, H. Vogel, F. Fiedler, B. Vogel: Photosmog - Möglichkeiten und Strategien zur Verminderung des bodennahen Ozons, ecomed-Verlagsgesellschaft, 86899 Landsberg 1995, ISBN 3-609-65320-5.

Panitz, H.J., K. Nester, F. Fiedler (1995) Bestimmung der Massenbilanzen chemisch reaktiver Luftschadstoffe in Baden-Württemberg, *FZKA-PEF* **130** (1995) 239–249.

Ruggaber, A., R. Dlugi, R. Forkel, W. Seidl, H. Hass, T. Nakajima, B. Vogel, M. Hammer: Modelling of Radiation Quantities and Photolysis Frequencies in the Troposphere, in: A. Ebel, N. Moussiopoulos (eds), *Air Pollution III, Volume 4: Observation and Simulation of Air Pollution*, Computational Mechanics Publications, Southampton 1995, pp. 111–118.

Vögtlin, R., M. Koßmann, M. Löffler-Mang, F. Fiedler, U. Corsmeier, N. Kalthoff, H.J. Binder, O. Klemm, H. Schlager: Verteilung und Transport atmosphärischer Spurengase über komplexem Gelände während TRACT, *Ann. Meteor.* **31** (1995) 151–152.

Vögtlin, R., F. Fiedler, U. Corsmeier: Tageszeitliche Entwicklung der konvektiven Grenzschicht über hügeligem Gelände, *Ann. Meteor.* **31** (1995) 400–401.

Vogel, B., F. Fiedler, H. Vogel: Influence of Topography and Biogenic Volatile Organic Compounds Emission in the State of Baden-Wuerttemberg on Ozone Concentrations During Episodes of High Air Temperatures, *J. Geophys. Res.* **100** (1995) 22907–22928.

Vogel, B., F. Fiedler, H. Vogel: Determination of Biogenic VOC and NO_x Emission Data using the High Resolution Mesoscale Model KAMM/DRAIS, in Proc. of The Emission Inventory : Programs and Progress, *Air & Waste Management Association,* Research Triangle Park, N.C., 1995.

Zimmermann, H.; Field phase report of the TRACT field measurement campaign, *EUROTRAC Special Publications*, EUROTRAC ISS, Garmisch-Partenkirchen 1995, 196 pp.

1996

Kanter, H.-J., F. Slemr, V. Mohnen, U. Corsmeier: Airborne Chemical and Meteorological Measurements Made during the 1992 TRACT Experiment: Quality Control and Assessment, *J. Air & Waste Management* **46** (1996) 710–724.

Löffler-Mang, M., M. Kunz, M. Koßmann: Eine nächtliche Kaltfront über orographisch gegliedertem Gelände und die Veränderung der Ozonkonzentration am Boden, *Meteor. Z., N.F.* **5** (1996) 308-317.

Obermeier, A., J. Seier, C. John, P. Berner, R. Friedrich, 1996, TRACT: Erstellung einer Emissionsdatenbasis für TRACT, *Forschungsbericht des Instituts für Energiewirtschaft und Rationelle Energieanwendung,* Band 27, Stuttgart, ISSN 0938-1228.

Panitz, H.-J., K. Nester, F. Fiedler: Bestimmung der Massenbilanzen chemisch reaktiver Luftschadstoffe in Baden-Württemberg, *FZKA-PEF* **142** (1996) 255-266.

Panitz, H.-J., K. Nester, F. Fiedler: Mass Balances and Interactions of Budget Components of Chemically Reactive Air Pollutants over the State of Baden-Württemberg, in: *Air Pollution Modeling and its Application XI*, Plenum Publishing Company, New York, London 1996, pp. 89-97.

1997

Anfossi, D., P. Gaglione, G. Graziani, K. Nodop, A. Stingele A. Marzorati, and G. Graziani; The TRANSALP tracer campaigns: lessons learned and possible project evolution, in: P.M. Borrell, P. Borrell, T. Cvitaš, K. Kelly, W. Seiler (eds), *Proc. EUROTRAC Symp 96 Vol. 1.* Computational Mechanics Publications, Southhampton 1997, pp. 751–756.

Berner, P., R. Friedrich, C. John, A. Obermeier, J. Seier; Generation of an emission data base for TRACT, in: P.M. Borrell, P. Borrell, T. Cvitaš, K. Kelly, W. Seiler (eds), *Proc. EUROTRAC Symp 96 Vol. 1.* Computational Mechanics Publications, Southhampton 1997, pp. 795–799.

DeWekker, S.F.J., M. Koßmann, F. Fiedler: The Behaviour of the Convective Boundary Layer Height over Orographically Complex Terrain., in: P.M. Borrell, P. Borrell, T. Cvitaš, K. Kelly, W. Seiler (eds), *Proc. EUROTRAC Symp 96 Vol. 1.* Computational Mechanics Publications, Southhampton 1997, pp. 789–793.

Koßmann, M., S.F.J. DeWekker, R. Vögtlin, N. Kalthoff, U. Corsmeier, H.J.Binder, F. Fiedler: Advective Venting - an Effective Process for the Handover of Trace Gases Between the Atmospheric Boundary Layer and the Free Troposphere over Complex Terrain, in: P.M. Borrell, P. Borrell, T. Cvitaš, K. Kelly, W. Seiler (eds), *Proc. EUROTRAC Symp 96 Vol. 1.* Computational Mechanics Publications, Southhampton 1997, pp. 693–697.

Koßmann, M., F. Fiedler, M. Löffler-Mang, R. Vögtlin (1997): Surface Layer Wind Field and Associated Transport of Air Pollutats over Complex Terrain during the TRACT Campaign in September, 1992, in: P.M. Borrell, P. Borrell, T. Cvitaš, K. Kelly, W. Seiler (eds), *Proc. EUROTRAC Symp 96 Vol. 1.* Computational Mechanics Publications, Southhampton 1997, pp. 771–775.

Koßmann, M., H. Vogel, B. Vogel, R. Vögtlin, U. Corsmeier, F. Fiedler, O. Klemm, H. Schlager (1997): The Composition and Vertical Distribution of Volatile Organic Compounds in Southwest Germany, Eastern France and Norther Switzerland during the TRACT Campaign in September 1992, in: P.M. Borrell, P. Borrell, T. Cvitaš, K. Kelly, W. Seiler (eds), *Proc. EUROTRAC Symp 96 Vol. 1.* Computational Mechanics Publications, Southhampton 1997, pp. 767–770.

Koßmann, M., R. Vögtlin, H.-J. Binder, F. Fiedler, U. Corsmeier, N. Kalthoff: Vermischungsvorgänge in der unteren Troposphäre über orographisch strukturiertem Gelände, Wissenschaftliche Berichte, *FZKA* 5897, 1997.

Koßmann, M., H. Vogel, B. Vogel, R. Vögtlin, U. Corsmeier, F. Fiedler, O. Klemm, H. Schlager: The composition and Vertical Distribution of Volatile Organic Compounds in Southwest Germany, Eastern France and Northern Switzerland during the TRACT Campaign in September 1992, *Physics and Chemistry of the Earth* **21** (1997) 429–433.

Kuntze, K., M. Löffler-Mang, K. Nester, G. Adrian, F. Fiedler: Modeling of Transport, Mixing and Deposition over Inhomogeneous Terrain Including Chemical Reactions, in: P.M. Borrell, P. Borrell, T. Cvitaš, K. Kelly, W. Seiler (eds), *Proc. EUROTRAC Symp 96 Vol. 1.* Computational Mechanics Publications, Southhampton 1997, pp. 533–538.

Löffler-Mang, M., M. Koßmann, R. Vögtlin, F. Fiedler: Valley Wind Systems and their Influence on Nocturnal Ozone Concentrations, *Beitr. Phys. Atmos.* **70** (1997) 1–14.

Mohnen, V.: Quality Assurance in Atmospheric Measurements and Assessments, Transport and transformation of pollutants in the troposphere, in: P.M. Borrell, P. Borrell, T. Cvitaš, K. Kelly, W. Seiler (eds), *Proc. EUROTRAC Symp 96 Vol. 2.* Computational Mechanics Publications, Southhampton 1997, pp. 417–423.

Panitz, H.J., K. Nester, F. Fiedler: Bestimmung der Massenbilanzen chemisch reaktiver Luftschadstoffe in Baden-Württember. Untersuchungen für die Region Freudenstadt-Stuttgart, *FZKA-PEF* **153** (1997) 207–219.

Trini Castelli S., D. Anfossi; Intercomparison of 3D turbulence parameterizations for dispersion models in complex terrain derived from a circulation model, *Nuovo Cimento* **20** (1997) 287–313

Vögtlin, R., M. Koßmann, U. Corsmeier, M. Löffler-Mang, F. Fiedler, A.R. Marsh: Topographically Influenced Distributions of Air Pollutants in the TRACT Area, in: P.M. Borrell, P. Borrell, T. Cvitaš, K. Kelly, W. Seiler (eds), *Proc. EUROTRAC Symp 96 Vol. 1.* Computational Mechanics Publications, Southhampton 1997, pp. 783–787.

Vögtlin, R., M. Koßmann, H. Güsten, G. Heinrich, F. Fiedler, U. Corsmeier, N. Kalthoff, A. Pfeifle: Transport of Trace Gases from the Upper Rhine Valley to a Mountain Site in The Northern Black Forest, in: P.M. Borrell, P. Borrell, T. Cvitaš, K. Kelly, W. Seiler (eds), *Proc. EUROTRAC Symp 96 Vol. 1.* Computational Mechanics Publications, Southhampton 1997, pp. 777–781.

Vogel, B.: Ozonanstieg in der Troposphäre während der letzten 100 Jahre – Welchen Beitrag liefert das Klima? *Ann. Meteor.* **31** (1995) 27–28

Vogel, B.: The effect of climate change during the last 120 years on the ozone concentration in the south-western part of Germany, in: P.M. Borrell, P. Borrell, T. Cvitaš, K. Kelly, W. Seiler (eds), *Proc. EUROTRAC Symp 96 Vol. 1.* Computational Mechanics Publications, Southhampton 1997, pp. 939–942.

Vogel, B., M. Hammer, N. Riemer, H. Vogel, A. Ruggaber: A Method to Determine the Three-Dimensional Distribution of Photolysis Frequencies in the Regional Scale, in: P.M. Borrell, P. Borrell, T. Cvitaš, K. Kelly, W. Seiler (eds), *Proc. EUROTRAC Symp 96 Vol. 1.* Computational Mechanics Publications, Southhampton 1997, pp. 745–749.

Vogel, H., F. Fiedler, B. Vogel: NO_y as an Indicator Species for Ozone Sensitivity to Emission Reductions, in: P.M. Borrell, P. Borrell, T. Cvitaš, K. Kelly, W. Seiler (eds), *Proc. EUROTRAC Symp 96 Vol. 1.* Computational Mechanics Publications, Southhampton 1997, pp. 757–760.

1998

Anfossi, D., F. Desiato, G. Tinarelli, G. Brusasca, E. Ferrero, D. Sacchetti; TRANSALP 1989 experimental campaign - II. Simulation of a tracer experiment with Lagrangian particle models, *Atmos. Environ.* **32** (1998) 1157–1166.

Ambrosetti, P., D. Anfossi, S. Cieslik, G. Graziani, R. Lamprecht, A. Marzorati, K. Nodop, S. Sandroni, A. Stingele, H. Zimmermann; Mesoscale transport of atmospheric trace constituents across the central Alps: TRANSALP tracer experiments, *Atmos. Environ.* **32** (1998) 1257–1272.

Cieslik, S.: Energy and Ozone Fluxes in the Atmospheric Surface Layer Observed in Southern Germany Highlands, *Atmos. Environ.* **32** (1998) 1273–1282.

Desiato, F., S. Finardi, G. Brusasca, M.G. Morselli; TRANSALP 1989 experimental campaign - I. Simulation of 3D flow with diagnostic wind field models, *Atmos. Environ.* **32** (1998) 1141–1156.

Güsten, H., G. Heinrich D. Sprung: Nocturnal Depletion of Ozone in the Upper Rhine Valley, *Atmos. Environ.* **32** (1998) 1195–1202.

Horváth, L, Z. Nagy, T. Weidinger: Estimation of Dry Deposition Velocities of Nitric Oxide, Sulfur Dioxide, and Ozone by the Gradient Method Above Short Vegetation During the TRACT Compaign, *Atmos. Environ.* **32** (1998) 1317–1322.

Kalthoff, N., H.J. Binder, M. Kossmann, R. Vögtlin, U. Corsmeier, F. Fiedler, H. Schlager: Temporal Evolution and Spatial Variation of the Boundary Layer over Complex Terrain, *Atmos. Environ.* **32** (1998) 1179–1194.

Koßmann, M., R. Vögtlin, U. Corsmeier, B. Vogel, F. Fiedler, H.-J. Binder, N. Kalthoff, F. Beyrich (1998): Aspects of the Convective Boundary Layer Structure over Complex Terrain, *Atmos. Environ.* **32** (1998) 1323–1348.

Löffler-Mang, M., H. Zimmermann, F. Fiedler: Analysis of Ground Based Operational Network Data Acquired During the September 1992 TRACT Campaign, *Atmos. Environ.* **32** (1998) 1229–1240.

Martilli, A., G. Graziani; Mesoscale circulation across the Alps: preliminary simulation of the TRANSALP 1990 simulations, *Atmos. Environ.* **32** (1998) 1241–1256.

Pilegaard, K., P. Hummelshoj, N. O. Jensen: Fluxes of Ozone and Nitrogen Dioxide Measured by Eddy Correlation over a Harvested Wheat Field, *Atmos. Environ.* **32** (1998) 1167–1178.

Schmidt, R. W. H., F. Slemr, U. Schurath: Airborne peroxyacetyl nitrate (PAN) and peroxypropionyl nitrate (PPN) measurements during TRACT 1992, *Atmos. Environ.* **32** (1998) 1203–1228.

Varvayanni M., J.G. Bartzis, N.Catsaros N., G.Graziani and P. Deligiannis, 1998, Numerical simulation of daytime mesoscale flow over highly complex terrain: Alps case, *Atmos Environ.* **32** (1998) 1301–1316.

1999

Hasager, C. B., N. O. Jensen, 1999, Surface-Flux aggregation in heterogeneous terrain, *Quart. J. Roy. Meteorol. Soc.* **125** (1999) 2075–2102.

Theses

M. Sc. / Diploma

Binder, H.-J.: Tageszeitliche und räumliche Entwicklung der konvektiven Grenzschicht über stark gegliedertem Gelände. Dissertation, Inst. Meteorol. Klimaforsch., Universität Karlsruhe, 1997.

Bitzer A.: Untersuchung der Erdoberflächentemperatur in der 'nested area' während TRACT, Seminararbeit, Inst. Meteorol. und Klimaforschung Karlsruhe, Universität Karlsruhe/Forschungszentrum Karlsruhe 1996.

de Wekker, S.F.J.; The behaviour of the convective boundary layer height over orographically complex terrain. Institut fuer Meteorologie und Klimaforschung, Karlsruhe and Department of Meteorology, Wageningen Agricultural University, 1995.

Grabe, F.: Die Auswirkungen geänderter Emissionsverhältnisse an Wochenenden auf die Ozonverteilung: Beobachtungen und numerische Simulationen, Diplomarbeit, Inst. Meteorol. Klimaforsch., Universität Karlsruhe / Forschungszentrum Karlsruhe, 1995.

Hammer, M.: Numerische Untersuchungen zur Variation der Photolyserate im regionalen Bereich, Seminararbeit, Inst. Meteorol. Klimaforsch., Universität Karlsruhe / Forschungszentrum Karlsruhe, 1995.

Koßmann, M.: Einfluß orographisch induzierter Transportprozesse auf die Struktur der atmosphärischen Grenzschicht und die Verteilung von Spurengasen. Dissertation, Inst. Meteorol. Klimaforsch., Universität Karlsruhe 1998.

Kunz, M.: Änderung der Spurengaskonzentrationen bei Frontpassagen, Seminararbeit, Inst. für Meteorol. und Klimaforschung Karlsruhe, Universität Karlsruhe/Forschungszentrum Karlsruhe 1996.

Pfeifle A.: Ozonverteilung in einer konvektiven Grenzschicht, Diplomarbeit, Inst. für Meteorol. und Klimaforschung Karlsruhe, Universität Karlsruhe/Forschungszentrum Karlsruhe, 1996.

Riemer, N.: Ein Verfahren zur Berechnung der dreidimensionalen Verteilung der Photolyserate im regionalen Bereich, Seminararbeit, Inst. Meteorol. Klimaforsch., Universität Karlsruhe / Forschungszentrum Karlsruhe, 1995.

Uhrner, U.: Physikalische Ursachen für regionale Unterschiede in der nächtlichen Ozonverteilung, Diplomarbeit, Inst. Meteorol. Klimaforschung, Universität Karlsruhe / Forschungszentrum Karlsruhe 1997.

Subject Index

A

B

C

D

E

F

G

H

I

L

M

N

O

T

U

V

W

Production: Druckhaus Beltz, Hemsbach